Springer
Berlin
Heidelberg
New York
Hong Kong
London
Milan
Paris
Tokyo

Physics and Astronomy ONLINE LIBRARY

springeronline.com

SPRINGER SERIES IN SURFACE SCIENCES

Series Editors: G. Ertl, H. Lüth and D.L. Mills

This series covers the whole spectrum of surface sciences, including structure and dynamics of clean and adsorbate-covered surfaces, thin films, basic surface effects, analytical methods and also the physics and chemistry of interfaces. Written by leading researchers in the field, the books are intended primarily for researchers in academia and industry and for graduate students.

38 **Progress in Transmission Electron Microscopy 1**
Concepts and Techniques
Editors: X.-F. Zhang, Z. Zhang

39 **Progress in Transmission Electron Microscopy 2**
Applications in Materials Science
Editors: X.-F. Zhang, Z. Zhang

40 **Giant Magneto-Resistance Devices**
By E. Hirota, H. Sakakima, and K. Inomata

41 **The Physics of Ultra-High-Density Magnetic Recording**
Editors: M.L. Plumer, J. van Ek, and D. Weller

42 **Islands, Mounds and Atoms**
Patterns and Processes in Crystal Growth Far from Equilibrium
By T. Michely and J. Krug

43 **Electronic Properties of Semiconductor Interfaces**
By W. Mönch

Volumes 1–37 are listed at the end of the book

Winfried Mönch

Electronic Properties of Semiconductor Interfaces

With 146 Figures and 17 Tables

Springer

Professor em. Dr. Winfried Mönch
Universität Duisburg-Essen
Fakultät für Naturwissenschaften
Laboratorium für Festkörperphysik
47048 Duisburg, Germany
w.moench@uni-duisburg.de

Series Editors:

Professor Dr. Gerhard Ertl
Fritz-Haber-Institute der Max-Planck-Gesellschaft, Faradayweg 4–6,
14195 Berlin, Germany

Professor Dr. Hans Lüth
Institut für Schicht- und Ionentechnik
Forschungszentrum Jülich GmbH,
52425 Jülich, Germany

Professor Douglas L. Mills, Ph.D.
Department of Physics, University of California,
Irvine, CA 92717, USA

Library of Congress Cataloging-in-Publication Data: Mönch, Winfried. Electronic properties of semiconductor interfaces / Winfried Mönch. p. cm. – (Springer series in surface sciences ; 43) Includes bibliographical references and index. ISBN 3-540-20215-3 (alk. paper) 1. Diodes, Schottky-barrier. 2. Semiconductors–Junctions. I. Title. II. Series. TK7871.89.S35M66 2003 621.3815'2–dc22 2003066401

ISSN 0931-5195
ISBN 3-540-20215-3 Springer-Verlag Berlin Heidelberg New York

Springer-Verlag is a part of Springer Science+Business Media

springeronline.com

Typesetting: Data prepared by the author using a Springer Word macro package
Final layout: LeTeX, Leipzig
Cover concept: eStudio Calamar Steinen
Cover production: *design & production* GmbH, Heidelberg

Printed on acid-free paper SPIN: 10959480 57/3141/ba - 5 4 3 2 1 0

To Gisela
with many thanks
for your patience and encouragement

Preface

Semiconductor interfaces are a most instructive example of the intimate interconnections between fundamental and applied science. *Braun*'s discovery of the rectifying properties of metal–semiconductor contacts marks the beginning of semiconductor science in 1874. Although *Schottky*'s explanation of this phenomenon in 1938 was possible only after the quantum-theory of solids had been formulated cuprous oxide and selenium rectifiers were industrially produced in large quantities much earlier. Ironically, *Poganski*'s work revealed the technical selenium rectifiers to consist of *n*-CdSe/*p*-Se semiconductor heterostructures rather than to be Cd/*p*-Se metal–semiconductor contacts as initially inferred. Finally, it took another 60 years before experimental data and theoretical concepts of the band lineup at both metal–semiconductor contacts and semiconductor heterostructures could be brought into harmony. On the theoretical side, *Heine*'s concept of interface–induced gap states, which he published in 1965, plays the most prominent role. As "new" semiconductors, such as silicon carbide and the III-nitrides, emerged, the interplay between fundamental research and device-oriented development continues and keeps the fascinating field of semiconductor interfaces in vivid motion.

My thanks go to the many of my former students who have produced a real wealth of experimental data on metal–semiconductor interfaces. Their data were the basis not only of our intense internal discussions but served also for the many and sometimes controversial but in the end fruitful arguments with members and heads of other research groups around the world. In addition, I would like to thank Professor Axel Lorke, my successor, for providing me with the possibility to continue my scientific work. It is my great pleasure to thank quite a number of colleagues of mine who trustfully provided me with valuable experimental data, mostly effective barrier heights and ideality factors of Schottky diodes, which were hidden in their files but unfortunately not contained in their publications. Furthermore, I have to thank Professors Inder P. Batra, Marvin L. Cohen, and Max G. Lagally, who gave permission to reprint figures which appeared in articles of theirs.

Mülheim an der Ruhr, October 2003 *Winfried Mönch*

Contents

1. Introduction 1
1.1 Metal–Semiconductor Contacts 1
1.1.1 Rectification of Metal–Semiconductor Contacts 1
1.1.2 Explanation of Rectification by Depletion Layers 4
1.1.3 Metal-Induced Gap States 7
1.1.4 Rules, Correlations, and Models 9
1.1.5 MIGS-and-Electronegativity Theory 14
1.2 Semiconductor Heterostructures and the IFIGS Concept 18

2. Depletion Layer 21
2.1 The Schottky Barrier 21
2.2 Capacitance of Schottky Barriers 25
2.3 Image-Force or Schottky Effect 27

3. Determination of Barrier Heights and Offsets 33
3.1 Current Transport across Schottky Contacts 33
3.2 Effective Barrier Height of Ideal Schottky Contacts 38
3.3 Effective Richardson Constant 39
3.4 I/V Characteristics of Real Schottky Contacts 42
3.5 Barrier Heights of Real Schottky Contacts 1: I/V Characteristics 47
3.6 Ballistic-Electron-Emission Microscopy 57
3.7 Barrier Heights of Real Schottky Contacts 2: BEEM I_{coll}/V_{tip} Characteristics 61
3.8 Barrier Heights of Real Schottky Contacts 3: Internal Photoemission Yield Spectroscopy 63
3.9 Core-Level Photoemission Spectroscopy 67
3.9.1 Chemical Shifts and Interface Band-Bending Determination by PES 67
3.9.2 Growth Modes of Metals on Semiconductors 70
3.10 Barrier Heights of Real Schottky Diodes 4: XPS 74
3.11 Metal-Induced Core-Level Shifts and the Intrinsic Interface Electric-Dipole Layer 75
3.12 Determination of Band Offsets at Heterostructures 79
3.12.1 Internal Photoemission Yield Spectroscopy 79
3.12.2 Core-Level X-Ray Photoemission Spectroscopy 81

4. Laterally Inhomogeneous Schottky Contacts ... 83
4.1 Potential Distribution ... 83
4.2 Current Transport in Schottky Contacts with One Circular Patch ... 91
4.3 Current Transport in Schottky Contacts with Many Circular Patches ... 93
4.4 Test of the Patch Concept with Modified Contacts ... 97
4.5 *I/V* and BEEM Test of the Patch Concept ... 98
4.6 Origin of Lateral Barrier-Height Inhomogeneities ... 103
4.6.1 Natural or Intrinsic Nonuniformities ... 103
4.6.2 Extrinsic Nonuniformities ... 104

5. The IFIGS-and-Electronegativity Theory ... 107
5.1 Band-Structure Lineup and Electronegativity: A Chemical Approach ... 107
5.1.1 Barrier Heights of Schottky Contacts ... 107
5.1.2 Interface Dipoles at Semiconductor Heterostructures ... 109
5.2 Barrier Heights and Interface Dipoles: A Phenomenological Approach ... 111
5.3 Interface-Induced Gap States ... 115
5.3.1 Metal-Induced Gap States ... 115
5.3.2 Semiconductor Heterostructures ... 118
5.3.3 Valence-Band Offsets at Semiconductor Heterostructures ... 119
5.4 Virtual Gap States: One-Dimensional Model ... 119
5.5 Virtual Gap States: Three-Dimensional Model ... 122
5.5.1 Mean-Value $\underline{k}$-Points and the Dielectric Band Gap ... 122
5.5.2 Branch-Point Energy in the ViGS Continuum ... 125
5.5.3 Decay Length of the ViGS at their Branch Point ... 129
5.6 Slope Parameter of Barrier Heights ... 131

6. The MIGS-and-Electronegativity Concept: Experiment and Theory ... 135
6.1 Schottky Contacts ... 135
6.1.1 Introductory Remarks ... 135
6.1.2 Silicon Schottky contacts ... 136
6.1.3 GaAs Schottky Contacts ... 139
6.1.4 GaP Schottky Contacts ... 140
6.1.5 GaN Schottky Contacts ... 141
6.1.6 SiC Schottky Contacts ... 142
6.1.7 ZnSe Schottky Contacts ... 146
6.1.8 Schottky Contacts on Layered Ga–Chalcogenides ... 146
6.1.9 $CuInSe_2$ Schottky Contacts ... 148
6.1.10 Schottky Contacts on Ternary III–V Alloys ... 149
6.1.11 Metal–Insulator Contacts ... 155
6.1.12 Direct Observations of MIGS ... 159
6.2 Heterostructures ... 160
6.2.1 Introductory Remarks ... 160
6.2.2 Non-Polar Heterostructures ... 162

6.2.3 Lattice-Matched Ternary and Quaternary III–V Alloys....... 163
6.2.4 I–III–VI_2 Chalcopyrites........ 168
6.2.5 Layered Semiconductors........ 170
6.2.6 Metamorphic Heterostructures........ 172
6.2.7 Semiconductor–Insulator Interfaces........ 173
6.2.8 *Langer–Heinrich* Rule........ 176

7. First-Principles Calculations of Barrier Heights and Valence-Band Offsets........ 181
7.1 Introductory Remarks........ 181
7.2 Schottky Barrier Heights........ 181
7.3 Valence-Band Offsets at Heterostructures........ 186
7.3.1 Isovalent and Lattice-Matched AlGaAs/GaAs........ 186
7.3.2 SiC Heterostructures........ 187

8. Temperature and Pressure Effects........ 189
8.1 Temperature Effects........ 189
8.2 Pressure Effects........ 193
8.2.1 Schottky Barrier Heights........ 193
8.2.2 Valence-Band Offsets........ 196
8.3 Pseudomorphic Heterostructures........ 197

9. Barrier Heights and Extrinsic Interface Defects........ 203
9.1 Defect-Induced Changes of Barrier Heights........ 203
9.2 Application to *p*-GaP(110) Schottky Contacts........ 206
9.3 Unified Defect Model........ 208

10. Extrinsic Interface Dipoles........ 209
10.1 Interface Doping of Schottky Contacts........ 209
10.2 Interface Structure........ 214
10.2.1 Metal/Si(111)-$(7\times7)^i$ Contacts........ 214
10.2.2 Epitaxial Silicide/Silicon Interfaces........ 217
10.2.3 Polar Heterostructures........ 224

11. Ohmic Contacts........ 227

Appendix........ 231

References........ 235

Subject Index........ 257

1. Introduction

1.1 Metal–Semiconductor Contacts

1.1.1 Rectification of Metal–Semiconductor Contacts

Semiconductor science began in 1874. At the end of this year, *Ferdinand Braun* described his discovery of the rectifying properties of metal–semiconductor contacts in an article entitled *Über die Stromleitung durch Schwefelmetalle*.[1] This publication deals with an interface-controlled device and, in this, it is an excellent example of the intimate interaction between fundamental research of bulk, surface, as well as interface properties and device physics that has been a characteristic of semiconductor physics since then. The expression *Halbleiter* or *semi-conductor* was, most probably, first used by *Koenigsberger* and *Weiss* in 1911.

"Bei einer großen Anzahl natürlicher und künstlicher Schwefelmetalle und sehr verschiedenen Stücken, sowohl Krystallen von so vollkommener Ausbildung, wie ich überhaupt bekommen konnte, als derben Stücken habe ich gefunden, daß der Widerstand derselben verschieden war mit Richtung, Intensität und Dauer des Stromes. Die Unterschiede betragen bis zu 30pCt. des ganzen Werthes. ... Ich habe benutzt Quecksilbercontact, stark gegen gepreßte Kupfer-, Platin- und Silberdrähte und endlich bei einem Stück eine bereits vorhandene Fassung mit dicken Neusilberbügeln."[2] *Braun* verified this summary of his observations with two tabulated series of experimental data. One displays the current flow through a metal resistor and a chalcopyrite sample, while the other one compares the current through the chalcopyrite sample when the applied voltage and by this the direction of the current is reversed. Figure 1.1 shows these data of *Braun*. The chalcopyrite crystal clearly exhibits *unipolar* conduction, as the rectifying behavior was called then

---

[1] *On the Current Transport in Metal Sulfides*

[2] "In a large number of natural and synthetic sulfides and with very different samples, in crystals as perfectly shaped as I could get them as well as rough pieces, I have found that their resistance varied by up to 30%, depending on direction, intensity, and duration of the current. ... I have used mercury contacts, wires of copper, platinum, and silver, which were heavily pressed against the samples, and an already existing mounting made of German silver with another piece."

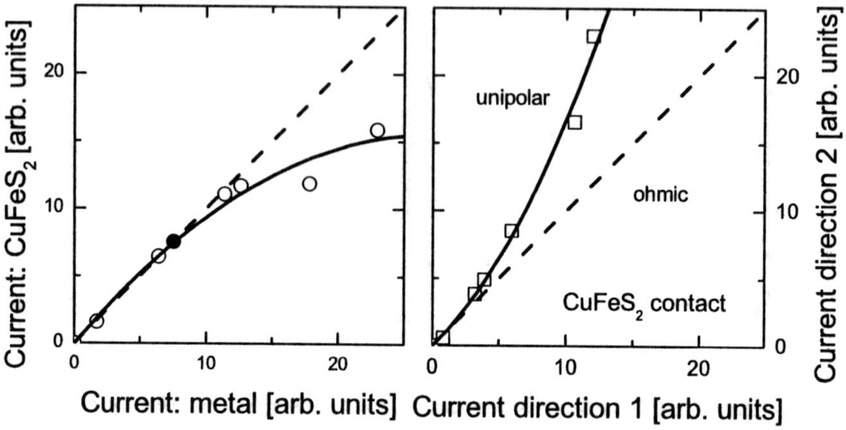

Fig. 1.1. Current flow through a chalcopyrite crystal and a metal resistor and through the forward- and reverse-biased chalcopyrite crystal. The maximum applied voltage was 1 Bunsen (1.8 to 1.9 V). Data from *Braun* [1874]

[*Schuster* 1874], in contrast to the familiar ohmic behavior of metals. In addition to chalcopyrite *Braun* also used such compounds as iron pyrite, galena, and fahlore and, later on, also cuprous oxide.

Braun's observations of unipolar conduction attracted much interest and shortly afterwards *Siemens* [1875] and *Adams* and *Day* [1876] confirmed them with metal–selenium contacts. Although they also met skepticism by other workers who were unsuccessful in reproducing their results, they rapidly found widespread technical applications. As *Braun* already realized in 1901, but only reported five years later, point contacts showing unipolar conduction should act as valves for alternating currents and, he successfully used psilomelane crystals for the effective detection of electromagnetic radiation. In the same year, 1901, *Bose* applied for a patent on the use of galena rectifiers as receivers of electromagnetic radiation. The patent was granted in 1904. At almost the same time, *Pickard* filed and obtained a patent on silicon point-rectifiers in 1904 and 1906, respectively. Apart from these "cat's whisker" rectifiers, which consisted of a metal point pressed against a semiconductor crystal, large-area or plate rectifiers were developed. *Fritts* [1883] and *Grondahl* [1925, 1926] made and studied selenium and cuprous oxide plate-rectifiers, respectively. Such units were also connected in parallel and in series for technical applications requiring larger currents or higher voltages, respectively.

Although copper–cuprous oxide and selenium rectifiers were commercially manufactured and widely used, their production methods were of a purely empirical nature. The physical understanding of rectification at metal–semiconductor contacts developed very slowly. In his monograph *Rectifying Semi-Conductor Contacts* of 1949, *Henisch* has described the many early attempts in detail and they will be not reviewed here. A huge step forward was made by *Schottky* and his collaborators in 1929. They scanned the potential gradient along Cu/Cu_2O rectifiers with electrostatic probes. With the rectifiers biased in the reverse direction they demonstrated that the voltage drop occurs across a narrow *Sperrschicht* (blocking layer) right at the interface itself. A few years later, in 1933, *Grondahl* confirmed these experimental results. *Schottky* and *Deutschmann* [1929] estimated the width of this layer from the differential capacitance of Cu/Cu_2O rectifiers measured as a function of the applied voltage. They found the differential capacitance to decease with reverse bias but to increase with forward bias. They modeled the blocking layer by a parallel-plate capacitor, i.e.,

$$C(V_a) = \varepsilon_b \varepsilon_0 \frac{A}{d(V_a)} , \tag{1.1}$$

where A and d are the area and the separation, respectively, of the two capacitor plates and ε_b is the static bulk dielectric constant of the semiconductor. From the capacitance measured they obtained a zero-bias width $d(0) = 3\times10^{-5}$ cm of the blocking layer. They further concluded that as a function of the applied voltage V_a the width of the blocking layer increases in reverse but decreases in forward current direction. Already at that time, they definitely excluded current transport by quantum-mechanical tunneling since the blocking layer was too wide for this mechanism to be effective and they concluded that the conduction mechanism in the blocking layer has to be similar to the one in bulk Cu_2O. *Schottky* and *Deutschmann* had already discussed the possible existence of a space charge in the blocking layer. However, they could not arrive at the correct model for several reasons.

First of all, *Wilson* presented his *Theory of Electronic Semi-Conductors* only in 1931, and it was then that the concept of defect electrons and, consequently, of *p*-type semiconductors was introduced. Second, the two most important semiconductors at that time, cuprous oxide and selenium, were only poorly characterized. Already at the end of 1931, Schottky and Peierls corresponded on the depletion-layer concept of the blocking layer at rectifying metal–semiconductor interfaces, as *Schottky* [1939] mentioned in one of this later articles. However, their considerations remained inconclusive, since the sign of the Hall coefficient of cuprous oxide was erroneously determined to be negative. It took some time, until in 1935 *Fritsch* eventually arrived at the correct, positive sign and established Cu_2O as a *p*-type semiconductor. The *p*-type character of selenium was later inferred from the decrease of the barrier heights of metal–selenium rectifiers with the increase of the work function of the metal [*Schweickert* 1939, *Schottky* 1940]. Third, the difference between *chemical* and *physical* blocking layers had to be understood. At Cu/Cu_2O contacts, the most important type of crystal rectifiers at that time, the oxygen concentration is reduced at the interface between the copper and the thermally grown oxide layer. This chemical depletion layer has to be considered in addition to the physical depletion or band bending caused by the Cu/Cu_2O barrier height. *Schottky* [1942] determined the spatial distribution of the acceptors at Cu/Cu_2O interfaces from the aforementioned capacitance of such rectifiers measured as a function of the blocking voltage [*Schottky* and *Deutschmann* 1929]. With selenium rectifiers, on the other hand, only physical depletion layers were thought to be of importance.

1.1.2 Explanation of Rectification by Depletion Layers

At the end of 1938, *Schottky* eventually published his *Halbleitertheorie der Sperrschicht.*[3] He explained the rectification of metal–semiconductors contacts by a band bending in the semiconductor right at the interface. He assumed the metal–semiconductor work function or, as it is called now, the barrier height to be so great that in thermal equilibrium the boundary layer is depleted of the mobile majority carriers. Consequently, the remaining charged static dopants make up a space charge. Figure 1.2 explains *Schottky*'s depletion-layer concept of metal–semiconductor contacts. An externally applied voltage will either reduce or enlarge the band bending and, correspondingly, the depletion-layer width and accordingly decrease and increase, respectively, the resistance of the boundary layer.

The rectifying properties of Schottky contacts – as metal–semiconductor contacts are generally labeled in order to honor *Schottky*'s many contributions to their understanding – are determined by the *work function of the majority carriers*, as *Schottky* originally called the barrier height. This quantity equals the distance in energy between the Fermi level and the edge of the respective majority carrier band, i.e., the conduction-band minimum for *n*-type semiconductors and the valence-band maximum for *p*-type semiconductors. *Mott* [1938] and *Schottky* [1940] were the first to propose a simple rule for calculating barrier heights. Figure 1.3 explains their concept.

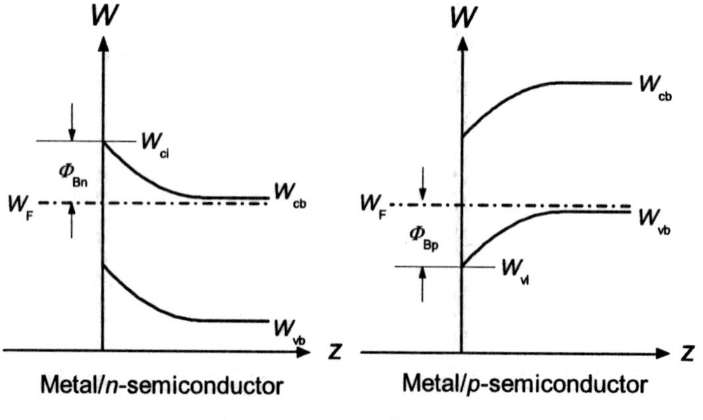

Fig. 1.2. Schematic band diagram of intimate, abrupt, and rectifying metal contacts on *n*- and *p*-type semiconductors

[3] *Semiconductor Theory of the Blocking Layer*

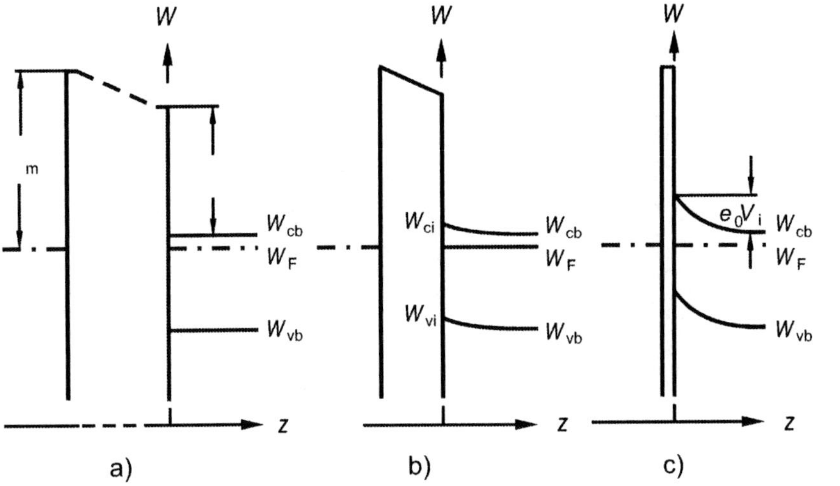

Fig. 1.3. Development of a Schottky barrier as a function of decreasing metal-to-semiconductor distance. The semiconductor and the interface are assumed to be free of surface and interface states, respectively

In a Gedanken experiment, Schottky contacts may be formed by gradually reducing the distance between a metal and a semiconductor until finally an intimate, clean, and abrupt interface exists. Most importantly, the semiconductor and the finished contact are assumed to have no surface and interface states, respectively, within the band gap. The bands of the semiconductor are thus flat up to the surface for infinite separation between metal and semiconductor (see Fig. 1.3a).

The work functions of the metal and of the bare semiconductor generally differ so that in thermal equilibrium an electric field exists in the vacuum gap between the two solids. Both metal and semiconductor thus carry surface charges of equal density but of opposite sign. The electric field enters the metal as well as the semiconductor (see Fig. 1.3b). However, the characteristic lengths are quite different. For metals, the Thomas–Fermi screening length determines the field penetration. Due to the large electron densities of metals, their screening lengths are typically less than a tenth of a nanometer. The penetration of electric fields may thus be neglected in metals. Screening in semiconductors doped non-degenerately is described by their Debye length (see Sect. 2.1). For a doping level of, for example, 10^{17} cm^{-3} and at room temperature, the Debye length typically amounts to 13.4 nm (see Fig 2.1). Electric fields thus enter non-degenerately doped semiconductors and, as a consequence, extended space-charge layers exist.

In the example of Fig. 1.3, the metal is assumed to have a larger work function than the semiconductor, which is assumed to be non-degenerately doped *n*-type. Hence, the semiconductor space-charge has a positive sign and, consequently, the surface is depleted of mobile electrons. This means that the bands bend upward. With decreasing distance between the metal and the semiconductor, the electric-field strength in the vacuum gap reduces while the band bending at the semicon ductor surface increases. In the limit of an intimate contact, the vacuum levels of the metal and the semiconductor align. Hence, the barrier height of electrons

$$\Phi_{Bn}^{SM} = W_{ci} - W_F = \phi_m - \chi_s \qquad (1.2)$$

equals the difference of the metal work function ϕ_m and the electron affinity $\chi_s = W_{vac} - W_{cs}$ of the semiconductor. This is the famous *Schottky–Mott* rule. When the semiconductor is doped *p*-type the barrier height of holes

$$\Phi_{Bp}^{SM} = W_F - W_{vi} = I - \phi_m \qquad (1.3)$$

equals the difference of the ionization energy $I = W_{vac} - W_{vs}$ of the semiconductor and the metal work-function.

Schweickert had already performed a first test of the *Schottky–Mott* rule in 1939 and definitely demonstrated that his experimental results did not comply with this sim- ple rule. He determined the initial resistance of metal–selenium rectifiers, i.e., the zero-bias slope of their current–voltage characteristics. The early diffusion theory of *Schottky* and *Spenke* [1939] as well as *Bethe*'s [1942] thermionic-emission theory find the zero-bias resistance of metal–semiconductor contacts to depend exponentially on the barrier height (see Sect 3.1). Figure 1.4 displays

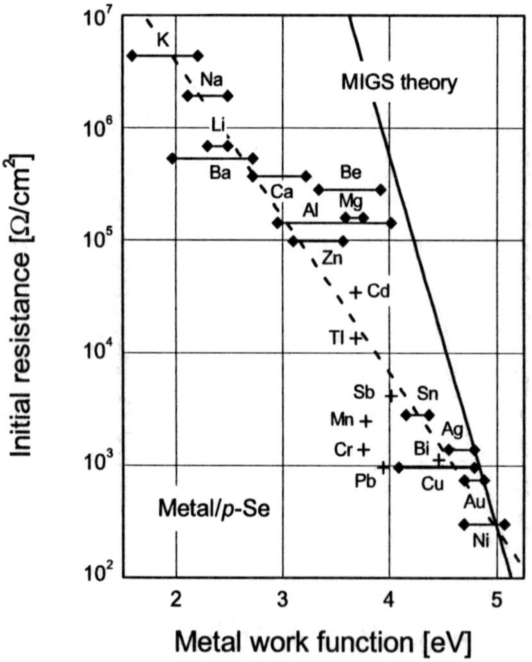

Fig. 1.4. Zero-bias resistance of selenium Schottky contacts as a function of the metal work-function. The dashed line is a linear least-squares fit to the data, which gives a slope parameter $S_X = -0.08$. Data from *Schweikert* [1939] as reported by *Schottky* [1940]

Schweickert's data as published by *Schottky* [1940]. From these observations several important conclusions were drawn. First, the observed decrease of the barrier heights with increasing work function of the metals proved selenium to be a *p*-type semiconductor. Second, the barrier heights of selenium Schottky contacts correlate linearly with the metal work-function but the slope parameter $S_\phi = -\mathrm{d}\Phi_{Bp}/\mathrm{d}\phi_\mathrm{m}$ has the magnitude of only 0.08 rather than of 1 as demanded by the *Schottky–Mott* rule (1.3). Alreadyat that time, it was generally accepted that this simple rule is incompatible with the experimental results. Therefore, neither *Shive* [1940] nor *Brattain* [1940] nor *Welker* [1943], who made similar observations with Se, Cu_2O, and Ge Schottky contacts, published their data.

1.1.3 Metal-Induced Gap States

Some years later, in 1947, *Bardeen* finally resolved this slope-parameter dilemma. He argued that in addition to the space-charge layer on the semiconductor side of the contact interface states might also absorb charge. The charge density Q_m on the metal side of the contact is then balanced by a charge density $Q_s = Q_{sc} + Q_{is}$ of equal size but of opposite sign on the semiconductor side, which distributes between the space-charge layer (sc) and interface states (is). The condition of charge neutrality may be written as

$$Q_\mathrm{m} + Q_s = Q_\mathrm{m} + Q_{sc} + Q_{is} = 0 \quad . \tag{1.4}$$

Interface states will screen the space-charge layer and, therefore, vary the slope parameter. For the two limiting examples of very large and almost vanishing densities of states this influence is obvious. In the latter case the interface states will have no significant effect, i.e., the charge densities on the metal side and in the space-charge layer will change to the same extent. This was exactly the assumption that led to the *Schottky–Mott* rule and the slope parameter $S_\phi = 1$. On the other hand, if their density of states is very large, the interface states will absorb most of the charge on the semiconductor side. Consequently, different charge densities on the metal side will cause only slight variations of the space-charge density and, what is equivalent, of the interface band-bending. Hence, the slope parameter will be small. Such behavior was called the *Bardeen* case.

As simple as the concept of interface states at semiconductor interfaces seems in retrospect, just as pioneering was *Bardeen*'s concept in 1947. The possible existence of surface states was theoretically discussed by *Tamm* [1932], *Maue* [1935], and *Goodwin* [1939a,b,c], and, most critically, by *Shockley* [1939]. However, *Shockley* and *Pearson* [1948], *Brattain* and *Bardeen* [1953], and *Morrison* [1953] were among the first who reported experimental studies on surface properties of semiconductors and gave experimental proof of the existence of semiconductor surface-states. The reasons for their work were the huge interest in semiconductor surface properties that originated from the discovery and the further developments of the transistor and, on the other hand, the invention of gauges for measuring gas pressures below 10^{-5} Pa [*Bayard* and *Alpert* 1950]. The latter progress specifically stimulated the development of the ultrahigh-vacuum technique, an essential prerequisite for studies of clean and intentionally modified surfaces and interfaces.

In 1965, *Cowley* and *Sze* quantified *Bardeen*'s pioneering concept of interface states at metal–semiconductor contacts. They derived the dependence of barrier heights on interfacial gap states and the metal work-function. They assumed a continuum of interface states which is characterized by a constant density of states and a charge-neutrality level. Most importantly, the interfacial gap states were thought to be a property of the semiconductor only and the corresponding dipole layer was postulated to have a thickness of a few Ångstroms.

The derivations performed by *Cowley* and *Sze* required no specific assumptions about the physical nature of the interface states. Right at the same time, *Heine* [1965] concluded that *for energies in the semiconductor band gap the volume states of the metal all have tails in the semiconductor.* These metal-induced gap

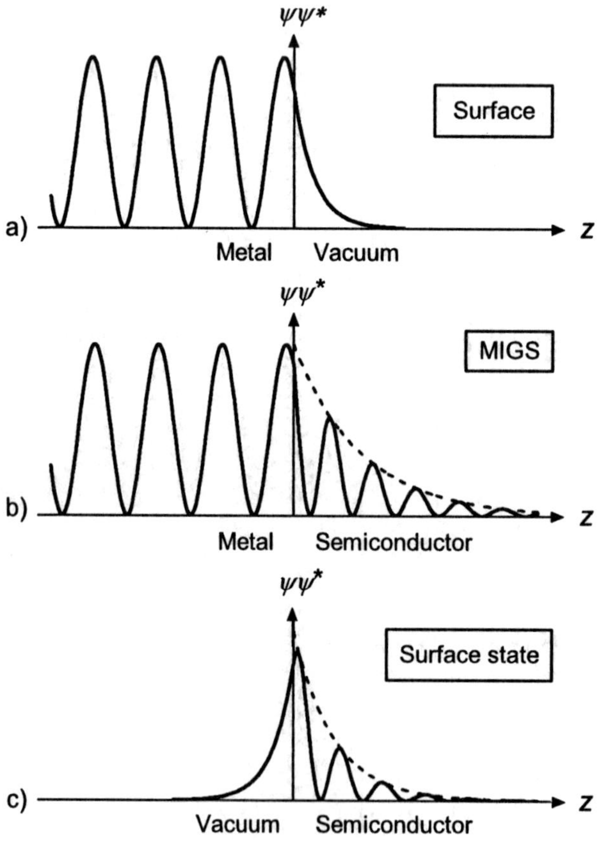

Fig. 1.5. Wave-functions at clean surfaces (a), at metal–semiconductor interfaces (b), and of surface states (c) (schematically)

states (MIGS), as they were called later [*Louie* and *Cohen* 1976], are an *intrinsic* property of the semiconductor and they are the fundamental mechanism that determines the barrier heights of Schottky contacts. The MIGS derive from the complex band structure of the semiconductor or, in other words, the continuum of virtual gap states (ViGS). This tailing behavior of the MIGS (see Fig. 1.5b) is quite similar to what happens at surfaces or, in other words, at solid–vacuum interfaces.

Due to the quantum-mechanical tunnel effect, the wave functions of conduction and valence electrons at metal and semiconductor surfaces tail exponentially into the vacuum (see Fig. 1.5a). Without a doubt, surface states decay exponentially into the vacuum but also towards the bulk and within the solid their charge distributions oscillate (see Fig. 1.5c). The MIGS derive from the valence- and the conduction-band states of the semiconductor and they will carry a net charge, the sign of which depends on the Fermi- level position relative to their branch point where their character changes from predominantly valence-band-like or donor-like to mostly conduction-band-like or acceptor-like. Hence, the MIGS give rise to an *intrinsic* interface dipole. Intuitively, the branch point of the MIGS is also described as a charge neutrality level (CNL).

However, *Heine*'s MIGS concept as simple, as it sounds, was very slowly adopted by the experimentalists. Theoreticians, on the other hand, appreciated the new model at once. In a series of articles, *Flores* and collaborators calculated the charge neutrality levels of the interface-induced gap states [*Tejedor* and *Flores* 1978] and the slope parameters of Schottky contacts of some of the III–V and II–VI compound semiconductors [*Louis* et al. 1976, *Tejedor* et al. 1977] that almost equal the meanwhile accepted values. While *Flores* and co-workers pursued more general concepts, *Cohen* and collaborators developed and applied a slab method. Using self-consistent pseudopotentials for the semiconductors, they calculated the electronic structure of Al–jellium contacts on ideally terminated Si(111) and GaAs-, ZnSe-, and ZnS(110) surfaces [*Louie* and *Cohen* 1976, *Louie* et al. 1977]. Their case studies clearly confirmed the existence of metal-induced gap states.

One of the reasons for the initially poor acceptance of *Heine*'s MIGS concept was that it predicts a linear dependence of the barrier heights on the metal workfunction while the experimental data available were arranged in *elongated clouds*. As an example, Fig. 1.6 displays data for GaAs Schottky contacts. The dashed line is a linear least-squares fit that indicates the *trend* of these data. It took another 35 years and additional conceptual input before *Heine*'s MIGS idea was finally verified by a wealth of experimental data [*Mönch* 1999, 2001]. Until then, the search for rules and correlations prevailed.

1.1.4 Rules, Correlations, and Models

The work function is a true surface property but not an interface property. In addition to a volume term it contains a surface-dipole contribution which depends on the charge distribution or, in other words, on the positions of the atoms at the surface. At metal–semiconductor interfaces the charge distribution will certainly dif-

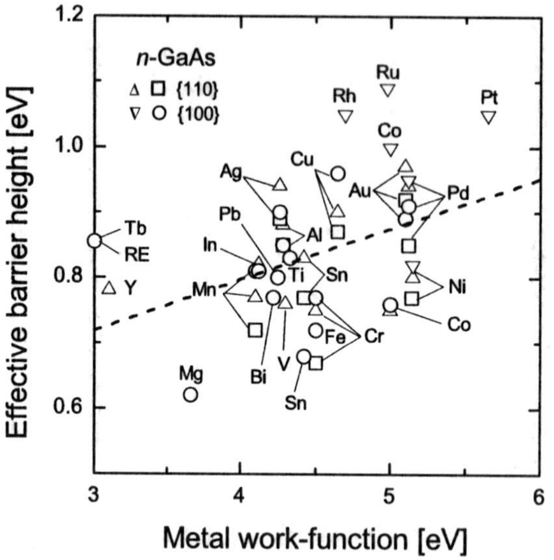

Fig. 1.6. Barrier heights of GaAs Schottky contacts as a function of the metal work-function: O data from *Waldrop* [1984], □ data from *Newman* et al. [1986a], Δ data from *McLean* et al. [1986b] and *McLean* and *Williams* [1988], ∇ data from *Ludwig* et al. [1986] and *Allongue* and *Souteyrand* [1987]. The dashed line is a linear least-squares fit to the data points

fer from the one at both the clean metal and the semiconductor surfaces. Therefore, it was pro-posed to consider the metal electronegativities rather than their work functions in describing the chemical trends of the barrier heights of Schottky contacts. However, *Gordy* and *Thomas* [1956] had already demonstrated a linear relationship between the work functions and the electronegativities of metals. Hence, it is not surprising that exchanging electronegativities for work functions did not reduce the scatter of the data. As an example, Figs. 1.6 and 1.10 display the same set of Schottky barrier heights as a function of the work functions and electronegativities of the metals, respectively. The later MIGS-and-electronegativity concept, on the other hand, reintroduced the use of electronegativities in explaining the chemical trends of Schottky barrier heights.

Gold–semiconductor contacts were most widely studied. *Mead* [1966] suggested that at such interfaces the Fermi level should be pinned at ⅓ of the fundamental band gap above the valence-band maximum. *McCaldin* et al. [1976] proposed that gold contacts on compound semiconductors with the same anion but different cations possess very similar values of the Schottky barrier of holes. Figure 1.7 displays the energy distance between the valence-band maximum and the Fermi level or, what is the same, the barrier heights of holes against the width W_g of the fundamental band gap of the semiconductors. No $W_g/3$ rule and no com-

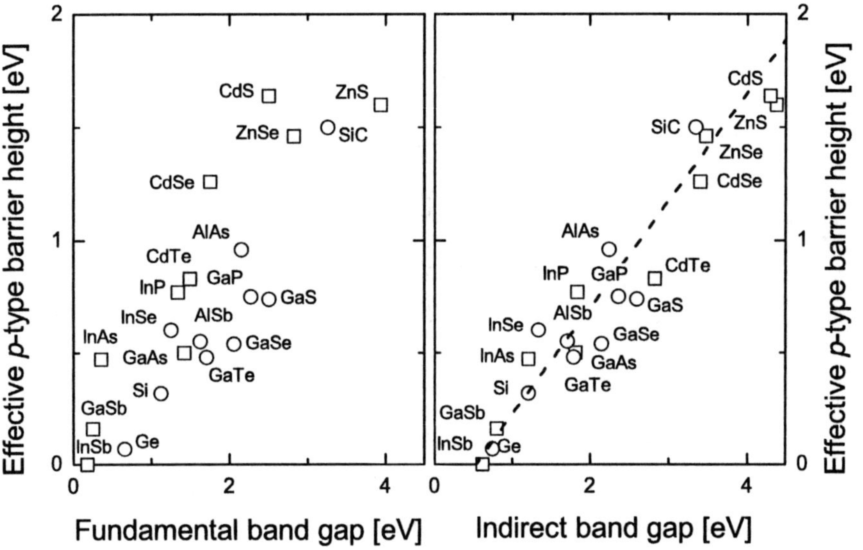

Fig. 1.7. Barrier heights of gold–semiconductor contacts as a function of the width of the fundamental band gap (left-hand plot) and of the indirect band gap (right-hand plot): ○ data for indirect and □ data for direct semiconductors from *Sze* [1981], *Kurtin* and *Mead* [1969], *Mamy* et al. [1990]. After *Mönch* [1986]

mon-anion rule are discernible. However, a remarkably good correlation is obtained when the same data are plotted against the width of the indirect band gaps [*Mönch* 1986].

For some 20 different semiconductors, *Kurtin* et al. [1969] plotted barrier heights versus the electronegativities X_m of the metals. Then they searched for a possible dependence of the slope parameters $S_X = d\Phi_{Bn}/dX_m$ on the ionicity of the semiconductors which they characterized by the electronegativity difference of their constituents. They claimed a step-like increase of the S_X values as the electronegativity difference exceeds ≈ 0.8 Pauling-units. *Schlüter* [1978, 1982] reexamined the original data but found no such correlation. His revised S_X values are displayed in Fig. 1.8, where the dashed line reproduces the trend asserted by *Kurtin* et al. The additional data point for metal–xenon contacts was obtained later by *Jacob* et al. [1987] and it clearly rules out the S-shaped trend that was initially inferred.

The barrier heights were also correlated with other physical parameters. *Andrews* and *Phillips* [1975] and *Ottaviani* et al. [1980] found linear relationships between the barrier heights of some silicide–silicon contacts and the formation energy and the eutectic temperature, respectively, of the silicides involved. As a result of the reactions of metals evaporated on compound semiconductors, interface compounds may form. *Brillson* [1978] noticed a step-like increase of the barrier

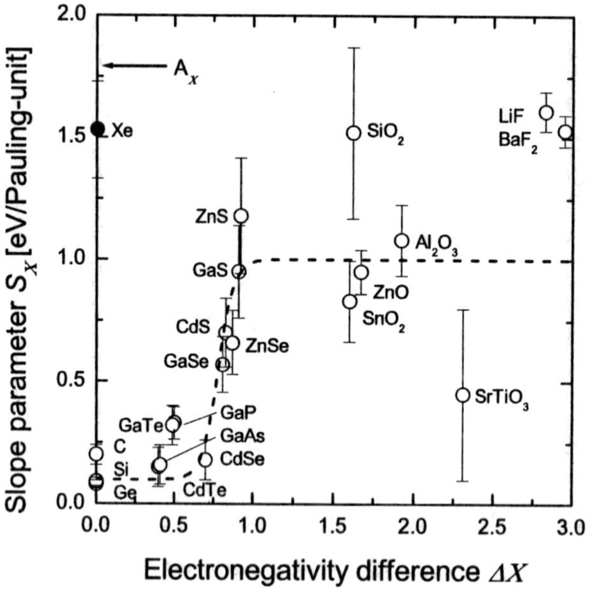

Fig. 1.8. Slope parameters $S_X = d\Phi_{Bn}/dX_m$ as a function of the electronegativity difference of the constituent atoms of the semiconductors. The S_X values were obtained from linear least-squares fits to plots of effective barrier heights versus the metal electronegativities. After *Kurtin* et al. [1969] and *Schlüter* [1978, 1982]

heights of III–V and II–VI Schottky contacts as a function of the heat of interface chemical reaction.

Freeouf and *Woodall* [1981], on the other hand, supposed that chemical reactions during the metallization of compound semiconductors produce microclusters of substrate anions at the interface. They still assumed the *Schottky–Mott* rule to be valid and suggested the substitution of the anion work function for the one of the metal. This *effective work-function model* is another common-anion rule which was also inconsistent with the experimental data.

A most widely recognized and interesting attempt to provide insight in the physical origin of Schottky barrier formation from quite a different viewpoint was *Spicer*'s *Unified Defect Model*. In the course of their photoelectron spectroscopy (PES) studies during the evaporation of different metals on, for example, cleaved GaAs(110) surfaces, *Spicer* et al. [1979] observed unique pinning positions of the Fermi level at 0.5 and 0.75 eV above the valence-band maximum with substrates doped *p*- and *n*-type, respectively. They concluded that discrete defect states of donor and acceptor character, respectively, are responsible for the observed Fermi-level pinning. *Spicer* and co-workers attributed these levels to missing Ga and As atoms, respectively, while *Weber* et al. [1982] ascribed them to As_{Ga} antisite defects. As they are double donors, As_{Ga} antisite defects alone could not explain Fermi-level pinning at both *p*- and- *n*-type Schottky contacts, so that acceptor-type defects had to be considered in addition.[4] The barrier heights shown in Fig. 1.6, on the other hand, span the wide range from 0.6 to 1.1 eV. Even more importantly, for one and the same metal–semiconductor pair the electron and hole barrier heights add up to the width of the band gap, i.e., no different Fermi-level pinning is observed with Schottky contacts on *n*- and *p*-type semiconductors. The data displayed in Fig. 1.6 were all determined from the current–voltage characteristics of Schottky contacts with thick metal films. The PES measurements, on the other hand, required very thin metal films since the escape depth of the photoelectrons is smaller than some 20 Å. At such low coverages, metals are known to grow as islands rather then to wet the semiconductor substrates. Only cesium shows layer-by-layer growth right from the beginning, and a distinct Fermi edge and the work function of massive cesium are observed already during the deposition of the second monolayer [*Linz* et al. 1993, *van Elsbergen* et al. 1996, *Eyckeler* et al. 1998, *Weyers* et al. 1999]. Therefore, it is difficult to obtain barrier heights of Schottky contacts other than Cs contacts using photoemission spectroscopy.

Instead of discrete defect levels *Hasegawa* and *Ohno* [1986] and *Walukiewicz* [1988] proposed a continuum of defect levels in the band gap to determine the barrier heights of Schottky contacts. *Walukiewicz* argued that in heavily radiation-damaged Si, Ge, GaP, GaAs, GaSb, InP, InAs, and InSb the Fermi level stabilizes at energy positions that are close to the corresponding charge-neutrality or branch-point levels of the MIGS, where their character changes from predominantly acceptor- or conduction-band-like to mostly donor- or valence-band-like. This agreement is not merely coincidental but has a physical explanation. All gap states, and this holds for not only for MIGS but also for defects, derive from the complex band structure of the semiconductor. Provided there are high densities of both donor- and acceptor-type defects present, then the Fermi level will indeed become stabilized at the branch point of the ViGS. However, there are no indications that a continuum of defect-induced gap states (DIGS) might determine the barrier heights of Schottky contacts. Some of the counterevidence shall be mentioned. Grazing-incidence X-ray diffraction demonstrated that under thick Ag and Pb films evaporated on Si(111)-7×7 surfaces the 7×7 reconstruction persists [*Hong* et al. 1992, *Howes* et al. 1995]. Medium-energy ion-channeling experiments gave upper limits of 1×10^{13} and 3×10^{13} Si atoms per cm^2, i.e., less than 1% and 4% of the atoms in a bulk Si{111} layer being displaced from lattice sites in type-A and type-B epitaxial $NiSi_2$/Si(111) contacts, respectively [*Vrijmoeth* et al. 1990].

[4] *Mönch* [1983] analyzed this behavior in detail.

1.1.5 MIGS-and-Electronegativity Theory

In 1984/86, *Tersoff* had calculated the ViGS branch points of Si, Ge, and a total of 12 of the III–V and II–VI compound semiconductors[5] and in this way revitalized the interest in *Heine*'s MIGS concept. Support also came from the experimental side. *Mönch* [1986] re-examined the slope parameters displayed in Fig. 1.8. In one dimension, the density of states as well as the decay length of the ViGS at their branch point vary inversely proportionally with the width of the band gap [*Maue* 1935]. This result cannot be directly transferred to three dimensions. But *Penn*'s [1962] average or dielectric band gap depends inversely on the square-root of the optical susceptibility $\varepsilon_\infty - 1$ of the semiconductor. Therefore, *Mönch* plotted the slope parameters displayed in Fig. 1.8 as a function of the optical susceptibility $\varepsilon_\infty - 1$ (see Fig. 1.9). As large as the margins of experimental error and the scatter of the data are, they nevertheless exhibit a well-pronounced quadratical trend which even includes the Xe data point [*Mönch* 1988b]. This correlation was immediately thought to strongly support the MIGS concept. Later on, this conclusion was also theoretically justified [*Mönch* 1996a].

In generalizing *Pauling*'s intuitive but very successful concept [1939], the partial ionic character of the covalent bonds between interfacial metal and semiconductor atoms may be described by the difference of the metal and semiconductor electronegativities. This *chemical* description is equivalent to the *physical* picture of the MIGS dipole. In combining the physical and the chemical concept, the barrier heights of Schottky contacts were plotted versus the difference of the metal and semiconductor electronegativities rather than the metal work function [*Mönch* 1987, 1988a]. This concept divides the barrier heights inro two contributions: the zero-charge-transfer barrier height and a dipole term. The dipole term equals the

[5] *Tersoff* used a linearized augmented plane wave method (APW) and treated exchange and correlation in the local-density approximation (LDA). The latter approximation notoriously underestimates the band gaps, which are then usually enlarged to the experimental values by rigid shifts of the conduction bands or, as this procedure is called, by applying the *scissors operation*. Later on, *Mönch* [1996b] avoided this shortcoming. First, he applied *Baldereschi*'s [1973] concept of mean-value $\underline{k}$-points in the Brillouin zone and demonstrated that the band gap at this special $\underline{k}$-point equals *Penn*'s [1962] dielectric or average band gap. Second, he found *Tersoff*'s [1986c] branch points of the ViGS continuum close to the middle of the dielectric band gap at the mean-value $\underline{k}$-point. This finding makes it possible to calculate the ViGS branch-point energies of all the tetrahedrally coordinated semiconductors not considered by *Tersoff*. *Cardona* and *Christensen* [1987] were the first to take advantage of *Baldereschi*'s concept of mean-value $\underline{k}$-points in calculating branch-point energies. They computed the energy bands within the linear muffin-tin orbital method and also used the local-density approximation. However, they increased the band gaps at the mean-value $\underline{k}$-point by taking the energy distance between the average of the two top valence and the two lowest conduction bands. These averaged gaps are remarkably close to the dielectric band gaps of the III–V and II–VI compounds considered. Surprisingly, *Cardona* and *Christensen*'s so-called dielectric mid-gap energies are close to *Tersoff*'s branch-point energies.

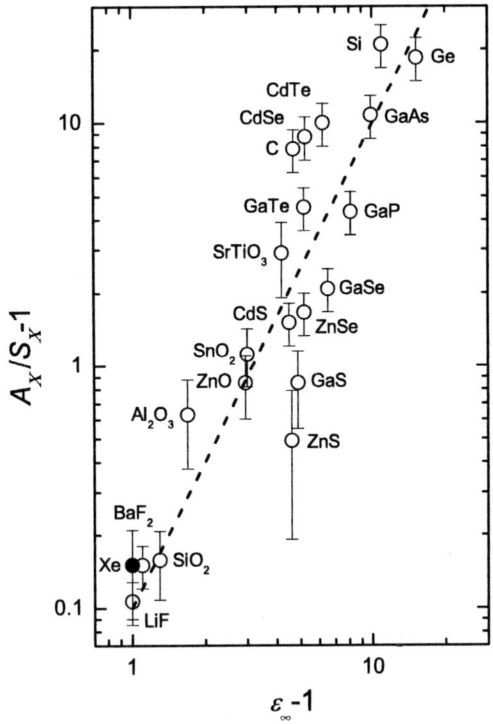

Fig. 1.9. Slope parameters $S_X = d\Phi_{Bn}/dX_m$ as a function of the optical susceptibility $\varepsilon_\infty - 1$ of the semiconductors. The data are the same as in Fig. 1.8 and the dashed line is a linear least-squares fit to the data. From *Mönch* [1986]

product of the slope parameter S_X and the difference $X_m - X_s$ of the metal and the semiconductor electronegativities. The slope parameter is determined by the density of states and the charge decay-length of the MIGS at their branch point. If the metal and the semiconductor have the same electronegativities, then no charge transfer occurs at the interface, and the Fermi level will coincide with the branch point of the MIGS which, for this reason, is also referred to as their charge neutrality level. The zero-charge-transfer barrier height thus equals the energy distance from the MIGS branch point to the edge of the corresponding majority-carrier band. In Fig. 1.10 the *n*-GaAs Schottky barriers shown in Fig. 1.6 as well as the prediction of the MIGS theory are plotted versus the electronegativity difference $X_m - X_s$ rather than the metal work-function. The figure clearly reveals that the MIGS line marks an upper limit to the data. Consequently, *Mönch* [1988a] concluded that the MIGS are the fundamental mechanism of the barrier formation and attributed the lowering of the barrier heights to fabrication-induced defects.

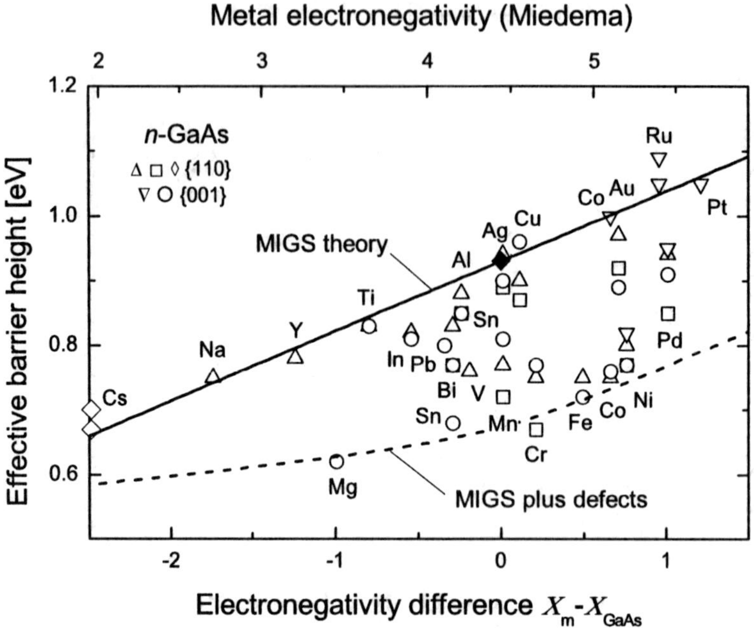

Fig. 1.10. Barrier heights of GaAs Schottky contacts as a function of the difference of the metal and GaAs electronegativities. The data are the same as in Fig. 1.6 plus ◊ data from *Spicer* et al. [1975] and *Grunwald* [1987]. From *Mönch* [1988a]

The dilemma why no barrier heights larger than those predicted by the MIGS theory were observed remained unsolved until *Schmitsdorf* et al. [1995] measured current–voltage characteristics of Ag/*n*-Si(111) contacts. They observed a linear decrease of the effective barrier heights with increasing ideality factors. The ideality factor describes the dependence of the barrier height on the applied voltage. *Schmitsdorf* et al. immediately attributed this behavior to the existence of patches with decreased barrier height and lateral dimensions smaller than the depletion layer width. They concluded that the barrier height of the laterally homogeneous contact is obtained by extrapolation to the ideality factor that is determined by the image-force or *Schottky* effect [1914] only.

As early as in 1982, *Freeouf* et al. studied to what extent patches of lowered barrier height and of lateral dimensions smaller than the depletion layer width change the potential distribution normal to metal–semiconductor interfaces and influence the current transport in Schottky contacts. They found the potential distribution underneath such patches to exhibit a saddle point. Their heights and, therefore, the *effective* barrier heights of patchy Schottky diodes are sensitive to the applied voltage. *Freeouf* et al. simulated the current transport in diodes with one patch only and found the effective barrier height to decrease and the ideality factor to become larger as they reduced the lateral dimensions of the patch. Figure 1.11 displays some of their results which were tabulated only. Unfortunately, *Freeouf*

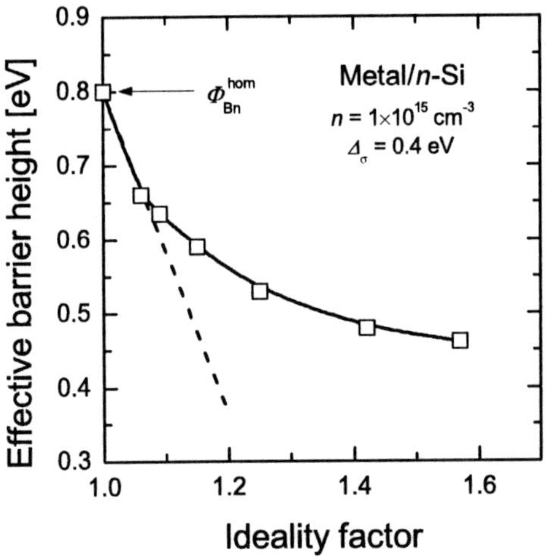

Fig. 1.11. Effective barrier height as a function of the ideality factor obtained from simulated current–voltage characteristics of laterally inhomogeneous metal/*n*-Si contacts. The contacts possess one stripe with a reduced barrier height of 0.4 eV and lateral widths of 0.0313, 0.0625, 0.125, 0.25, 0.5, and 1 μm. The □ data are from *Freeouf* et al. [1982a] and the dashed line is a result of *Tung*'s [1992] analytical solution

et al. omitted to note that the barrier heights of laterally homogeneous contacts may be obtained from such plots. This was left to *Schmitsdorf* et al. [1995], although much earlier *Ohdomari* et al. [1978] had already observed the effective barrier heights of Ir/*n*-Si contacts to become smaller with increasing ideality factor. It is worth mentioning that already *Schottky* [1942] and *Bethe* [1942] realized the most likely existence of patches with reduced barrier height due to the unavoidable fluctuations in the spatial distribution of dopants.

The MIGS are the origin of the *intrinsic* interface dipole at metal–semiconductor interfaces. *Schmitsdorf* et al. [1995] clearly observed *extrinsic* dipoles at intimate, abrupt, clean and laterally homogeneous Schottky diodes. They studied Ag/*n*-Si(111) contacts with both $(1\times1)^i$ and $(7\times7)^i$ interface structures.[6] The reduced barrier height observed with the $(7\times7)^i$ interface structure was qualitatively and quantitatively explained by the electric dipole associated with the stacking fault of the $(7\times7)^i$ reconstruction. The earlier observed differences in barrier heights of type-A and type-B $NiSi_2$/*n*-Si(111) contacts [*Tung* 1994a,b] were also explained by extrinsic interface dipoles. However, the results of theoretical calculations carried out by *Das* et al. [1989] and *Fujitani* and *Asano* [1990] required a very thorough analysis [*Mönch* 2001].

[6] Following a proposal of *Wood*'s [1964], the dimensions of surface and interface unit-meshes are expressed as multiples, m and n, of respective basis vectors in equivalent bulk {*hkl*} planes. Surface structures are then classified as (*hkl*)-m×n and interface structures (*hkl*)-$(m\times n)^i$ are distinguished by a superscript i.

Ballistic electron emission microscopy (BEEM), invented by *Kaiser* and *Bell* [1988], applies scanning tunneling microscopy to spectroscopic investigations of buried semiconductor interfaces with high spatial resolution. *Fowell* et al. [1990] were the first who carried out lateral mapping of Schottky contacts on the nanometer scale. They investigated Au/*n*-CdTe contacts. The local BEEM barrier heights generally exhibited Gaussian distributions and, therefore, demonstrated the existence of patches with reduced and increased barrier heights. The centers of the Gaussian distributions then equal the barrier heights of the laterally homogeneous contacts.

The existence of patches of reduced barrier height with lateral dimensions smaller than the depletion layer width easily explains the observation that the MIGS line in, for example, Fig. 1.10 gives an upper limit of the barrier heights of *real* Schottky contacts. *Ideal* contacts as treated in theory are intimate, abrupt, free of any impurities, and, of course, laterally homogeneous. Hence, *Mönch* [1999] plotted the barrier heights of well-prepared and laterally homogeneous contacts obtained from current-voltage and BEEM measurements versus the difference of the metal and semiconductor electronegativities. He found the Si, Ge, GaAS, GaN, and CdTe as well as the 3*C*-, 6*H*-, and 4*H*-SiC Schottky barrier heights to agree excellently with the predictions of the MIGS-and-electronegativity theory.

1.2 Semiconductor Heterostructures and the IFIGS Concept

The electronic properties of contacts between two semiconductors are determined by the lineup of their band structures across the interface and the existence of space-charge layers. The band gaps of semiconductors generally differ, so that in intimate and abrupt heterostructures the edges of the valence and/or conduction bands will be off-set. At such interfaces, these band-edge discontinuities are established within a few atomic layers, while the regions of band bending depend on the doping profiles on both sides of the contact and will extend over a few Debye lengths. Here, only the band-edge offsets shall be considered.

Anderson [1962] made a first attempt to explain band offsets at semiconductor heterostructures. In the spirit of the *Schottky–Mott* model for metal–semiconductor contacts, he aligned the vacuum levels of the two semiconductors in contact and, consequently, assumed no interface states to exist. Hence, the conduction-band offset

$$\Delta W_c = W_{cil} - W_{cir} = \chi_{sr} - \chi_{sl} \tag{1.5}$$

resulted as the difference of the electron affinities and the valence-band offset

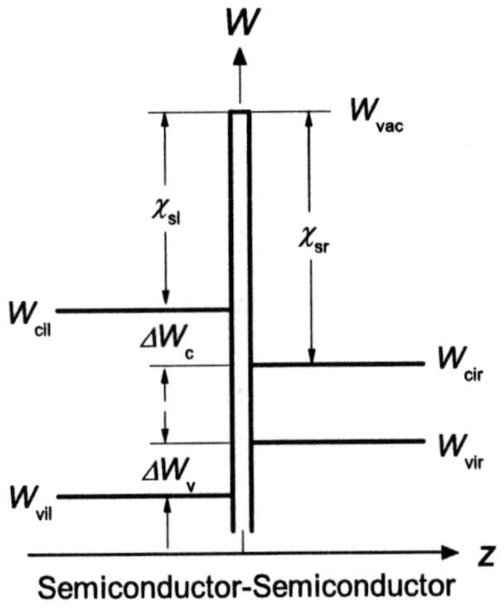

Fig. 1.12. Schematic band diagram of an intimate and abrupt heterostructure of an *n*-type semiconductor with larger band gap and a *p*-type semiconductor with smaller band gap for the case of no interface states

$$\Delta W_v = W_{vir} - W_{vil} = I_l - I_r \tag{1.6}$$

as the difference of the ionization energies of the semiconductors on the right-hand (r) and the left-hand side (l) (see Fig. 1.12). Again, it was correctly argued that electron affinities and ionization energies are surface rather than interface properties. A possible solution was *Harrison*'s [1977] atomic orbital theory. He used an empirical tight-binding approach and calculated the energy positions of the valence-band maxima below the vacuum level or, in other words, the ionization energies without any surface contributions. In 1983, *Krömer* [1985] very thoroughly inspected the experimental data available and identified a reliable data set of seven different heterostructures. He proclaimed excellent agreement between the differences of *Harrison*'s ETB ionization energies and the experimental valence-band offsets of the selected reference systems. However, a big problem came up when *Arnold* et al. [1984] and *Wang* and *Stern* [1985] reported new experimental valence-band offsets of lattice-matched epitaxial $Al_xGa_{1-x}As$–GaAs(100) heterostructures. Their values of 0.55 and 0.45 eV, respectively, for AlAs/GaAs heterostructures were considerably larger than the 0.19 eV reported earlier. However, a theoretical approach of *Tejedor* and *Flores* [1978], which had been overlooked but was taken up by *Tersoff* [1984b], immediately solved the dilemma.

The band-edge offsets in semiconductor heterostructures may be also viewed as a partial overlap of the valence and/or the conduction band of one of the semiconductors with the band gap of the other one. Within the band-edge discontinuities the wave-functions of the corresponding valence or conduction electrons will tail across the interface. As in metal–semiconductor contacts, these tailing states originate from the continuum of the virtual gap states of the semiconductor and constitute the intrinsic semiconductor-induced or, more generally speaking, interface-induced gap states (IFIGS). *Tejedor* and *Flores* then argued that the difference of the branch-point energies of the semiconductors in contact and an intrinsic IFIGS dipole determine the band lineup in semiconductor heterostructures. The IFIGS dipole contribution will again be proportional to the difference of the electronegativities of the two semiconductors forming the heterostructure. The atoms of the elemental and the compound semiconductors are all in the middle of the Periodic Table of the Elements and, hence, the electronegativities of the semiconductors differ on the average by only 5% [*Mönch* 1986]. Therefore, the contributions of the IFIGS dipole to the band-edge offsets in semiconductor heterostructures are small and may be safely neglected.

The IFIGS concept immediately explained the large experimental value of the valence-band discontinuity of AlAs/GaAs heterostructures. Excellent agreement was also found between the predictions of the IFIGS theory and experimental valence-band offsets, not only of lattice-matched binary, ternary, and quaternary III–V alloys but also of metamorphic heterostructures [*Mönch* 2001]. The interface-induced gap states thus explain the origin of the band-structure lineup at semiconductor interfaces of all types, and the IFIGS-and-electronegativity concept correctly predicts the barrier heights of Schottky contacts and the band-edge offsets of semiconductor heterostructures.

2. Depletion Layer

2.1 The Schottky Barrier

Spatial variations of band bending in space-charge layers are described by an electrostatic potential

$$V(z) = \frac{1}{e_0}[W_{cb} - W_c(z)] \ . \qquad (2.1)$$

In describing interfacial space-charge layers, subscripts b and i refer to properties in the bulk and at the interface, i.e., for $z \to \infty$ and $z = 0$, respectively. Relation (2.1) assumes the potential to vanish in the bulk, i.e., for large values of z. The Poisson equation

$$\mathrm{d}^2V/\mathrm{d}z^2 = -\rho(z)/\varepsilon_b\varepsilon_0 \qquad (2.2)$$

relates the potential $V(z)$ to the space charge $\rho(z)$ per unit volume, where ε_b is the static dielectric constant of the semiconductor. For $V(z) < 0$ the bands bend upward and the related space charge has a positive sign, while positive values $V(z) > 0$ mean a downward bending of the bands and, correspondingly, a negative space charge. Such band bending characterizes depletion layers at surfaces and interfaces of semiconductors doped n- and p-type, respectively. The interface band-bending

$$e_0V_i = W_{cb} - W_{ci} \qquad (2.3)$$

uniquely determines both the total space charge Q_{sc} per unit area and the shape $V(z)$ of the potential in space-charge layers. In this section the semiconductors are assumed to be semi-infinite so that a simple one-dimensional model has to be solved. Edge effects are discussed in Sect. 4.1.

The following treatment[1] will consider depletion layers at interfaces of semiconductors non-degenerately doped n-type so that the classical Maxwell–Boltzmann statistics

$$n(z) = N_c \exp\{-[W_c(z) - W_F]/k_BT\} = n_b \exp[eV(z)/k_BT] \qquad (2.4)$$

[1] A more general treatment of space-charge layers at semiconductor surfaces and interfaces may be found in *Mönch* [2001].

applies. The effective density of states and the density of states effective mass in the conduction band are defined as

$$N_c = 2M_c(2\pi m^*_{cd} k_B T / h)^{3/2} \tag{2.5}$$

and

$$m^*_{cd} = (m^*_{c1} m^*_{c2} m^*_{c3})^{1/3} \quad , \tag{2.6}$$

respectively, where m^*_{c1}, m^*_{c2}, and m^*_{c3} are the effective masses along the three principal axes of the ellipsoidal energy surfaces of the M_c equivalent conduction-band minima. The electron density in the bulk is given by

$$n_b = N_c \exp[-(W_{cb} - W_F)/k_B T] = N_c \exp(-W_n / k_B T) \quad , \tag{2.7}$$

provided that the donors are homogeneously distributed.

If the shallow donors are assumed to be completely ionized and no acceptors and deep traps are present, then the space-charge density

$$\rho_n(z) = e_0[N_d - n(z) + p(z)] \tag{2.8}$$

is composed of positively charged static donors (d) as well as mobile electrons and holes. The minority carriers, the holes in the present case of *n*-type doping, can be neglected and the space-charge density may be written as

$$\rho_n(z) \approx e_0 n_b \{1 - \exp[e_0 V(z)/k_B T]\} \quad . \tag{2.9}$$

The exponential in the braces accounts for the mobile electrons in the depletion layer.

It is convenient to introduce the reduced potential

$$v \equiv e_0 V / k_B T \tag{2.10}$$

and the reduced electric-field strength

$$-\mathrm{d}v/\mathrm{d}z \equiv F / L_D \quad , \tag{2.11}$$

where the extrinsic Debye length is defined by

$$L_D^2 = \varepsilon_b \varepsilon_0 k_B T / e_0^2 n_b \quad . \tag{2.12}$$

Figure 2.1 shows the variation of the Debye length as a function of the carrier density for a *typical* semiconductor ($\varepsilon_b = 12.5$) at room temperature. Using the reduced potential and electric-field strength defined above, Poisson's equation (2.2) may be rewritten as

$$\frac{\mathrm{d}^2 v}{\mathrm{d}z^2} = \frac{F}{L_D^2}\frac{\mathrm{d}F}{\mathrm{d}v} = -\frac{1 - \exp[v(z)]}{L_D^2} \quad . \tag{2.13}$$

Integration from the bulk to the interface yields

$$F_i^2 = -2\{v_i - [\exp(v_i) - 1]\} = 2[|v_i| + \exp(-|v_i|) - 1] \quad . \tag{2.14}$$

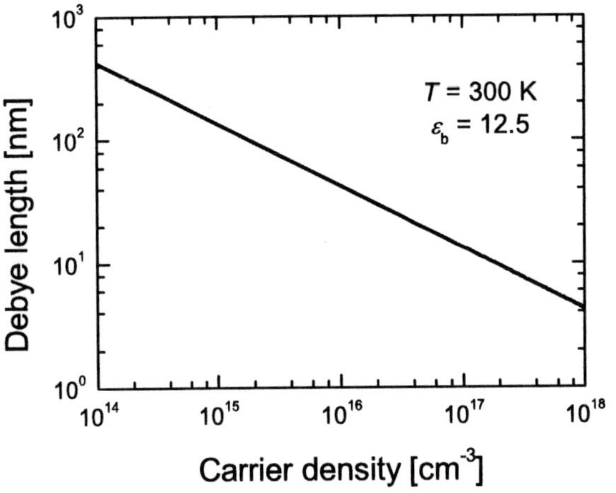

Fig. 2.1. Extrinsic Debye length as a function of carrier density for a typical semiconductor ($\varepsilon_b = 12.5$) at room temperature

Provided that the reduced band bending v_i at the interface is much larger than unity, i.e., if $e_0 V_i >> k_B T$ then the exponential term in (2.14) may be neglected.

Hence, the electric-field strength at the interface is given by

$$E_i = -\left.\frac{\mathrm{d}V}{\mathrm{d}z}\right|_{z=0} = \frac{k_B T}{e_0} F_i \approx -\sqrt{\frac{2N_d}{\varepsilon_b \varepsilon_0}\left(e_0 |V_i| - k_B T\right)} \quad . \tag{2.15}$$

Figure 2.2 displays the interfacial electric-field strength as a function of the interface band-bending for a *typical* semiconductor ($\varepsilon_b = 12.5$) at room temperature and the donor density as a parameter. Gauss's law relates the total charge Q_{sc} per unit interface area with the corresponding interfacial electric-field strength. For depletion layers one obtains

$$Q_{sc} = -\varepsilon_b \varepsilon_0 E_i \approx \sqrt{2\varepsilon_b \varepsilon_0 N_d (e_0 |V_i| - k_B T)} \quad . \tag{2.16}$$

Thus, the interface band-bending $e_0|V_i|$ uniquely determines the total space charge per unit interface area.

The two terms in brackets under the square roots on the right-hand side of the electric-field strength (2.15) and of the space-charge density (2.16) have different origins. The first one is simply obtained if the completely ionized static donors make up the space charge only, i.e., the space-charge density is approximated as

$$\rho_n \approx e_0 N_d = e_0 n_b \quad . \tag{2.17}$$

This is the basic assumption of *Schottky*'s [1942] depletion-layer model. The approximation (2.17) means that not only the minority holes but also the majority electrons are completely neglected within the space-charge layer. The space-charge layer then has a sharp edge toward the bulk and its width measures

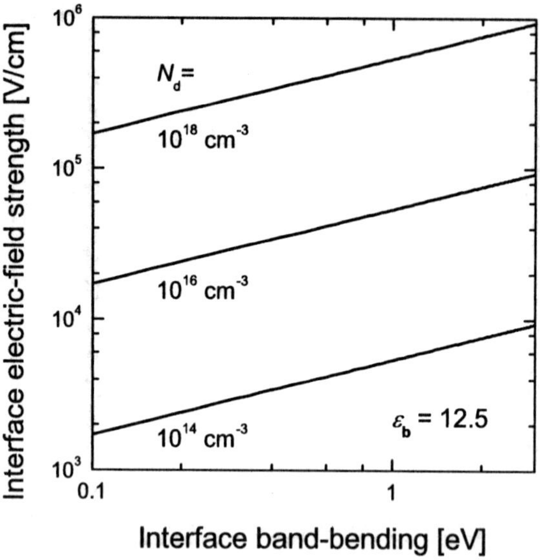

Fig. 2.2. Interface electric-field strength as function of interface band-bending for a typical semiconductor ($\varepsilon_b = 12.5$) with 10^{14}, 10^{16}, and 10^{18} donors per cm^3 at room temperature

$$z_{dep} = \frac{Q_{sc}}{e_0 N_d} = L_D \left(2 \frac{e_0 |V_i|}{k_B T} \right)^{1/2} . \tag{2.18}$$

Applying Gauss's law one obtains the interface electric-field strength

$$E_i = \frac{e_0 N_d}{\varepsilon_b \varepsilon_0} z_{dep} . \tag{2.19}$$

Twice repeated integration of Poisson's equation (2.2) finally yields the shape of the depletion layer

$$W_c(z) - W_{cb} = e_0 |V(z)| = e_0 |V_i| (1 - z / z_{dep})^2 \tag{2.20}$$

in *Schottky*'s approximation of depletion layers. The interface band-bending $e_0|V_i|$ uniquely determines the total space-charge (2.16) per unit area, the depletion-layer width (2.18), and the electric-field strength (2.19), as well as the shape (2.20) of the band bending.

The second term, k_BT, under the square roots of (2.15) and (2.16) accounts for the gradual transition of the electron density from its bulk value n_b to a value negligibly smaller than the donor density N_d within the depletion layer. It is worth mentioning that in this more accurate approach the depletion layer has no well-

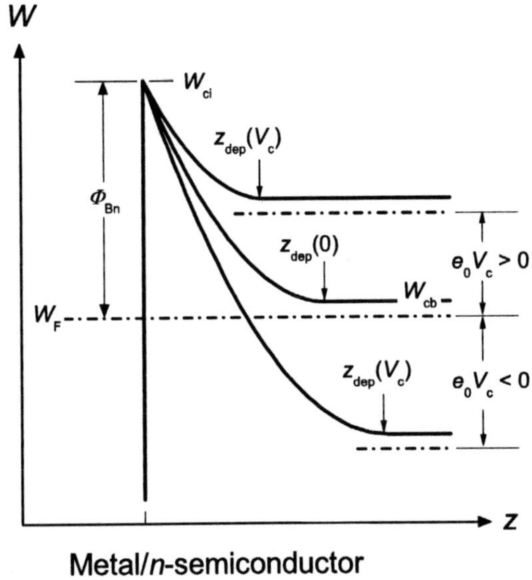

Fig. 2.3. Band bending at the interface of a metal contact and an n-type semiconductor in thermal equilibrium ($V_c = 0$) and with forward ($V_c > 0$) and reverse bias ($V_c < 0$) applied (schematically)

defined width but the band edges rather asymptotically approach their positions in the bulk.

With an external bias applied to Schottky contacts, i.e., in non-equilibrium, the interface band-bending may be written as $e_0(|V_i^0| - V_c)$ where $e_0|V_i^0|$ is the zero-bias or thermal-equilibrium interface band-bending. The externally applied voltage V_a divides up into a voltage drop V_c across the depletion layer of the contact and an IR drop at the series resistance R_s of the diode, i.e., $V_c = V_a - R_sI$. Figure 2.3 shows schematically the conduction-band bending at metal/n-semiconductor interfaces in thermal equilibrium and with an externally applied bias in reverse and in forward direction, respectively.

2.2 Capacitance of Schottky Barriers

The space charge and the width of the depletion layer at a metal–semiconductor contact both vary as a function of the externally applied voltage. The depletion layer thus behaves like a parallel-plate capacitor. Using the approximation (2.17), *Schottky* [1942] obtained the differential capacitance per unit area of depletion layers at metal–semiconductor contacts as

$$C_{dep} = \partial Q_{sc} / \partial V_c = \left[e_o^2 \varepsilon_b \varepsilon_0 N_d / 2e_0(|V_i^0| - V_c)\right]^{1/2} \ . \tag{2.21}$$

Hence, the inverse square $1/C_{dep}^2$ of the depletion-layer capacitance per unit area varies proportionally to the voltage drop across the barrier layer, and the slope

$$\frac{\mathrm{d}(1/C_{dep}^2)}{\mathrm{d}V_c} = -\frac{2}{e_0 \varepsilon_b \varepsilon_0 N_d} \tag{2.22}$$

is determined by the distribution $N_d(z)$ of the donors within the depletion layer.

The current through a Schottky contact biased in the reverse direction is small, so that the *IR* drop due to the series resistance of the diode may be neglected. The voltage drop V_c across the depletion layer then equals the externally applied voltage V_a. Consequently, the extrapolated intercepts on the abscissa of $1/C_{dep}^2$ versus V_a plots give the band bending $e_0 \mid V_i^0 \mid$ at the interface.

A somewhat more precise derivation of the differential capacitance of depletion layers considers the total space charge per unit area (2.16). The extrapolated intercept V_a^{ex} on the abscissa of a $1/C_{dep}^2$ versus V_a plot then equals $(e_0 \mid V_i^0 \mid - k_B T)/e_0$ rather than the interface potential $\mid V_i^0 \mid$ only. The extrapolation is equivalent to flat bands up to the interface, so that the intercept provides the flat-band barrier height

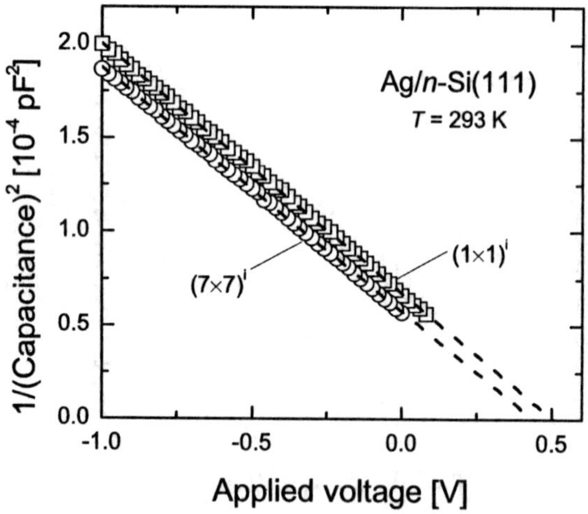

Fig. 2.4. Differential capacitance of an Ag/n-Si(111)-(7×7)^i diode and an Ag/n-Si(111)-(1×1)^i diode, each as a function of the applied bias in the reverse direction. The diodes had an area of 7.3×10^{-3} cm^2. The dashed lines are linear least-squares fits to the data points. From *Schmitsdorf* [1993]

$$\Phi_{Bn}^{fb} = e_0 \,|\, V_i^0 \,|+(W_{cb} - W_F) = e_0 V_a^{ex} + k_B T + (W_{cb} - W_F) \tag{2.23}$$

of the diode.

Schottky [1942] applied his model to both selenium and copper oxide rectifiers and determined the barrier heights as well as the spatial acceptor distributions from their capacitance–voltage (*C/V*) characteristics. The *C/V* data of the Cu/Cu_2O rectifiers [*Schottky* and *Deutschmann* 1929] showed the existence of a *chemical* blocking layer close to the interface which is characterized by a very low density of acceptors. As a more recent example, Fig. 2.4 displays the inverse square $1/C^2$ of the capacitance of two Ag/*n*-Si(111) diodes which differ in their interface structures versus the applied reverse bias [*Schmitsdorf* 1993]. The data points are well fitted by straight lines, i.e., the donor densities are constant in the depletion layers of both diodes. Linear least-squares fits give abscissa intercepts of 0.44 V and 0.51 V for the $(7\times7)^{i}$ diode and the $(1\times1)^{i}$ diode, respectively, and the same donor density $N_d = 1.7\times10^{15}$ cm^{-3} for both diodes. The donors are completely ionized at room temperature and relation (2.7) gives the corresponding energy position $W_{cb} - W_F = 0.24$ eV of the Fermi level below the conduction-band edge in the bulk. Hence, the flat-band barrier heights of the Ag/*n*-Si(111)-$(7\times7)^{i}$ diode and the -$(1\times1)^{i}$ diode measure 0.71 eV and 0.78 eV, respectively.

2.3 Image-Force or Schottky Effect

The presence of a conductor heavily distorts the radially symmetric Coulomb field of a point charge in front of it. The surface of the conductor is an equipotential surface since no current flows parallel to it, i.e., the electric field must be perpendicular to it. This requirement is easily met by a mirror image of the original point charge which is of equal size but has opposite sign. As a consequence, there exists an attractive interaction between the real point charge and its image induced in the conductor which lowers the potential energy of the electron.

Schottky [1914] applied this image-force concept to the thermal electron emission of hot cathodes in vacuum tubes. As a function of the applied voltage in the forward direction the electron current increases to above its saturation value. *Schottky* explained this observation by the lowering of the work function due to the image-force effect and to honor this important contribution of his it is called the *Schottky* effect.

The image-force concept not only applies to metal–vacuum interfaces but also to metal–semiconductor contacts. An electron in the depletion layer of the semiconductor at distance z in front of the metal experiences an image force

$$F_{if}(z) = -\frac{e_0^2}{4\pi\varepsilon_{if}\varepsilon_0}\frac{1}{(2z)^2} \ . \tag{2.24}$$

Consequently, its potential energy is lowered by

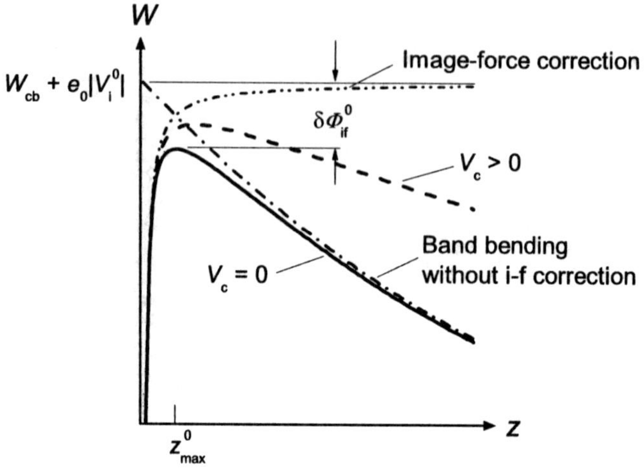

Fig. 2.5. Image-force effect at metal–semiconductor interfaces (schematic)

$$e_0 V_{if}(z) = \int_{\infty}^{z} F_{if} \mathrm{d}z = \frac{e_0^2}{16\pi\varepsilon_{if}\varepsilon_0 z} \tag{2.25}$$

relative to an electron at infinite distance from the interface. The only difference to bare metal surfaces is the insertion of the image-force dielectric constant ε_{if} in the denominator. Its exact value is discussed in a later paragraph of this section.

Including the image-force effect, the total potential energy of electrons in the conduction band of the semiconductor may be written as

$$W_c(z) = W_{cb} + e_0 \,|\, V(z) \,|\, - e_0 V_{if}(z) \quad , \tag{2.26}$$

where W_{cb} is the energy position of the conduction-band bottom in the bulk and $e_0|V(z)|$ accounts for the band bending (2.20) in the depletion layer. Figure 2.5 illustrates this behavior. To a first approximation, the electric-field-strength in the depletion layer may be approximated by its value E_i right at the interface and the total potential energy is then given by

$$W_c(z) \approx W_{ci} - e_0 E_i z - e_0 V_{if}(z) \quad . \tag{2.27}$$

As Fig. 2.5 explains, the potential energy passes through a maximum at a distance z_{max} in front of the interface where the band bending $e_0|E_i|z_{max}$ equals the image-force potential energy $e_0 V_{if}(z_{max})$. Using relations (2.15) and (2.25), one obtains the position of the potential maximum

$$z_{\max} = \frac{1}{4}\sqrt{\frac{e_0}{\pi\varepsilon_{if}\varepsilon_0 E_i}} = \frac{1}{4}\left[\frac{e_0^2 \varepsilon_b}{2\pi^2 \varepsilon_{if}^2 \varepsilon_0 N_d (e_0 \,|\, V_i \,|\, - k_B T)}\right]^{1/4} \tag{2.28}$$

and the image-force lowering

$$\delta\Phi_{if} = 2e_0 V_{if}(z_{\max}) = e_0 \left[\frac{e_0^2 N_d}{8\pi^2 \varepsilon_{if}^2 \varepsilon_b \varepsilon_0^3} (e_0 |V_i| - k_B T) \right]^{1/4} \tag{2.29}$$

of the barrier height at the position z_{max}.

To a first approximation, the position z_{max} of the maximum potential in front of metal–semiconductor interfaces will be estimated by assuming the image-force dielectric constant ε_{if} to equal the bulk value ε_b. Figure 2.6 displays the variation of z_{max} as a function of the potential energy $e_0|V_i| - k_B T$ for a typical semiconductor ($\varepsilon_b = 12.5$) having different donor densities. Obviously, a reasonable value of z_{max} is 5 nm. While the electrons travel from the interface to the potential maximum, they will polarize the semiconductor. At room temperature they have a mean thermal velocity of 1×10^7 cm/s if their effective mass is set equal to the free electron mass, so that it takes a time of approximately 5×10^{-14} s for them to travel a distance of 5 nm. The corresponding frequency corresponds to light in the infrared spectral region. Therefore, it appears to be reasonable to equate the image-force dielectric constant ε_{if} with the optical dielectric constant ε_∞ [*Sze* et al. 1964].

The distance z_{max} of the potential maximum from the metal–semiconductor interface is by far smaller than the width z_{dep} of the depletion layer, which scales with the Debye length L_D; see relations (2.12) and (2.18) and Fig. 2.1. This finding justifies the approximation of the electric-field strength in (2.27) by its value E_i right at the interface.

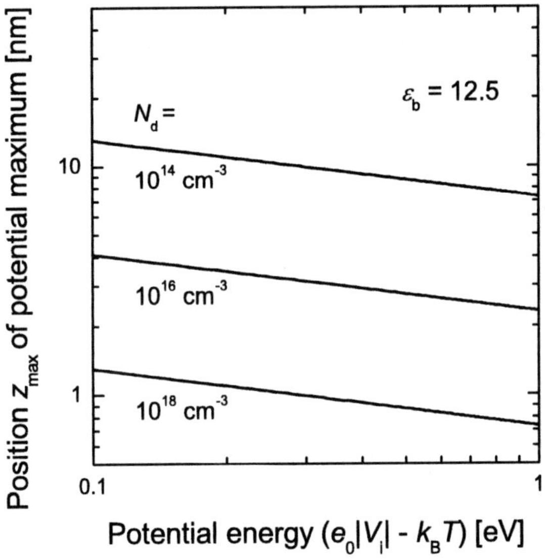

Fig. 2.6. Position of the potential maximum in depletion layers at metal–semiconductor contacts due to the image-force effect as a function of the interface band-bending for donor densities of 10^{14}, 10^{16}, and 10^{18} per cm^3 in the bulk of a typical semiconductor ($\varepsilon_b = 12.5$)

The application of an external bias to a Schottky contact alters the band bending in its depletion layer and, consequently, the image-force lowering depends on the voltage drop V_c across the depletion layer. Figure 2.5 illustrates this behavior and also shows the correlated spatial shift of the potential maximum. By explicitly considering the externally applied voltage, (2.29) may be rewritten as

$$\delta\Phi_{if} = e_0 \left\{ \frac{2e_0^2 N_d}{(4\pi)^2 \varepsilon_\infty^2 \varepsilon_b \varepsilon_0^3} [e_0(|V_i^0| - V_c) - k_B T] \right\}^{1/4} . \tag{2.30}$$

Hence, the image-force effect makes the barrier heights even of *ideal* metal–semiconductor contacts voltage-dependent. Quite generally, the bias dependence of the barrier heights may be expressed as

$$\Phi_{Bn}(V_c) = \Phi_{Bn}^0 + \beta e_0 V_c + \ldots = \Phi_{Bn}^0 + (1 - 1/n) e_0 V_c + \ldots , \tag{2.31}$$

where Φ_{Bn}^0 is the zero-bias barrier height

$$\Phi_{Bn}^0 = \Phi_{Bn}^{hom} - \delta\Phi_{if}^0 \tag{2.32}$$

and Φ_{Bn}^{hom} is the barrier height of the Schottky contact if the zero-bias image-force lowering

$$\delta\Phi_{if}^0 = e_0 \left[\frac{2e_0^2 N_d}{(4\pi)^2 \varepsilon_\infty^2 \varepsilon_b \varepsilon_0^3} e_0(|V_i^0| - k_B T) \right]^{1/4} \tag{2.33}$$

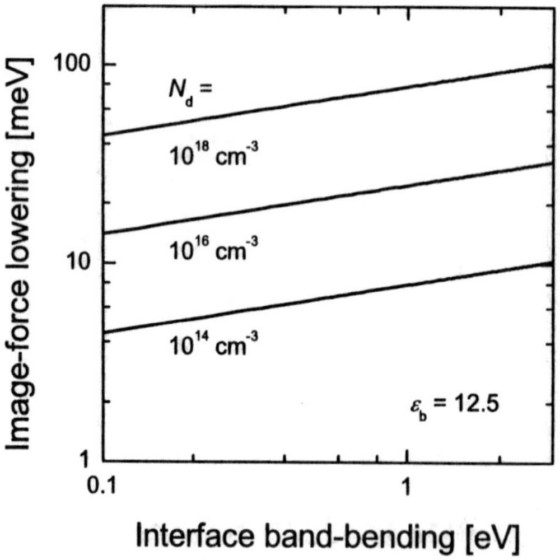

Fig. 2.7. Image-force lowering of the barrier height at metal–semiconductor contacts as a function of interface band-bending for donor densities of 10^{14}, 10^{16}, and 10^{18} per cm^3 in the bulk of a typical semiconductor ($\varepsilon_b = 12.5$)

is neglected. Usually, the zero-bias derivative $\beta = \partial\Phi_{Bn}/\partial e_0 V_c$ is replaced by an ideality factor n, as indicated in relation (2.31). Thus, the ideality factor

$$n_{if} = \left(1 - \frac{\delta\Phi_{if}^0}{4e_0 \,|\, V_i^0 \,|}\right)^{-1} \tag{2.34}$$

describes the voltage-dependence of the barrier heights of *ideal* Schottky contacts due to the image-force effect.

Figures 2.7 and 2.8 display the zero-bias image-force lowering $\delta\Phi_{if}^0$ as a function of the zero-bias interface band-bending $e_0 \,|\, V_i^0 \,|$ with the donor density N_d as a parameter and the ideality factor n_{if} as a function of the reduced image-force lowering $\delta\Phi_{if}^0 / e_0 \,|\, V_i^0 \,|$, respectively. It is worth mentioning that even for large values of the reduced image-force lowering, the value of the ideality factor n_{if} remains smaller than 1.1. As a numerical example, silicon Schottky contacts with a density N_d of 10^{17} donors per cm^3 and a zero-bias interface band-bending $e_0 \,|\, V_i^0 \,|$ of 0.7 eV will be considered. Relation (2.33) gives a lowering $\delta\Phi_{if}^0$ of the barrier height by approximately 40 meV and relation (2.34) an ideality factor $n_{if} = 1.02$ for such contacts.

The image-force lowering of the barrier height was also studied experimentally. *Sze* et al. [1964] investigated Au/n-Si(111) diodes using internal photoemission yield spectroscopy (IPEYS) while *Davies* and *Craighead* [1994] performed ballistic electron emission microscopy (BEEM) studies at Au/n-Si(001) contacts. Figure 2.9 displays their data. The dashed lines are least-squares fits of the relationship

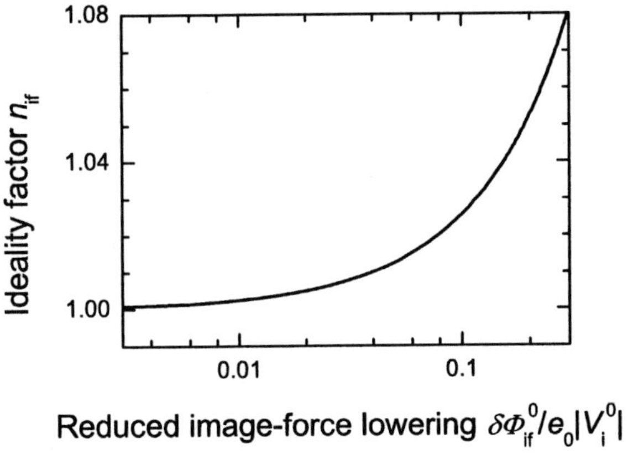

Fig. 2.8. Zero-bias ideality factor as the function of the zero-bias barrier-height lowering divided by the zero-bias interface band-bending

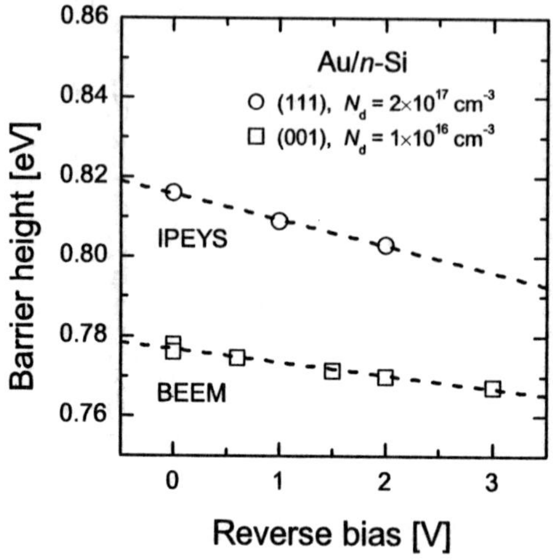

Fig. 2.9. Image-force lowering of the barrier heights of Au/*n*-Si(111) and Au/*n*-Si-(001) contacts observed as a function of reverse bias using internal photoelectron yield spectroscopy (IPEYS) and ballistic electron emission microscopy (BEEM), respectively. Data from *Sze* et al. [1964] and *Davies* and *Craighead* [1994], respectively

(2.30) to the data. The difference of 40 meV between the zero-bias barrier heights is not due to the different orientations of the Si substrates but rather is caused by the experimental techniques used. As is explained in Sect. 3.7, BEEM measures local barrier heights which vary laterally along the contact. IPEYS, on the other hand, is a large-area technique and the extrapolation of the photoelectron yield versus photon energy curves to zero yield provides the barrier height of the laterally homogeneous contact (see Sect. 3.8).

3. Determination of Barrier Heights and Offsets

3.1 Current Transport across Schottky Contacts

As in the bulk of semiconductors the current flow through space-charge regions of metal semiconductor contacts occurs via drift and diffusion. The semiconductors will be assumed to be doped non-degenerately *n*-type so that Maxwell-Boltzmann statistics applies. The temperature shall be chosen such that all donors are ionized, i.e., the semiconductors are in the exhaustion (or saturation) range. For low-level injection and moderate electric fields the current density is then given by[1]

$$J = e_0 n \mu_n E + e_0 D_n \frac{\mathrm{d}n}{\mathrm{d}z} \quad , \tag{3.1}$$

where μ_n and D_n are the electron mobility and diffusion constant, respectively. The electric-field strength $E(z)$ in the barrier region may be written as $e_0E(z) = \mathrm{d}[W_c(z) - W_F]/\mathrm{d}z$, where the conduction-band bottom $W_c(z)$ is referenced to the energy position W_F of the Fermi level in the metal bulk.

In (3.1) the mobility and the diffusion constant are both assumed not to depend on the electric-field strength. This requirement is met provided the electric-field strength is below the value at which the drift velocity of the mobile carriers reaches the saturation regime. This critical value amounts to approximately 10^4 V cm^{-1}. As the data displayed in Fig. 2.2 indicate, the above simple assumptions that the carrier mobilities and the diffusion constants are independent of the electric-field strength may become invalid at the top of the barriers in metal–semiconductor contacts.

Assuming low-level injection, i.e., small deviations from the thermal equilibrium and steady-state condition, the distributions of the mobile carriers in semiconductors are usually described by quasi-equilibrium distributions. Correspondingly, the Fermi level in the Maxwell–Boltzmann distribution (2.4) is replaced by a quasi-Fermi level W_{Fn} or imref, i.e.,

$$n(z) = N_c \exp\{-[W_c(z) - W_{Fn}(z)]/k_B T)\} \quad . \tag{3.2}$$

Using Einstein's relation

---

[1] The theory of current flow in metal–semiconductor contacts was developed by *Schultz* [1954], *Gossick* [1963], and *Crowell* and *Sze* [1966].

$$\frac{D_n}{k_B T} = \frac{\mu_n}{e_0} \quad , \tag{3.3}$$

the current density (3.1) may be written in the form

$$J = \mu_n n(z) \frac{\mathrm{d}(W_{Fn} - W_F)}{\mathrm{d}z} \quad , \tag{3.4}$$

i.e., the driving force of the current transport is the gradient of the quasi-Fermi level. If in (3.4) the definition (3.2) is substituted for the electron density $n(z)$ one obtains

$$J = \mu_n N_c k_B T \exp\{-[W_c(z) - W_F]/k_B T\} \frac{\mathrm{d}(\exp\{[W_{Fn}(z) - W_F]/k_B T\})}{\mathrm{d}z} \quad . \tag{3.5}$$

Integration of (3.5) from the metal–semiconductor interface to the edge of the depletion layer, i.e., between $z = 0$ and $z = z_{dep}$, gives

$$\frac{J}{\mu_n N_c k_B T} \int_0^{z_{dep}} \exp\{[W_c(z) - W_F]/k_B T\} \mathrm{d}z = \\ \exp\{[W_{Fn}(z_{dep}) - W_F]/k_B T\} - \exp\{[W_{Fn}(0) - W_F]/k_B T\} \ . \tag{3.6}$$

The value of $[W_{Fn}(z_{dep}) - W_F]/e_0$, of course, equals the voltage drop V_c across the space-charge layer.

In thermal equilibrium, no net current flow exists at metal–semiconductor contacts. This may be described as a compensation of two currents, one from the semiconductor to the metal and the other in the opposite direction. For the s→m current the metal acts as a collector and the corresponding current density may be written as

$$J_{s \to m} = e_0 n_m(0) v_{coll} \quad , \tag{3.7}$$

where v_{coll} is the so-called collection velocity and

$$n_m(0) = N_c \exp[-(W_{ci} - W_F)/k_B T] = N_c \exp(-\Phi_{Bn}/k_B T) \tag{3.8}$$

is the electron density at the interface in thermal equilibrium. For non-equilibrium conditions, i.e., with an externally applied voltage and a corresponding voltage drop V_c across the depletion layer, definition (3.2) gives the interfacial electron density as [*Crowell* and *Sze* 1966, *Crowell* and *Beguwala* 1971]

$$n_m(V_c) = N_c \exp\{[W_{Fn}(0) - W_F - \Phi_{Bn}]/k_B T\} \quad . \tag{3.9}$$

The usual assumption is that no recombination effects exist in the depletion layers of *p-n* junctions [*Shockley* 1949]. Consequently, the current density is constant through the depletion layer. If this assumption is applied to metal–semiconductor contacts one obtains the boundary condition

$$J(z) = J(0) = J(z_{dep}) = e_0[n_m(V_c) - n_m(0)]v_{coll} \quad . \tag{3.10}$$

The combination of (3.8) through (3.10) yields

$$J = e_0 N_c v_{coll} \exp(-\Phi_{Bn} / k_B T)[\exp\{[W_{Fn}(0) - W_F] / k_B T\} - 1] \quad . \tag{3.11}$$

If the second term on the right-hand side of (3.6) is eliminated by using (3.11) one finally obtains the current–voltage relation

$$J = \frac{e_0 N_c v_{coll}}{1 + v_{coll} / v_{diff}} \exp(-\Phi_{Bn} / k_B T)[\exp(e_0 V_c / k_B T) - 1] \quad , \tag{3.12}$$

where

$$v_{diff} = D_n \left\{ \exp(-\Phi_{Bn} / k_B T) \int_0^{z_{dep}} \exp\{[W_c(z) - W_F] / k_B T\} \mathrm{d}z \right\}^{-1} \tag{3.13}$$

is an effective diffusion velocity. Obviously, the current–voltage characteristics (3.12) has the two limiting cases

$$J_{te} = e_0 N_c v_{coll} \exp(-\Phi_{Bn} / k_B T)[\exp(e_0 V / k_B T) - 1] \quad \text{for } v_{coll}/v_{diff} << 1 \tag{3.14}$$

and

$$J_{diff} = e_0 N_c v_{diff} \exp(-\Phi_{Bn} / k_B T)[\exp(e_0 V / k_B T) - 1] \quad \text{for } v_{coll}/v_{diff} >> 1. \tag{3.15}$$

The differentiating parameters are the collection and the diffusion velocities, v_{coll} and v_{diff}, respectively.

The maximum value the collection velocity v_{coll} can attain equals the average thermal velocity

$$v_{coll}^{\max} \equiv \frac{< v_{th} >}{4} = \frac{1}{4} \left(\frac{8 k_B T}{\pi m_n^*} \right)^{1/2} \tag{3.16}$$

in the z-direction where $<v_{th}>$ is the average thermal velocity of Maxwell–Boltzmann-distributed electrons. This assumption means that the electrons at the interface are in equilibrium with the bulk or, in other words, the quasi-Fermi level is constant throughout the depletion layer. Inserting (3.16) into (3.14) then reduces the current–voltage characteristics to

$$J_{te} = A_R^* T^2 \exp(-\Phi_{Bn} / k_B T)[\exp(e_0 V_c / k_B T) - 1] \quad , \tag{3.17}$$

with the effective Richardson constant

$$A_R^* = \frac{4 \pi e_0 k_B^2 m_n^*}{h^3} = A_R \frac{m_n^*}{m_0} \quad , \tag{3.18}$$

where $A_R = 4\pi e_0 k_B^2 m_0 / h^3 = 120$ A cm^{-2} K^{-2} is the Richardson constant for thermionic emission of nearly free electrons into vacuum. The current–voltage characteristics (3.18) thus corresponds to thermionic emission over the barrier. This approach was first treated by *Bethe* [1942].

The effective diffusion velocity v_{diff}, on the other hand, is obtained by inserting the shape of the barrier (2.20) and the depletion-layer width (2.18) in the integral on the right-hand side of (3.13). One then gets

$$\int_0^{z_{dep}} \exp[(W_c(z) - W_F)/k_B T]dz =$$

$$= \sqrt{2}L_D \exp(\Phi_{Bn}/k_B T)\exp\{-eV_i/k_B T\} \int_0^{\sqrt{e_0 V_i/k_B T}} \exp(t^2)dt \quad (3.19)$$

$$= \sqrt{2}L_D \exp(\Phi_{Bn}/k_B T)D(y) \ ,$$

where $D(y)$ is the Dawson function

$$D(y) = \exp(-y^2)\int_0^y \exp(t^2)dt \quad (3.20)$$

with

$$y^2 = e_0 V_i/k_B T = (\Phi_{Bn} - W_n - e_0 V_c)/k_B T \ .$$

Hence, the effective diffusion velocity (3.13) may be rewritten as

$$v_{diff} = \frac{D_n}{\sqrt{2}L_D D(y)} \ . \quad (3.21)$$

Figure 3.1 shows that the Dawson function[2] is well approximated by

$$D(y) \approx 1/2y = 1/2\sqrt{e_0 V_i/k_B T} \quad (3.22)$$

for $y^2 < 4$, i.e., for $e_0 V_i < 4k_B T$. This approximation is equivalent to the assumption that the electric-field strength is constant throughout the depletion layer and equals its maximum value (2.19) right at the interface. The diffusion velocity then results as

$$v_{diff} \approx \frac{e_0 D_n}{k_B T} E_i = \mu_n E_i \quad (3.23)$$

so that the current–voltage relation (3.15) reduces to

$$J_{diff} = e_0 N_c \mu_n E_i \exp(-\Phi_{Bn}/k_B T)[\exp(e_0 V_c/k_B T) - 1] \ . \quad (3.24)$$

This current–voltage characteristics was first obtained by *Schottky* and *Spenke* [1939] in their diffusion theory of the current transport in metal–semiconductor contacts.

The current transport in metal–semiconductors is dominated by thermionic emission over the barrier if the condition $v_{coll}/v_{diff} << 1$ is satisfied. This condition

---

[2] Numerical tables of the Dawson function may be found in *Karpov* [1965].

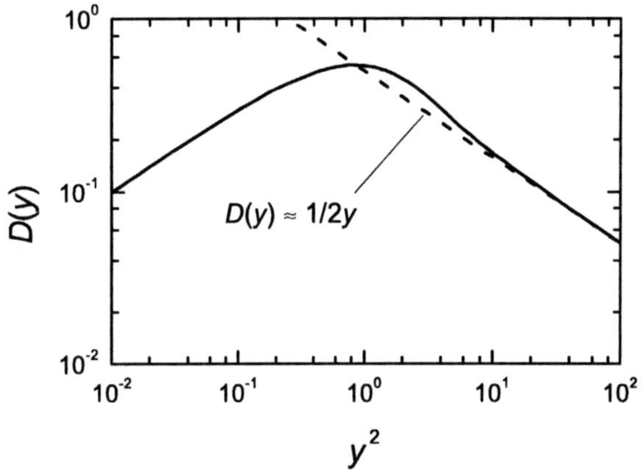

Fig. 3.1. Dawson function $D(y)$ as a function of argument y. The approximation $D(y) \approx y/2$ (dashed line) fits to an accuracy of better than 1% for $y > 10$

may be reformulated using relations (3.16) and (3.21). In the relaxation-time approximation the electron mobility may be written as

$$\mu_n = \frac{e_0 \overline{\tau_n}}{m_n^*} \quad . \tag{3.25}$$

The energy-averaged relaxation time $\overline{\tau_n} = \langle v^2 \tau_n(v) \rangle / \langle v^2 \rangle$ and the mean free path $\overline{\lambda_n}$ are related by the definition

$$\overline{\tau_n} = \overline{\lambda_n} / \langle v_{th} \rangle \quad . \tag{3.26}$$

The condition $v_{coll}/v_{diff} \ll 1$ then reduces to

$$e_0 E_i \overline{\lambda_n} \gg \frac{2}{\pi} k_B T \quad , \tag{3.27}$$

i.e., the thermionic-emission theory applies if at the top of the depletion layer the potential energy changes by more than $k_B T$ within a distance of one mean free path. This criterion was first put forward by *Bethe* [1942]. He was also the first to propose that the current flow in metal–semiconductor contacts occurs via thermionic emission over the top of the barrier.

Figure 3.2 illustrates criterion (3.27) for n-type silicon. It shows the decrease of the barrier from its maximum value $e_0 E_i$ right at the interface within one mean free path $\overline{\lambda_n}$ of the conduction electrons as a function of the height $\Phi_{Bn} - e_0 V_c$ of the barrier. Criterion (3.27) is certainly satisfied for donor densities above 10^{16} cm^{-3} and $\Phi_{Bn} - e_0 V_c > 0.4$ eV, i.e., under these conditions the current transport in silicon Schottky contacts definitely occurs via thermionic emission.

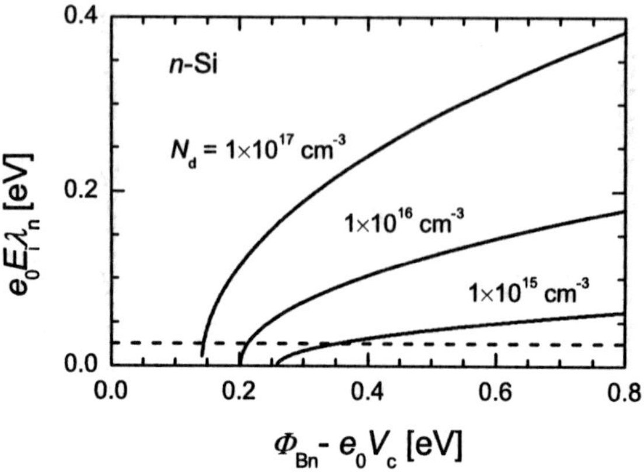

Fig. 3.2. Illustration of criterion (3.27) for *n*-silicon Schottky contacts with donor densities of 10^{15}, 10^{16}, and 10^{17} cm^{-3}

The physical meaning of *Bethe*'s criterion may be explained by considering the conditions within one mean free path in front of the interfacial potential maximum. At this distance, the electrons experience their last collisions before they overcome the interfacial potential maximum and reach the metal. However, only those electrons will pass the barrier which have kinetic energies in excess of the potential rise within one mean free path. All other electrons will be back-reflected into the semiconductor. Directly after the collisions the electrons obey a Maxwell–Boltzmann distribution, i.e., for motion normal to the interface the portion of their average kinetic energy amounts to $k_BT/2$. If the interfacial potential barrier rises by more than a few k_BT within one mean free path in front of its maximum, only very few electrons will reach the metal and most of them will be back-reflected toward the semiconductor bulk. Thus, neither the density nor the energy distribution of the electrons close to the interface will significantly differ from the equilibrium distribution in the semiconductor bulk. These are the necessary conditions for applying thermionic-emission theory to current transport in metal–semiconductor contacts.

3.2 Effective Barrier Height of Ideal Schottky Contacts

The image-force effect causes the barrier heights even of ideal Schottky contacts to vary as a function of the external bias. This is discussed in Sect. 2.3. However, the derivation of the general current–voltage or *I*/*V* relationship (3.12) for emission over the barrier assumed the barrier height not to depend on the applied voltage. Hence, the voltage-dependent barrier height

$$\Phi_{Bn}(V_c) = \Phi_{Bn}^0 + (1 - 1/n_{if})e_0V_c = \Phi_{Bn}^{hom} - \delta\Phi_{if}^0 + (1 - 1/n_{if})e_0V_c \tag{3.28}$$

has to be substituted for Φ_{Bn}. In (3.28) Φ^0_{Bn} is the zero-bias barrier height. Provided thermal emission is dominant the corresponding *I/V* characteristics (3.17) may then be rewritten as

$$J_{te} = A_R^* T^2 \exp[-(\Phi_{Bn}^{hom} - \delta\Phi_{if}^0)/k_B T]\exp(e_0 V_c / n_{if} k_B T)\times \\ [1-\exp(-e_0 V_c / k_B T)] \ , \tag{3.29}$$

where $\delta\Phi^0_{if}$ is the zero-bias image-force lowering (2.33) of the barrier height Φ^{hom}_{Bn} and n_{if} is the ideality factor (2.34) which is determined by the image-force effect only.

3.3 Effective Richardson Constant

For Schottky contacts, the Richardson constant A_R derived for thermal emission into vacuum is modified in several ways. First of all, an appropriate effective mass $m^*_{n,p}$ of the respective majority carriers is substituted for the mass m_0 of the free electrons, see (3.18). Direct semiconductors are the easiest case to discuss. Their lowest conduction bands are isotropic and have parabolic shape with the minimum at the Γ point in the middle of the Brillouin zone. The effective mass to be considered in the effective Richardson constant (3.18) then is the usual isotropic effective mass, i.e.,

$$m_n^* = m_e^* \ . \tag{3.30}$$

For holes, both the light and the heavy holes have to be considered and one obtains

$$m_p^* = m_{hh}^* + m_{lh}^* \ . \tag{3.31}$$

Indirect semiconductors have several conduction-band valleys. The corresponding ellipsoidal constant-energy surfaces are characterized by the components m^*_{ex}, m^*_{ey}, and m^*_{ez} of the effective-mass tensor. *Crowell* [1965] calculated the appropriate effective mass for a single minimum as

$$m_{n1}^* = (l_1 m_{ey}^* m_{ez}^* + l_2 m_{ez}^* m_{ex}^* + l_3 m_{ex}^* m_{ey}^*)^{1/2} \ , \tag{3.32}$$

where l_1, l_2, and l_3 are the direction cosines of the normal to the interface relative to the principal axes of the ellipsoid. For semiconductors with six conduction-band minima in the *X* direction – as in, for example, Si and GaP – one obtains

$$m_n^* = 2m_{et}^* + 4(m_{el}^* m_{et}^*)^{1/2} \quad \text{for } \{100\}\text{-oriented interfaces} \tag{3.33}$$

Table 3.1 Effecive Richardson constants of differently oriented interfaces in n-Si and n-Ge Schottky contacts

Semiconductor	A_R^* / A_R	
	{111}	{100}
n-Si	2.15	2.05
n-Ge	1.07	1.19

and

$$m_n^* = 6\left(\frac{m_{et}^{*2} + 2m_{el}^* m_{et}^*}{3}\right)^{1/2} \quad \text{for } \{111\}\text{-oriented interfaces.} \tag{3.34}$$

For semiconductors with four conduction-band minima in the L direction – as in, for example, Ge – the appropriate values of the effective electron-masses are

$$m_n^* = 4\left(\frac{m_{et}^{*2} + 2m_{el}^* m_{et}^*}{3}\right)^{1/2} \quad \text{for } \{100\}\text{-oriented interfaces} \tag{3.35}$$

and

$$m_n^* = m_{et}^* + (m_{et}^{*2} + 8m_{el}^* m_{et}^*)^{1/2} \quad \text{for } \{111\}\text{-oriented interfaces.} \tag{3.36}$$

As prototypical examples, Table 3.1 shows the Richardson constants of differently oriented n-Si and n-Ge Schottky contacts. For both n-Si and n-Ge contacts, the effective Richardson constants have almost the same values irrespective of the interface orientation. Table 3.2 summarizes the values of the effective Richardson constant for most of the semiconductors studied.

The image force effect places the maximum of the potential barrier at metal–semiconductor interfaces approximately a few nanometers into the semiconductor. Therefore, electrons that crossed the barrier may absorb and excite optical phonons and, consequently, be back-scattered over the barrier into the semiconductor. This effect will reduce the net current over the barrier. Furthermore, two quantum-mechanical effects have to be considered in addition. First, electrons that have enough kinetic energy to pass the barrier will experience quantum-mechanical reflection. Second, close to the potential maximum quantum-mechanical tunneling through the barrier will occur. *Crowell* and *Sze* [1966] investigated these processes for Ge, Si, and GaAs Schottky contacts. They characterized the phonon-scattering and the quantum-mechanical corrections by two factors, f_p and f_q, respectively. They replaced the Richardson constant by

$$A_R^{**} = f_p f_q A_R^* \tag{3.37}$$

and estimated the product $f_p f_q$ to have a value of approximately 0.5. Their data are shown in Table 3.2. *Andrews* and *Lepselter* [1970] re-analyzed these effects and

Table 3.2. Effective and corrected effective Richardson constants, A_R^* and A_R^{**}, respectively, in A cm^{-2} K^{-2}.

Semiconductor	*n*-type		*p*-type	
	A_R^*	A_R^{**}	A_R^*	A_R^{**}
Diamond			84	
Si	258	96[a] 112[b]	79	32[b]
Ge	138	50[a]	43	
3*C*-SiC				
4*H*-SiC	146			
6*H*-SiC	194			
AlN	57.6			
AlAs	94		95	
2*H*-GaN	26.4		71	
GaP	151		90	
GaAs	8.2	4.4[a]	74	
GaSb	5.1		84	
InP	9.4		75	
InAs	2.8		51	
InSb	1.7		52	
ZnS	48			
ZnSe	19		72	
ZnTe			72	
CdS	25			
CdSe	16			
CdTe	13			
InSe	29			

[a]: *Crowell* and *Sze* [1966].
[b]: *Andrews* and *Lepselter* [1970]

obtained slightly larger values of the corrected effective Richardson constants A_R^{**} of Si than previously reported by *Crowell* and *Sze* [1966]. The A_R^{**} values calculated by *Andrews* and *Lepselter* are also displayed in Table. 3.2.

The correction term $f_p f_q$ decreases the effective Richardson constant A_R^* by a factor of approximately 2. However, this reduction is of no great significance when experimental barrier heights are evaluated from measured *I*/*V* curves. At room temperature, the barrier height becomes larger by approximately 17 meV, i.e., by less than $k_B T$ if the estimated correction term is not considered. For typical barrier heights of 0.7 eV, this difference corresponds to an error of only 2.5%.

3.4 *I/V* Characteristics of Real Schottky Contacts

A numerical estimate of *Bethe*'s criterion (3.27) showed that thermionic-emission theory should describe the current transport in *n*-Si Schottky contacts provided the donor density is larger than 10^{16} per cm^3 and interface band-bending exceeds 0.4 eV. Figure 3.3 displays the current–voltage characteristics of two Ag/*n*-Si(111) diodes with $(7\times7)^i$ and $(1\times1)^i$ interface structure[3], respectively [*Schmitsdorf* et al. 1995].

Both types of Ag/*n*-Si(111) diode contacts were fabricated by evaporation of silver at room temperature onto clean Si(111)-7×7 surfaces Figure 3.4 shows the dimer-adatom-stacking fault (DAS) model of the famous Si(111)-7×7 reconstruction [*Takayanagi* et al. 1985a,b]. Each 7×7 unit mesh contains

- a stacking fault in one of its triangular subunits,
- a corner hole,
- 9 dimers along the boundary of its faulted, triangular subunit,
- 12 adatoms in 2×2-like arrangement and, as a consequence of this,
- 6 restatoms between the adatoms in the layer below them.

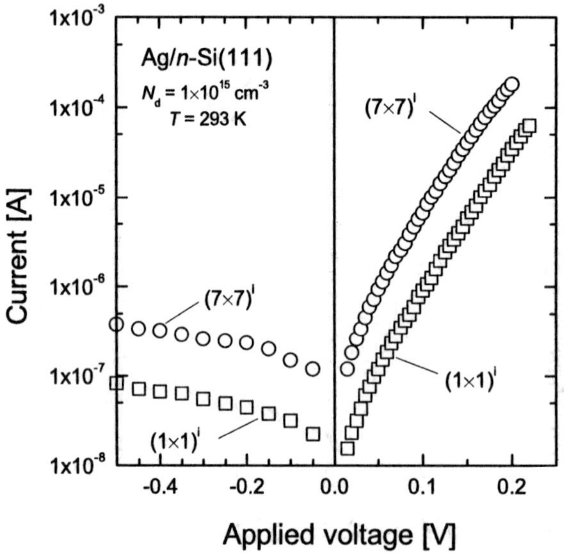

Fig. 3.3. Current–voltage characteristics of forward- and reverse-biased Ag/*n*-Si(111)-$(7\times7)^i$ and Ag/*n*-Si(111)-$(1\times1)^i$ Schottky diodes. The diode areas are 6.3×10^{-3} and 6.4×10^{-3} cm^2, respectively. From *Schmitsdorf* [1993]

[3] Following a proposal of *Wood*'s [1964], the dimensions of surface and interface unit-meshes are expressed as multiples, m and n, of respective basis vectors in equivalent bulk {*hkl*} planes. Surface structures are then classified as (*hkl*)-m×n and interface structures (*hkl*)-$(m\times n)^i$ are distinguished by a superscript i.

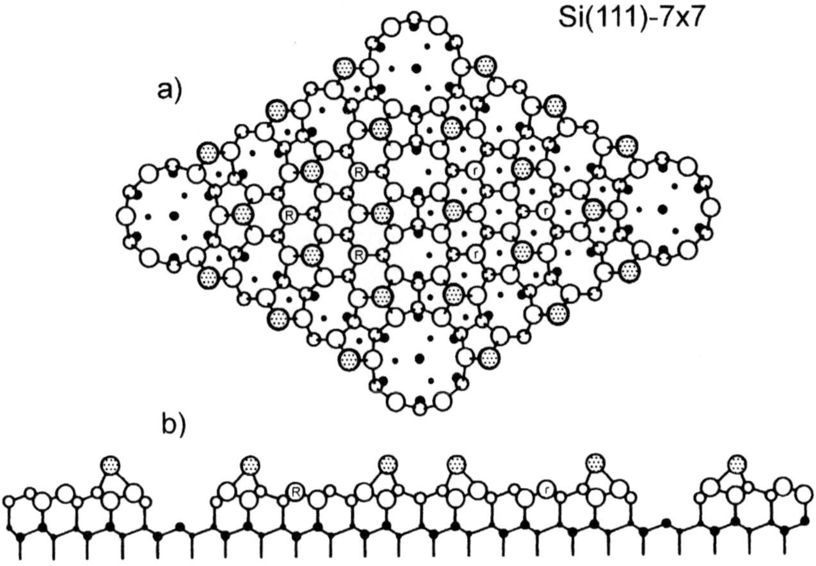

Fig. 3.4 Atom arrangement on clean Si(111)-7×7 surfaces according to the dimer-adatom-stacking fault (DAS) model; a) top view, b) side view. Adatoms are shaded; restatoms in the faulted and non-faulted triangles of the 7×7 unit mesh are marked R and r, respectively. After *Takayanagi* et al. [1985b]

Clean and 7×7-reconstructed Si(111) surfaces may be obtained by thermal desorption of hydrogen from Si(111):H-1×1 surfaces prepared by wet chemical etching in buffered HF solutions. Using grazing incidence X-ray diffraction (GIXD) *Hong* et al. [1992] demonstrated that under thick Ag films the stacking faults and, as a consequence of this, the associated dimers and corner holes of the 7×7 reconstruction persist, while the Si adatoms are missing. The $(7\times7)^{\mathrm{i}}$ interface reconstruction of Ag/Si(111) contacts is unstable against thermal treatments at temperatures above 250°C and converts to a $(1\times1)^{\mathrm{i}}$-unreconstructed structure [*Hong* et al. 1992].

The thermionic-emission current–voltage relation (3.29) predicts a linear dependence of $\ln[J_{te}/(1-\exp(-e_0V_c/k_BT)]$ as a function of the voltage drop V_c across the depletion layer [*Missous* and *Rhoderick* 1986]. The slope parameter and the intercept should then equal $n_{if}k_BT$ and $A_R^*T^2\exp(-\Phi_{Bn}^0/k_BT)$, respectively. Figure 3.5 displays the corresponding plots of the *I/V* data shown in Fig. 3.3. The dashed lines are linear least-squares fits to the data points. The linear regression coefficients $r_1 = 0.99996$ demonstrate the very good quality of the *I/V* characteristics. The same good agreement between thermionic-emission theory and experimental data was obtained earlier with Al/*n*-GaAs diodes [*Missous* and *Rhoderick*

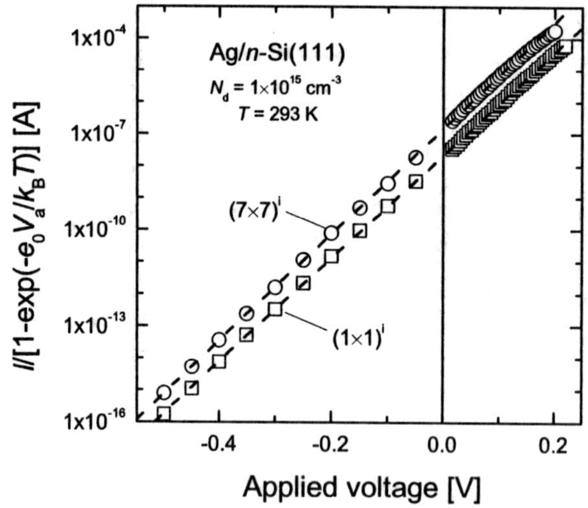

Fig. 3.5. Logarithmic plot of $I/[1-\exp(-e_0V_a/k_BT)]$ as a function of the applied voltage V_a for a Ag/n-Si(111)-(1×1)^i and a Ag/n-Si(111)-(7×7)^i Schottky diode. Data are the same as in Fig. 3.3

1986]. The slight deviations from the straight lines for the largest voltages applied in the forward direction are due to the series resistance of the diodes. Hence, the error is small when the applied voltage is assumed to drop completely across the depletion layer.

The slope parameters and the intercept values of the linear least-squares fits to the experimental data plotted in Fig. 3.5 yield ideality factors of 1.08 ± 0.001 and 1.05 ± 0.001 and zero-bias barrier heights of 0.74 and 0.69 eV for the diode with (1×1)^i and (7×7)^i interface structure, respectively. The ideality factors of the diodes are definitely larger than the image-force value n_{if} = 1.01 obtained from relation (2.34). The histogram on the left-hand side of Fig. 3.6 shows the ideality factors of other Ag/n-Si(111)-(7×7)^i contacts, which were fabricated under *experimentally identical conditions*, to vary up to 1.2 [*Schmitsdorf* et al. 1995]. This finding indicates that the zero-bias barrier heights of *real* diodes depend more strongly on the applied bias than predicted by the image-force effect only. Not only the ideality factors but also the zero-bias barrier heights differ from one diode to the next even if they were fabricated under *experimentally identical conditions*. The histogram on the right-hand side of Fig. 3.6 displays the zero-bias barrier heights of the same set of Ag/n-Si(111)-(7×7)^i diodes. Furthermore, there exists a pronounced correlation between the zero-bias barrier heights and the ideality factors of real Schottky diodes. This behavior is discussed in Sect. 3.5.

Since the ideality factors and the zero-bias barrier heights of Schottky diodes *fabricated under experimentally identical conditions* vary from one specimen to the next the I/V relation (3.29) derived for *ideal* contacts is now replaced by the *standard* relation

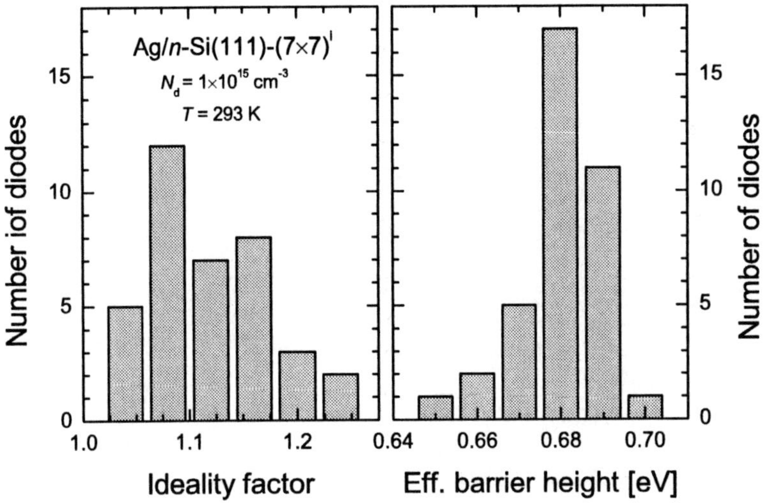

Fig. 3.6. Histograms of ideality factors and effective barrier heights determined from the *I/V* characteristics of identically fabricated Ag/*n*-Si(111)-(7×7)^{i} diodes at room temperature. Data from *Schmitsdorf* et al. [1995]

$$\begin{aligned} J_{te}^{stan} &= A_R^* T^2 \exp[-\Phi_{Bn}^{eff} / k_B T] \exp(e_0 V_c / n k_B T)[1 - \exp(-e_0 V_c / k_B T)] \\ &= J_0^{stan} \exp(e_0 V_c / n k_B T)[1 - \exp(-e_0 V_c / k_B T)] \end{aligned} \tag{3.38}$$

for the thermionic-emission *I/V* characteristics of *real* Schottky diodes, where

$$J_0^{stan} = A_R^* T^2 \exp[-\Phi_{Bn}^{eff} / k_B T] \tag{3.39}$$

is the saturation current. The transport properties of real Schottky contacts are thus characterized by two *phenomenologically* defined parameters: their *effective* barrier heights Φ_{Bn}^{eff} and ideality factors n.

The quality of the *I/V* characteristics of Schottky diodes strongly depends on details of the fabrication process. Clean Si(111)-7×7 surfaces, to name an example, may be obtained by decomposition of an native oxide layer or by thermal desorption of hydrogen from Si(111):H-1×1 surfaces prepared by HF-dips, both in ultrahigh vacuum (UHV). Figure 3.7a displays the *I/V* characteristics of two Schottky diodes made by evaporation of silver in UHV onto clean Si(111)-7×7 surfaces processed by the two methods mentioned. Although low-energy diffraction (LEED) shows sharp 7×7 spots in both cases, the *I/V* characteristics differ considerably. Only the diode fabricated on the Si(111)-7×7 surface prepared by thermal H-desorption from a Si(111):H-1×1 surface may be described by the standard thermionic-emission *I/V* relation (3.38), as Fig. 3.7b demonstrates. This means that fits to sections of such strongly distorted *I/V* curves of Schottky contacts are highly questionable and should be avoided.

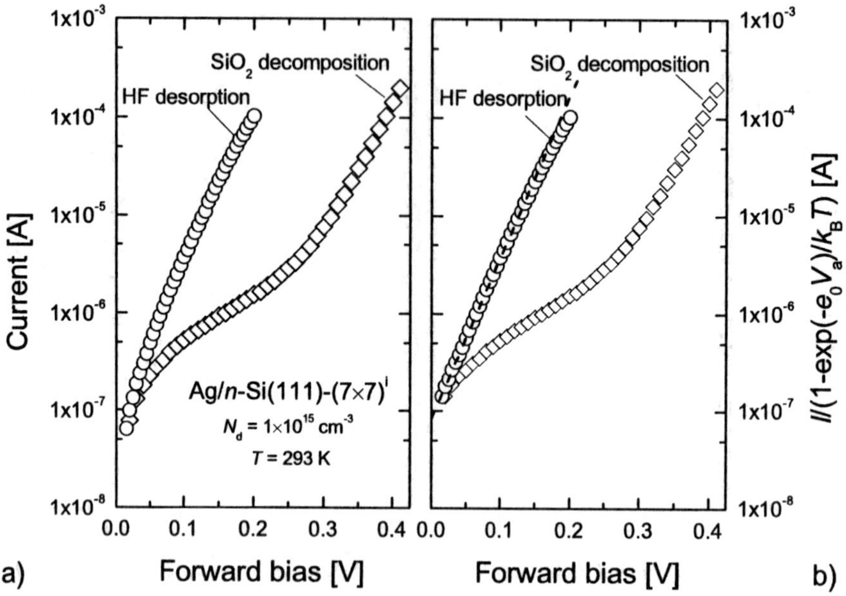

Fig. 3.7. Current–voltage characteristics of differently prepared Ag/*n*-Si(111)-(7×7)^i Schottky diodes. The clean Si(111)-7×7 surfaces prior to Ag evaporation were obtained by H-desorption from Si(111)-(111):H surfaces (○ data) and by SiO_2 decomposition on etched surfaces (◇ data). The dashed line is a linear least-squares fit to the corresponding data. Data from *Schmitsdorf* et al. [1995]

Further support that thermal-emission theory describes the current transport in Schottky diodes comes from the energy position of the quasi-Fermi level in the depletion layer [*Rhoderick* 1972]. The energy difference $W_{Fn}(0) - W_{Fn}(z_{dep})$ of the quasi-Fermi level at the interface and at the edge of the depletion layer may be obtained from relation (3.6) and corresponding experimental data. The expression on the left-hand side of (3.6) may be rewritten as

$$\begin{aligned}\exp\{[W_{Fn}(z_{dep}) - W_F]/k_BT\} - \exp\{[W_{Fn}(0) - W_F]/k_BT \\ = \exp(e_0V_c/k_BT)[1 - \exp\{[W_{Fn}(0) - W_{Fn}(z_{dep})]/k_BT\}] \; .\end{aligned} \tag{3.40}$$

If the shape (2.20) of the barrier is inserted in the integral on the left-hand side of (3.6) and the integration is performed, one then obtains

$$\begin{aligned}\exp\{[W_{Fn}(0) - W_{Fn}(z_{dep})]/k_BT\} \\ = 1 - \frac{J}{e_0\mu_n N_c}\sqrt{\frac{\varepsilon_b\varepsilon_0}{2N_d e_0 V_i}}\exp\left(\frac{\Phi_{Bn}^{eff}}{k_BT}\right)\exp\left(-\frac{e_0V_c}{k_BT}\right) ,\end{aligned} \tag{3.41}$$

where the approximation (3.22) of the Dawson function is also considered. As an example, the *I/V* characteristics of the Ag/*n*-Si(111)-(1×1)^i diode which is dis played in Fig. 3.3 will be considered. The maximum current density amounts to 55 A/cm^2 at a forward bias of 0.2 V. Taking the effective barrier height $\Phi_{Bn}^{eff} = 0.74$ eV obtained for this diode and the mobility $\mu_n = 1500$ cm^2 V^{-1} s^{-1}, relation (3.41) yields $W_{Fn}(0) - W_{Fn}(z_{dep}) = 0.005$ eV, i.e., the quasi-Fermi level is virtually flat throughout the depletion layer in forward-biased Ag/*n*-Si(111)-(1×1)^i contacts. *Rhoderick* [1972] arrived at the same conclusion for Au/*n*-GaAs, Au/*n*-Si, and Au/*n*-CdS Schottky diodes. These findings thus justify the application of thermionic-emission theory to describe the current transport in Schottky diodes of common semiconductors.

3.5 Barrier Heights of Real Schottky Contacts 1: *I/V* Characteristics

The effective barrier heights and the ideality factors differ from one diode to the next, even if they were fabricated under *experimentally identical conditions*. As an example, Fig. 3.6 displays histograms of the ideality factors and effective barrier heights that were determined from the *I/V* characteristics of Ag/*n*-Si(111)-(7×7)^i diodes [*Schmitsdorf* et al. 1995]. The effective barrier heights of the Ag/*n*-Si(111)-(7×7)^i diodes vary between 0.65 and 0.695 eV while the ideality factors span the range from 1.03 to 1.2 and are thus much larger than $n_{if} = 1.01$, the value controlled by the image-force effect. This finding indicates that the potential barriers at *real* metal–semiconductor interfaces depend much more strongly on the applied voltage than predicted by the image-force effect for *ideal* contacts.

However, the ideality factors n and the effective barrier heights Φ_B^{eff} do not vary at random. Figure 3.8 displays the same data shown in Fig. 3.5 and additional results obtained with Ag/*n*-Si(111)-(1×1)^i diodes. The latter (1×1)^i-unreconstruc ted interfaces were obtained by annealing of Ag/*n*-Si(111)-(7×7)^i contacts above 250°C. Figure 3.8 reveals a pronounced correlation between effective barrier heights and ideality factors in that the effective barrier heights become smaller with increasing ideality factors. The dashed lines are linear least-squares fits to the data. The dependence of the effective barrier heights on the ideality factors may thus be written as

$$\Phi_B^{eff} = \Phi_B^{nif} - \varphi_p(n - n_{if}) \ , \tag{3.42}$$

where Φ_B^{nif} is the barrier height at the ideality factor n_{if} which is determined by the image-force-effect only. This linear correlation is generally observed with real metal–semiconductor contacts up to ideality factors $n \approx 1.4$ while for larger n values Φ_B^{eff} versus n curves tend to level off. Already *Ohdomari* et al. [1978], *Chin* et al. [1989], and *Werner* et al. [1993] noted the correlation between effective bar-

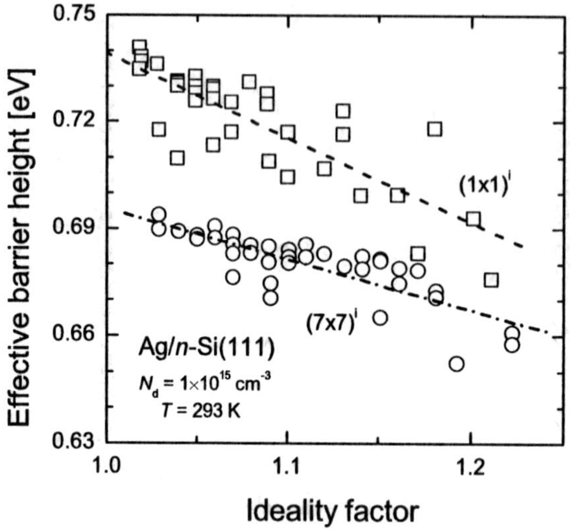

Fig. 3.8. Effective barrier heights versus ideality factors determined from *I/V* curves of Ag/*n-Si*(111)-(7×7)^{i} and Ag/*n*-Si(111)-(1×1)^{i} contacts at room temperature. The data of the diodes with (7×7)^{i}-reconstructed interfaces are the same as in Fig. 3.6. The dashed and dash-dotted lines are linear least-squares fits to the data. From *Schmitsdorf* et al. [1995]

rier heights and ideality factors but only *Schmitsdorf* et al. [1995] realized its significance.

Several conclusions may be immediately drawn from relation (3.42). First, the correlation between effective barrier heights and ideality factors demonstrates the existence of more than one physical mechanism that determines the barrier heights of *real Schottky* contacts. Second, the extrapolation of the Φ_B^{eff} versus *n* curves to n_{if}, the ideality factor controlled by the image-force effect only, leaves all effects out of consideration which cause a larger bias-dependence of the barrier height than the image-force effect itself. Third, the extrapolated barrier heights Φ_B^{nif} are equal to the zero-bias barrier height $\Phi_B^0 = \Phi_B^{hom} - \delta\Phi_{if}^0$. The superscript "*hom*" indicates that the barrier heights are laterally uniform or homogeneous. However, the Φ_B^{hom} values are not necessarily the barrier heights of the corresponding *ideal* contacts as the two data sets shown in Fig. 3.8 demonstrate. They differ in their interface structures: (7×7)^{i}-reconstructed and (1×1)^{i}-unreconstructed.

The extrapolated barrier height Φ_{Bn}^{nif} of the Ag/*n*-Si(111) contacts with (1×1)^{i}-unreconstructed interfaces is by 47 ± 20 meV larger than the corresponding value of the diodes that exhibit (7×7)^{i}-reconstructed interfaces. Similar observations were made with Pb/*n*-Si(111) contacts [*Schmitsdorf* and *Mönch* 1999]. Surface X-ray diffraction detected the stacking faults of the 7×7 reconstruction to persist not

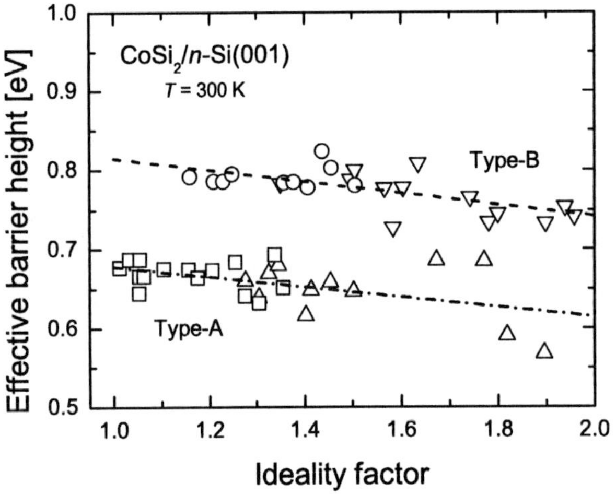

Fig. 3.9. Effective barrier heights versus ideality factors of $CoSi_2$/*n-Si*(001) contacts with two different co-ordinations of the Co interface atoms. The dashed and dash-dotted lines are linear least-squares fits to the data. Data from *Werner* et al. [1993]

only under thick Ag films[*Hong* et al. 1992] but also under thick Pb films evaporated onto clean Si(111)-7×7 surfaces at room temperature [*Howes* et al. 1995]. The extrapolated barrier heights Φ_{Bn}^{nif} of the $(7\times7)^i$-reconstructed contacts are lower by 74 ± 20 meV than those observed in Pb/*n*-Si(111)-$(1\times1)^i$ diodes [*Kampen* and *Mönch* 1995, *Schmitsdorf* and *Mönch* 1999]. The difference in barrier heights between the $(1\times1)^i$-unreconstructed and the $(7\times7)^i$-reconstructed interfaces is obviously caused by the different interface structures or, to be more specific, by the stacking fault. This effect is explained in Sect. 10.2.1.

Some of the epitaxial metal-silicides may be grown on silicon substrates with different interface structures. The two most prominent examples are $NiSi_2$/Si(111) and $CoSi_2$/Si(001) contacts. Figure 3.9 displays the effective barrier heights of two types of epitaxial $CoSi_2$/*n*-Si(001) contacts as a function of the ideality factors [*Werner* et al. 1993]. Again, the effective barrier heights decrease with increasing ideality factors and the linear relationship (3.42) applies. High-resolution transmission electron diffraction showed the interfaces to have different arrangements of the Co interface atoms. Their coordinations are six- and four-fold at the type-B and six- and eight-fold at type-A interfaces. The extrapolated barrier heights Φ_{Bn}^{nif} differ by 140 meV. A possible explanation of this difference is discussed in Sect. 10.2.2.

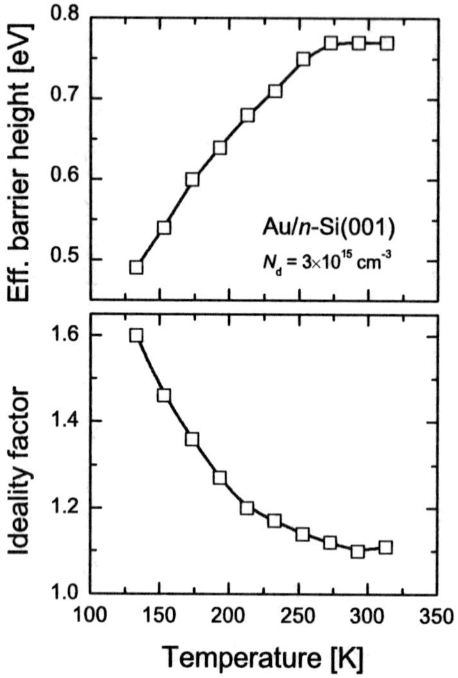

Fig. 3.10. Effective barrier height and ideality factor of a single Au/n-Si(001) diode as a function of temperature. The lines through the data points are meant to guide the eyes only. Data from *Maeda* and *Kitahara* [1998]

The linear $\Phi_B^{eff} - n$ correlation (3.42) is not only true when, as in Figs. 3.8 and 3.9, many diodes of one and the same kind are considered at room temperature but also when the temperature of one single Schottky contact is varied. As an example, Fig. 3.10 shows experimental data for one Au/n-Si(001) diode [*Maeda* and *Kitahara* 1998]. With decreasing temperature its effective barrier height becomes smaller, while the ideality factor gets larger. Prior to the evaporation of gold at room temperature, the Si wafers used were dipped into diluted HF and, thereby, became hydrogen-terminated. During the subsequent metal deposition the hydrogen most probably desorbed so that the interfaces were $(1\times1)^i$-unreconstructed. This conclusion is based on observations with Ag and Pb contacts prepared on hydrogen-terminated Si(111) surfaces, which are mentioned above.

Evaporation of Ag onto H-terminated n-Si(111):H-1×1 surfaces at room temperature produces rectifying contacts [*Kampen* et al. 1995]. The effective barrier heights again decrease linearly with increasing ideality factors and the extrapolation to n_{if} yields a barrier height Φ_{Bn}^{nif} of 0.74 ± 0.03 eV. Exactly the same extrapolated value is obtained from the data of the Ag/n-Si(111)-$(1\times1)^i$ diodes prepared by annealing of Ag/n-Si(111)-$(7\times7)^i$ contacts that are displayed in Fig. 3.8. The obvious conclusion that the evaporation of Ag removes the hydrogen-termination on Si(111):H-1×1 surfaces is confirmed by investigations using the resonance nuclear reaction $^{1}H(^{15}N,\alpha\gamma)^{12}C$ [*Tsunoda* et al. 1994]. Evidently, the Si(111) interfaces remain $(1\times1)^{i}$-unreconstructed under silver films evaporated onto Si(111):H-1×1 surfaces. Lead deposited on Si(111):H-1×1 surfaces, on the other hand, behaves quite differently.

Pb contacts that are prepared by room-temperature evaporation on Si(111):H-1×1 surfaces are ohmic on *n*-type but rectifying on *p*-type substrates [*Kampen* and *Mönch* 1995]. The effective barrier heights Φ_{Bp}^{eff} again become smaller with increasing ideality factors and the correlation is also linear. Relation (3.41) thus applies to rectifying metal contacts on both *n*- and *p*-type semiconductors. The Pb contacts prepared on Si(111):H-1×1 surfaces have an extrapolated barrier height Φ_{Bp}^{nif} of 0.71 eV. This observation gives an extrapolated Φ_{Bn}^{nif} value of 0.41 eV on *n*-type samples since for one and the same metal–semiconductor contacts the barrier heights Φ_{Bn}^{hom} and Φ_{Bp}^{hom} on *n*- and *p*-type substrates, respectively, add up to the width of the semiconductor band-gap. Simulated *I*/*V* curves that assume such low barrier heights and a reasonable series resistance of the diodes are indeed ohmic at room temperature, see Sect. 11. Contrary to what was observed with Ag evaporated onto Si(111):H-1×1 surfaces, the hydrogen persists under evaporated Pb films as again detected by the resonant nuclear reaction $^{1}H(^{15}N,\alpha\gamma)^{12}C$ [*Fukutani* et al. 1997]. *Clean* Pb/Si(111)-$(1\times1)^{i}$ contacts, on the other hand, are obtained by evaporation of lead onto Si(111):Pb-(√3×√3)R30° surfaces [*Howes* et al. 1995] or by annealing of Pb/Si(111)-$(7\times7)^{i}$ contacts at approximately 250°C [*Lucas* and *Loretto* 1995], as observed using surface X-ray diffraction. The extrapolated barrier height Φ_{Bn}^{nif} of Pb/*n*-Si(111)-$(1\times1)^{i}$ diodes is 0.72 eV [*Schmitsdorf* and *Mönch* 1999]. This value is larger by 0.31 eV than the extrapolated barrier height of Pb/H/Si(111)-$(1\times1)^{i}$ diodes with a monolayer of hydrogen at the interface. The effect of interface doping on the barrier heights is discussed in Sect 10.1.

Figure 3.11 displays the effective barrier heights of one Au/*n*-Si(001) and Au/*n*-Si(111) diode each [*Maeda* and *Kitahara* 1998; *Chen* et al. 1993] as a function of the ideality factors measured at the same temperature. The data of the Au/*n*-Si(001) contact are the same as in Fig. 3.10. Again, the dashed lines are linear least-squares fits to the data for ideality factors smaller than 1.4. Evidently, the linear correlation (3.42) is also valid for the $\Phi_{B}^{eff}(T)$ and $n(T)$ data of single Schottky contacts.

The two Au/*n*-Si diodes considered in Fig 3.11 have identical ideality factors $n_{if} = 1.01$. The slope parameters of the corresponding linear least-squares fits differ. Nevertheless, the extrapolations to the ideality factor n_{if} yield virtually identi-

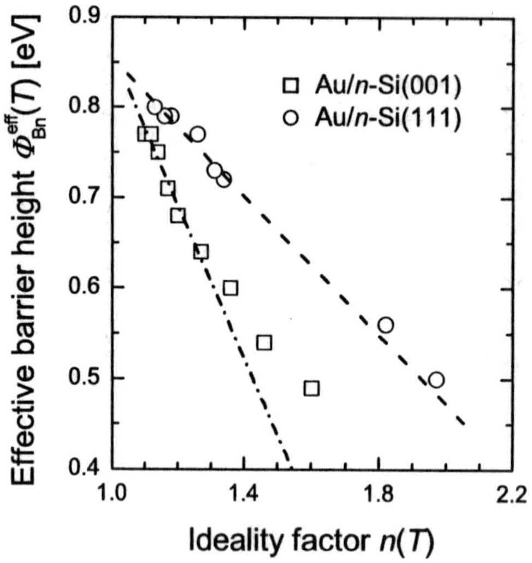

Fig. 3.11. Effective barrier heights versus ideality factors of one Au/n-Si(001) diode and one Au/n-Si(111) diode. The corresponding I/V curves were recorded in the temperature range between 100 and 350 K. The data of the Au/n-Si(001) diode are the same as in Fig. 3.10. The dashed and dash-dotted lines are linear least-squares fits to the data. Data from *Maeda* and *Kitahara* [1998] and *Chen* et al. [1993], respectively

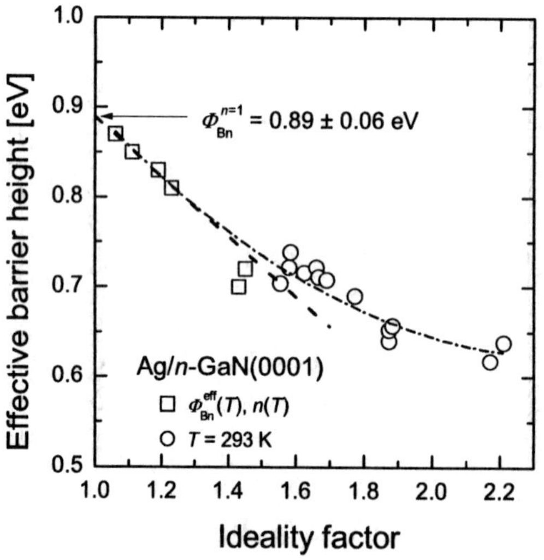

Fig. 3.12. Effective barrier heights versus ideality factors of Ag/n-GaN(0001) diodes. □ data of one diode at different temperatures between 150 and 400 K, from *Sawada* et al. [2000]; ○ data of 13 different diodes at room temperature from *Kampen* and *Mönch* [1997]. The dashed line is a linear least-squares fit to the data points up to n = 1.2 while the dash-dotted line is only meant to guide the eye.

cal barrier heights Φ_B^{nif} of 0.85 ± 0.07 eV on the (001)-oriented and of 0.86 ± 0.06 eV on the (111)-oriented silicon substrate. This observation obviously indicates that the surface orientation *per se*, even if at all, has only a minor influence on the barrier height of *Schottky* contacts. However, the examples shown in Figs. 3.8 and 3.9 evidence that the interface structure plays a crucial role.

Effective barrier heights and ideality factors determined with a large number of identically prepared diodes or measured with one and the same diode but at different temperatures behave alike. Figure 3.12 displays corresponding data of Ag/*n*-GaN(0001) diodes which exemplify this behavior. *Kampen* and *Mönch* [1997] investigated 13 different diodes at room temperature while *Sawada* et al. [2000] studied one diode at different temperatures between 150 and 400 K. Their data indeed merge smoothly.

Another example is PPy/*n*-InP(001) Schottky contacts studied by *Jones* et al. [1999]. They used the metallic conjugated polymer poly(pyrrole) (PPy) instead of a traditional metal. *Jones* et al. measured the *I*/*V* characteristics of nine diodes at room temperature and of one of them in the temperature range between 298 and 98 K and then determined the ideality factors and the saturation current densities. Figure 3.13 displays the effective barrier heights which were calculated from the saturation current densities using relation (3.39) over the corresponding ideality factors. The linear least-squares fits to the two data sets agree quantitatively. For illustration, the dashed line shown in Fig. 3.13 represents the fit to the data points obtained with the one diode studied at different temperatures. The extrapolation to

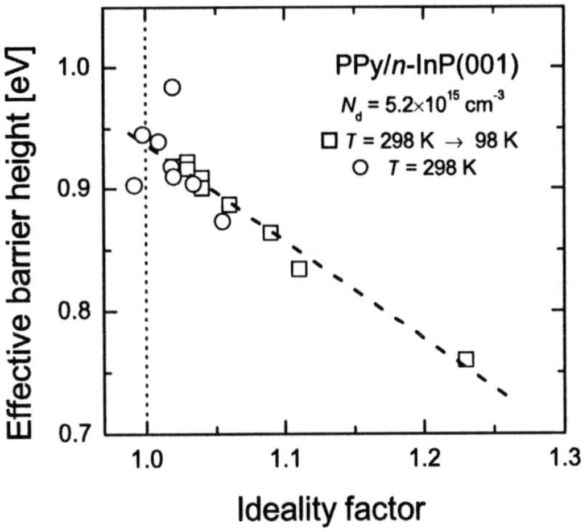

Fig. 3.13. Effective barrier heights versus ideality factors of PPy/*n*-InP contacts. The dashed line is a linear least-squares fit to the □ data. Data from Jones et al. [1999]

Table 3.3. Barrier heights $\Phi_B^{hom} = \Phi_B^{nif} - \delta\Phi_{if}^0$ of laterally homogeneous Schottky contacts, in eV, as obtained by linear extrapolation of Φ_B^{eff} versus n curves to n_{if} and by considering the image-force lowering $\delta\Phi_{if}^0$

Substrate	Metal	Interface	Φ_{Bn}^{hom}	Φ_{Bp}^{hom}	Reference
Si(111)	Ag	(1×1)^i	0.76		*Schmitsdorf* et al. [1995]
		(7×7)^i	0.70		*Schmitsdorf* et al. [1995]
(111)	Au	(1×1)^i	0.87		*Chen* et al. [1993]
(001)	Au	(1×1)^i	0.87		*Maeda & Kitahara* [1998]
(001)	Au		0.81		*Zhu* et al. [2001]
(111)	Al		0.79		*Jäger & Kassing* [1977] *Landmann & Jäger* [1975]
(111)	Al	(1×1)^i	0.76		*Miura* et al. [1994]
		(7×7)^i	0.69		*Miura* et al. [1994]
(111)	Pb	(1×1)^i	0.74		*Schmitsdorf & Mönch* [1999]
		(7×7)^i	0.67		*Heslinga* et al. [1990], *Kampen & Mönch* [1995]
(111)	Ir		0.90		*Ohdomari* et al. [1978]
(001)	Ir		0.92		*Ohdomari* et al. [1978]
(001)	Ir			0.25	*Nur* et al. [1995]
(111)	Pt		0.88		*Bäumer & Jäger* [1976]
(001)	Sm			0.80	*Eftekhari* 1991]
(001)	$TiSi_2$		0.64		*Aboelfotoh & Tu* [1986]
(001)	$TiSi_2$		0.60	0.55	*Aboelfotoh* [1988]
(001)	$TiSi_2$		0.62		*Oh & Nemanich* [2002]
(111)	$CrSi_2$		0.69		*Donoval* et al. [1998]
(111)	$CrSi_2$		066		*Turan & Akman* [1993]
(001)	$CrSi_2$		0.67	0.45	*Aniltürk & Turan* [1999]
(111)	$FeSi_2$		0.98		*Radermacher* et al. [1994]
(001)	$CoSi_2$	type-B	0.83		*Werner* et al. [1993]
		type-A	0.70		*Werner* et al. [1993]
(111)	$NiSi_2$	type-B	0.79	0.32	*Tung* [1984b],
		type-A	0.66	0.45	*Tung* et al. [1986]
(111)	$MoSi_2$		0.68		*Donoval* et al. [1998]
(001)	Pd_2Si		0.88		*Chand & Kumar* [1996]
(001)	WSi_2		0.93		*Shenai* [1991]
(001)	IrSi		0.98		*Wittmer* [1990]
(111)	IrSi		0.91		*Sanz-Maudes* et al. [1999]
(001)	PtSi		0.88		*Werner & Güttler* [1991]
(001)	PtSi		0.89		*Donoval* et al. [1991]
(001)	PtSi			0.25	*Chin* et al. [1989]
(001)	PtSi			0.24	*Chin* et al. [1990]
(001)	PtSi			0.21	*McCafferty* et al. [1996]
(001)	PtSi	epi.		0.25	*Wang* et al. [1993]
(001)	PtSi	poly.		0.23	*Wang* et al. [1993]

Table 3.3 (continued)

Substrate	Metal	Interface	Φ_{Bn}^{hom}	Φ_{Bp}^{hom}	Reference
(001)	$GdSi_2$†	epi.		0.52	*Horváth* et al. [1993]
(001)	$GdSi_2$‡	epi.		0.42	*Horváth* [1999]
(001)	$ErSi_{2-x}$		0.28		*Unewisse & Storey* [1993]
SiC					
3*C*-(001)	Au		0.84*		*Constantinidis* et al. [1998]
4H-(0001)	Au			1.63*	*Lee et al.* [2001]
4*H*-(0001)	Ti		1.26		*Defives* et al. [1999]
4*H*-(0001)	Ti		1.14		*Skromme* et al. [2000]
4*H*-(0001)	Ti			2.08*	*Lee* et al. [2001]
4*H*-(0001)	Ti			1.99*	*Lee* et al. [2002]
4*H*-(0001)	Ni		1.57*		*Khemka* et al. [1998], *Schoen* et al. [1998]
4*H*-(0001)	Ni		1.60		*Skromme* et al. [2000]
4*H*-(0001)	Ni			1.65*	*Lee* et al. [2001*]*
4*H*-(0001)	Pd		1.63*		*Im* et al. [1998]
4*H*-(0001)	Pt		1.68*		*Im* et al. [1998]
4*H*-(0001)	Pt		1.63		*Skromme* et al. [2000]
6*H*-0001)	Cu		1.26		*Suezaki* et al. [2001]
6*H*-(n.s.)	Ti		1.03*		*Schröder* et al. [1996]
6*H*-(0001)	Ti			1.99*	*Lee* et al. [2002*]*
6*H*-(000-1)	Ti		1.11*		*Shigiltchoff* et al. [2003]
6*H*-(000-1)	Ti		1.10*		*Bai* et al. [2003]
6*H*-(0001)	Pd		1.35*		*Im* et al. [1998]
6*H*-(0001)	Pd		1.24		*Im* et al. [2001]
6*H*-(0001)	Pt		1.31*		*Im* et al. [1998]
GaN{0001}	Ag		0.82*		*Kampen & Mönch* [1997]
{0001}	Ag		0.90		*Sawada* et al. [2000]
{0001}	Ag		0.95		*Maffeis* et al. [2000]
{0001}	Au		1.15		*Wang & He* [1998]
{0001}	Au		1.05		*Sawada* et al. [2000]
{0001}	Au		1.27		*Maffeis* et al. [2000]
{0001}	Pb		0.73*		*Kampen & Mönch* [1997]
{0001}	Ti		0.57		*Yu* et al. [1998]
{0001}	Ni		0.95		*Sawada* et al. [2000]
{0001}	Ni		0.84		*Yu* et al. [1998]
{0001}	Pd		1.15		*Karrer* et al. [1999]
{0001}	Pt		1.1		*Karrer* et al. [1999]
GaP(110)	Na			1.1	*v. d. Emde* et al. [1992]
(110)	In			1.04	*v.d. Emde* [1991]
(110)	Ti			1.03	*v.d. Emde* [1991]
(110)	Pt			0.84	*Evans* et al. [1992]
GaAs(001)	Au		0.94		*Hardikar* et al. [1999]
(001)	Al		0.90		*Bhuiyan* et al. [1988]

Table 3.3 (continued)

Substrate	Metal	Interface	Φ_{Bn}^{hom}	Φ_{Bp}^{hom}	Reference
(n.s.)	Ti		0.86		*DiDio* et al. [1995]
(n.s.)	Ti		0.87		*Arulkumaran* et al. [1996]
(001)	Ti		0.85		*Nuhoglu* et al. [1998]
(001)	Ni		0.89		*Hackam & Harrop* [1972]
(001)	Ni		0.91		*Nathan* et al. [1996]
(001)	Pd		0.95		*Dharmarasu* et al. [1998]
(001)	Pt		1.00		*Barnard* et al. [1996]
(001)	Pt		1.03		*Hübers & Röser* [1998]
(-1-1-1)	MnSb		0.94		*Manago* et al. [2000]
ZnSe(001)	Zn		0.85		*Rubini* et al. [2001]
CdTe(001)	Au		0.94*		*Sands* & Scott [1995]
(n.s.)	Fe		0.68*		*McLean* et al. [1986a]
(110)	Pd		0.69*		*Dharmadasa* et al. [1987]
GaTe(0001)	Al			0.72	*Bose & Pal* [1997]
	Cd			0.80	*Abay* et al. [2003]

* Data obtained by extrapolation to $n = 1$
† hexagonal
‡ orthorhombic
n.s. not specified

$n_{if} = 1.01$, the ideality factor determined by the image-force effect only, gives a barrier height of 0.947 eV for laterally homogeneous PPy/*n*-InP(001) diodes.[4]

Table 3.3 summarizes barrier heights Φ_B^{hom} of laterally homogeneous Si, GaN, GaAs, CdTe, and SiC Schottky contacts. The Si and GaAs data resulted from extrapolation of Φ_B^{eff} versus n curves to n_{if}, the ideality factor controlled by the image-force effect. The values given for GaN, CdTe, and SiC were determined by extrapolation to $n = 1$, since some of the parameters required as input for calculating the image-force effect were not specified or are even not known. The resulting error will be negligible because the difference $\Phi_B^{n=1} - \Phi_B^{nif}$ and the image-force correction $\delta\Phi_{if}^0$ are comparable in size. The data set of Table 3.3 is compared with the predictions of the MIGS-and-electronegativity concept in Sect 6.1.

[4] *Jones* et al. [1999] analyzed their data in a different way. They assumed the barrier heights which they determined from the characteristics of their PPy/*n*-InP(001) diodes to be more significant than the effective barrier heights obtained from the *I*/*V* curves. For the set of the nine diodes investigated at 298 K, they observed the *C*/*V* barrier heights to range from 0.76 to 0.81 eV. *Jones* et al. modeled the difference between the larger effective *I*/*V* and the smaller *C*/*V* barrier heights by introducing a transmission coefficient α in the *I*/*V* relation (3.38). They found α to vary between 0.0022 and 0.017.

3.6 Ballistic-Electron-Emission Microscopy

Scanning tunneling microscopy (STM), a technique pioneered and developed by *Binnig* et al. [1982a, b; 1983], images the geometrical and electronic structure of clean and adsorbate-covered surfaces. With the exception of large molecules adsorbed, atomic resolution is easily obtained. However, the properties of interfaces are not accessible to this technique. *Kaiser* and *Bell* [1988] overcame this obstacle. With metal–semiconductor contacts, as a first example, they employed the metal film as a third terminal in addition to the usual tip-and-substrate configuration in STM. In analogy to bipolar transistors, the tip, metal film, and semiconductor may be considered as the emitter, base, and collector, respectively. Figure 3.14 shows a schematic band diagram of this three-terminal BEEM configuration developed by *Kaiser* and *Bell.* The acronym BEEM stands for ballistic-electron-emission microscopy.[5] It stems from the idea that the electrons injected by the tip lose no energy on their way through the metal base and thus reach the semiconductor collector as ballistic electrons. Hence, the collector current is expected to set in if the ballistic electrons pass the metal–semiconductor barrier or, in other words., if the voltage applied between tip and metal film exceeds a threshold value that equals the barrier height. In favorable cases, BEEM makes it possible to determine local barrier heights of Schottky contacts with lateral resolutions of 1 nm. The BBEM

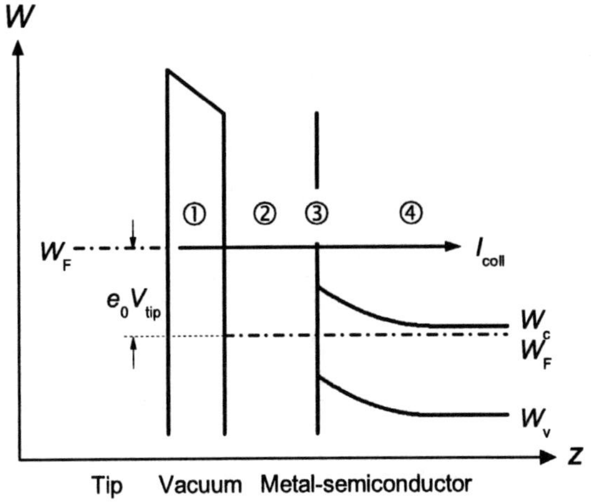

Fig. 3.14. Energy diagram and the tree-step model of Ballistic-Electron-Emission Microscopy (BEEM) (schematical). ①: Injection of electrons from the tip through the vacuum gap (emitter) into the metal film by tunneling; ②: Ballistic transport of injected electrons through the metal film (base); ③: Transmission of injected ballistic electrons across the metal–semiconductor interface into the semiconductor (collector)

[5] The BEEM technique has been reviewed by *Prietsch* [1995] and *Narayanamurti* and *Kozhevnikov* [2001].

technique is not limited to Schottky contacts but was also applied to studies of the band lineup in buried semiconductor heterostructures.

The BEEM transport process may be described by four consecutive steps. First, the tip injects almost monoenergetic electrons through the vacuum gap into the metal film by tunneling. Their energy ranges between the Fermi level and the vacuum level of the metal film. Second, these hot electrons will travel through the thin metal "base" to the metal–semiconductor interface. Provided they have sufficient energy to pass the Schottky barrier at the metal–semiconductor interface, they will, third, reach the semiconductor collector. This process determines the BEEM threshold voltage. Finally, and fourth, effects specific of hot electrons such as, for example, back-injection due to phonon back-scattering and impact ionization have to be considered within the semiconductor. Due to wave-vector conservation, impact ionization requires kinetic energies larger than approximately $1.5W_g$.[6] Such processes may thus be safely neglected for electrons close to the conduction-band minimum or, in other words, for tip voltages only slightly larger than the BEEM threshold. Therefore, the present discussions will be limited to simple one-electron models of the initial three BEEM steps.[7] The resulting relationship between collector current and tip voltage will be compared with experimental I_{coll}/V_{tip} characteristics to obtain experimental values of Schottky barrier heights, the most important electronic property of metal–semiconductor contacts in the present context, with high lateral resolution.

In BEEM experiments, the collector current I_{coll} is measured as a function of the tip voltage V_{tip} while the tunnel current I_{tip} is kept constant. In the following, the I_{coll}/V_{tip} characteristics close to the threshold voltage, i.e., tip voltages only slightly larger than the barrier height will be considered. In a rough approximation, the tunneling probability as well as the density of states of both the tip and the metal film may be assumed to be constant in the then narrow energy window $e_0V_{tip} - \Phi_{Bn}$. At these energies, the conduction-band states of the metal film may be safely assumed to be completely unoccupied since the Schottky barrier heights are generally larger than a few tenths of an eV. The energy distribution of the electrons injected by tunneling from the tip into the metal "base" may thus be described by the Fermi–Dirac distribution function

$$f_0(W - W_F - e_0V_{tip}) = 1/(\exp\{[W - (W_F + e_0V_{tip})]/k_BT\} + 1) \qquad (3.43)$$

of the electrons in the tip. Conveniently, the Fermi level W_F in the metal is taken as the reference.

The transport in the metal films used in BEEM was investigated by varying the thickness of the metal film. Such studies showed an exponential attenuation

$$I_{coll} = \mathrm{K}I_{tip}\exp(-d_{\mathrm{m}}/\Lambda_{\mathrm{m}}) \qquad (3.44)$$

of the collector current as a function of the thickness d_{m} of the metal films while the tunnel current I_{tip} was kept constant. When the thickness d_{m} of the metal film is

---

[6] See, for example, *Mönch* [1969].

[7] More detailed BEEM descriptions may be found in a recent review article by *de Andres* et al. [2001].

much smaller than the respective attenuation length or mean free path Λ_m, then the electrons injected into the metal film are described as ballistic electrons. It was this picture that named the BEEM technique. The attenuation length Λ_{Au} of electrons injected in, for example, Au films of Au(111)/*n*-Si(111) and Au(111)/*n*-Si(001) structures was observed to be larger than 23 nm [*Weilmeier* et al. 1999].

For ballistic electrons passing through the metal–semiconductor interface their energy and the parallel component $k_{\|}$ of their wave-vector are conserved. Assuming the conduction band of the semiconductor to be parabolic, the kinetic energy of the electrons right at the interface may then be written as

$$W - W_{ci} = W - W_F - \Phi_{Bn} = \frac{\hbar^2}{2m_n^*}(k_{\perp s}^2 + k_{\|}^2) \ , \tag{3.45}$$

where $k_{\perp s}$ is the component of the wave-vector normal to the interface. Only electrons with parallel wave-vector smaller than

$$k_{\|}^{\max} = \sqrt{2m_n^*(W - W_F - \Phi_{Bn})/\hbar^2} \tag{3.46}$$

will thus be able to get over the barrier and to enter the semiconductor.

With the above arguments the BEEM collector current may be approximated by

$$I_{coll} \propto I_{tip} \int_{e_0V_i+W_n}^{\infty} \mathrm{d}W \, f_0(W - W_F - e_0V_{tip}) \iint_{2\pi} \mathrm{d}\Omega \ , \tag{3.47}$$

where $\mathrm{d}\Omega$ is the solid-angle segment. Making use of the relation

$$\underline{k}^2 \mathrm{d}\Omega \cong \mathrm{d}(\pi k_{\|}^2) = 2\pi k_{\|} \mathrm{d}k_{\|} \ , \tag{3.48}$$

$\mathrm{d}\Omega$ may be replaced by the parallel component of the wave-vector. Relation (3.47) may then be rewritten as

$$I_{coll} \propto I_{tip} \int_{e_0V_i+W_n}^{\infty} \mathrm{d}W \, f_0(W - W_F - e_0V_{tip}) \int_0^{k_{\|}^{\max}} k_{\|} \mathrm{d}k_{\|} \ . \tag{3.49}$$

Carrying out the integration over the $k_{\|}$, one obtains

$$I_{coll} \propto I_{tip} \int_{e_0V_i+W_n}^{\infty} \mathrm{d}W \, f_0(W - W_F - e_0V_{tip})(W - W_F - \Phi_{Bn}) \ . \tag{3.50}$$

Partial integration yields

$$I_{coll} \propto I_{tip} \int_{e_0V_i+W_n}^{\infty} \mathrm{d}W \frac{-\mathrm{d}f_0(W - W_F - e_0V_{tip})}{\mathrm{d}W}(W - W_F - \Phi_{Bn})^2 \ . \tag{3.51}$$

Since $\mathrm{d}f_0/\mathrm{d}W$ is a Dirac δ-function, one finally obtains

$$I_{coll} = R^* I_{tip}(e_0V_{tip} - \Phi_{Bn})^2 \ , \tag{3.52}$$

which is the famous square-law of the BEEM I_{coll}/V_{tip} characteristics already proposed by *Bell* and *Kaiser* [1988].

Many different approaches have been tried in order to improve the descriptions of the four BEEM steps mentioned above. *Ludeke* et al. [1991], for example, considered a quantum-mechanical square-root energy-dependence of the transmission coefficient at the metal–semiconductor interface. They arrived at a power-law I_{coll}/V_{tip} relationship similar to (3.51) but with an exponent 5/2 rather than 2. *Cuberes* et al. [1994] performed least-squares fits of their experimental BEEM I_{coll}/V_{tip} characteristics measured with Au/*n*-Si(111)-(7×7)^{i} contacts to the power-law dependency

$$I_{coll} = R(V_{tip} - V_{BEEM})^{m} \tag{3.53}$$

using different values of m. Among the exponents tested, 1, 3/2, 2, and 5/2, they obtained the smallest χ^2-deviation for m = 2, the result of the above, most crude approximation. The threshold voltages V_{BEEM} differed by 0.03 V for m values of 2 and 5/2. This difference indicates the accuracy of the barrier heights which are determined from BEEM I_{coll}/V_{tip} characteristics.

Figure 3.15 explains the quality of the fits of the quadratic relation (3.52) to experimental BEEM data and the determination of the BEEM threshold voltage. As an example, two typical BEEM I_{coll}/V_{tip} curves are displayed which were recorded at two different but well-defined spots of one and the same $CoSi_2$/*n*-Si(001) con-

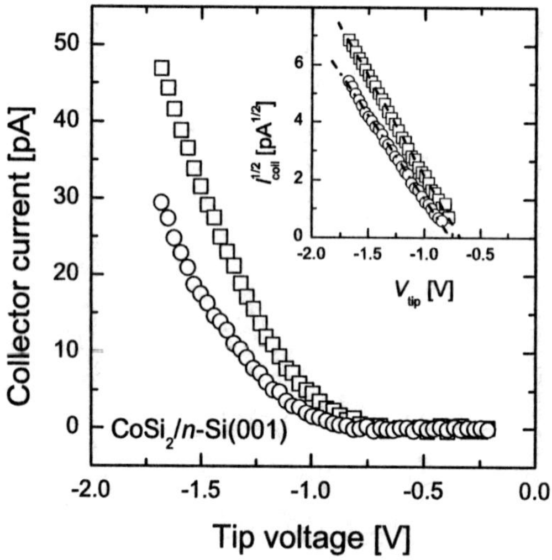

Fig. 3.15. BEEM collector current as a function of the tip voltage measured on top of a dislocation (squares) and in the neighboring undisturbed region (circles) of a $CoSi_2$/*n*-*Si*(001) contact. The inset shows the square root of the collector current plotted versus the tip voltage. The dashed lines are linear least-squares fits to the data. The BEEM threshold voltages were obtained as 0.76 ± 0.05 eV and 0.68 ± 0.04 eV, respectively. Data from *Sirringhaus* et al. [1996]

tact; see Sect. 4.5.2 for details [*Sirringhaus* et al. 1996]. The inset of the figure shows the square-root of the BEEM collector current as a function of the tip voltage. The dashed lines are linear least-squares fits to the data. They demonstrate that the experimental data are excellently described by the quadratic I_{coll}/V_{tip} relationship (3.52) first derived by *Bell* and *Kaiser* [1988]. This finding agrees with the observations of *Cuberes* et al. [1994] which are mentioned in the preceding section. The threshold voltages at the two different spots of the same contact were obtained as 0.76 ± 0 05 eV and 0.68 ± 0.04 eV.

3.7 Barrier Heights of Real Schottky Contacts 2: BEEM I_{coll}/V_{tip} Characteristics

Ballistic-electron-emission microscopy (BEEM) is an experimental tool for measuring spatial variations of the barrier height on the nanometer scale. For this purpose, the BEEM collector current I_{coll} is recorded as a function of the tip voltage V_{tip} at tip positions along line scans. The BEEM threshold energies e_0V_{BEEM} measure the *local* zero-bias barrier heights $\Phi_B^{loc} - \delta\Phi_{if}^0$ of the Schottky contact at the corresponding tip positions.

Figure 3.16 displays a typical lateral profile of BEEM threshold voltages or BEEM barrier heights on a PtSi/*n*-Si(001) contact [*Talin* et al. 1994]. The spacing between successive tip positions is approximately 7 nm. The local barrier heights vary between 0.83 and 0.92 eV. This spread is well beyond the margins of experi-

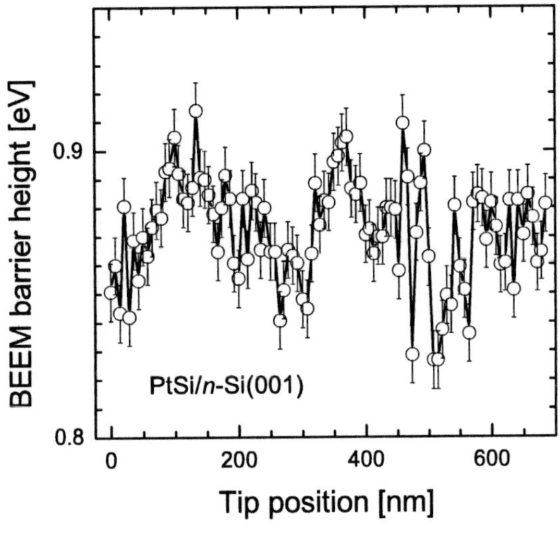

Fig. 3.16. Lateral profile of BEEM barrier heights on a PtSi/*n*-Si(001) contact. Data from *Talin* et al. [1994]

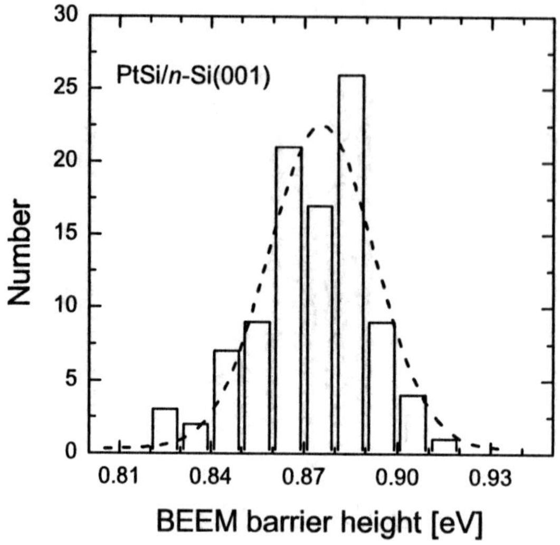

Fig. 3.17. Histogram of BEEM barrier heights along the line scan on the PtSi/*n*-Si(001) contact shown in Fig. 3.15. The dashed line is a Gaussian least-squares fit to the data. The center and the standard deviation of the Gaussian are 0.876 ± 0.002 eV and 0.0169 ± 0.0024 eV, respectively. Data from *Talin* et al. [1994]

mental error of ± 0.01 eV. Obviously, the contact possesses lateral inhomogeneities of the barrier height on the nanometer scale. Figure 3.17 displays the distribution of the BEEM barrier heights along the line scan shown in Fig. 3.16. The dashed Gaussian is a least-squares fit to the data. It describes the experimental results quite well. The fitting procedure gives the center value and the standard deviation of the Gaussian as 0.876 ± 0.002 eV and 0.0169 ± 0.0024 eV, respectively. Intuitively, the center value of the Gaussian distribution may be considered as the barrier height $\Phi_{Bn}^{hom} - \delta\Phi_{if}^{0}$ of the laterally homogeneous contact. For the linear scan on the PtSi/*n*-Si(001) diode shown in Fig. 3.16, the Gaussian fit gives $\Phi_{Bn}^{hom} - \delta\Phi_{if}^{0} = 0.876$ eV. Considering the margins of experimental error, this BEEM value agrees with the barrier heights of 0.88 to 0.91 eV that were obtained from linear extrapolations of the Φ_{Bn}^{eff} versus n curves to n_{if}, the ideality factor determined by the image-force effect only.

Table 3.4 summarizes barrier heights Φ_{B}^{hom} of laterally homogeneous Si, SiC, GaN, GaAs, GaInP, GaAsP, ZnSe, and CdTe, Schottky contacts which were obtained from histograms of BEEM threshold voltages.

Table 3.4. Barrier heights $\Phi_{Bn}^{\text{hom}} - \delta\Phi_{if}^{0}$ of laterally homogeneous Schottky contacts in eV as obtained from histograms of BEEM threshold voltages.

Substrate	Metal	Interface	$\Phi_{Bn}^{\text{hom}} - \delta\Phi_{if}^{0}$	Reference
Si(001)	Au		0.83	*Palm* et al. [1993]
(001)	Au		0.82	*Morgan* et al. [1996a]
(001)	Au		0.79	*Zhu* et al. [2001]
(001)	$CoSi_2$	epi-{001}	0.74	*Sirringhaus* et al. [1995]
		epi-{110}	0.58	*Sirringhaus* et al. [1995]
(111)	$CoSi_2$	epi.	0.68	*Sirringhaus* et al. [1994]
(001)	$NiSi_2$	type-B	0.84	*Fernandez* et al. [1991]
		type-A	0.71	*Fernandez* et al. [1991]
(001)	PtSi		0.87	*Niedermann* et al. [1992]
(001)	PtSi		0.87	*Talin* et al. [1994]
(001)	PtSi		0.88	*Zhu* et al. [1999]
(001)	PtSi		0.87	*Ru* et al. [2000]
(001)	PtSi		0.86	*Zhu* et al. [2001]
6*H*-SiC(0001)	Pd		1.30	*Im* et al. [1998]
(0001)	Pd		1.25	*Im* et al. [2001]
(0001)	Pt		1.30	*Im* et al. [1998]
2*H*-GaN(0001)	Au		1.04	*Bell* et al. [1998]
GaAs(110)	Au		0.98	*Stockmann & Kempen* [1998]
$Ga_{0.1}In_{0.9}P(001)$	Au		1.46	*Morgan* et al. [1996a]
$Ga_{0.51}In_{0.49}P(001)$	Au		1.14	*Morgan* et al. [1996a]
$GaAs_{0.67}P_{0.33}(001)$	Au		1.16	*Olbrich* et al. [1998]
(001)	Co		1.04	*Olbrich* et al. [1998]
ZnSe(001)	Au		1.38	*Morgan* et al. [1996b]
CdTe(001)	Au		0.83	*Fowell* et al. [1990]
(001)	Au		1.01	*Dharmadasa* et al. [1998]

3.8 Barrier heights of Real Schottky Contacts 3: Internal Photoemission Yield Spectroscopy

Metal–semiconductor contacts show a photoelectric response to optical radiation with photon energies far lower than the width of the bulk band gap. *Gyulai* [1926] measured the spectral dependence of photoconduction in rocksalt containing colloidal sodium metal. Much later, *Gilleo* [1953] studied photoelectric yield spectroscopy with silver films evaporated onto cleaned AgCl, KBr, and NaCl surfaces, i.e., he used more conventional metal–insulator contacts. Already *Mott* and *Gurney* [1940] had noticed that the low-energy part of *Gyulai*'s results is similar in shape to the photoemission current from clean sodium surfaces but is displaced by 0.5 eV towards lower photon energies. They interpreted these observations as the photoemission of electrons from the metallic sodium particles into the conduc-

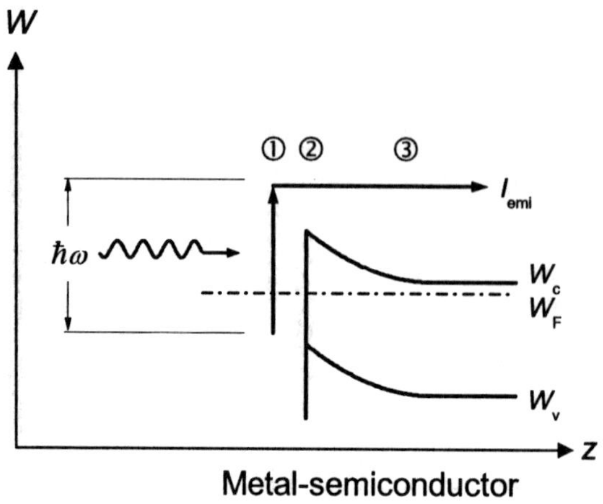

Fig. 3.18. Energy diagram of internal photoemission yield spectroscopy (IPEYS) (schematic)

tion band of the sodium-chloride insulator. This process is now called *internal photoemission* and Fig. 3.17 explains the basic model. Hot electrons are photoexcited in the metal. Provided their energy is sufficiently large they may surmount the interfacial barrier and reach the insulator or semiconductor. Schottky diodes thus have photothresholds which equal their barrier heights or, in other words, internal photoemission spectroscopy is another experimental tool for determining the barrier heights of metal–semiconductor contacts. In studying Cu/*n*-CdS rectifiers, *Williams* and *Bube* [1960] were the first to apply this method to Schottky contacts.

Internal photoemission experiments may be performed with Schottky diodes in two basically different operational modes, which are

- front illumination, where the light is incident on the metal film from the air-side so that most of the electrons are excited close to the metal–air interface and have to travel to the metal–semiconductor interface, and
- back-illumination, where the radiation enters from the semiconductor side so that the electrons are excited in the metal right at the interface.

Electron-electron and electron-phonon collisions reduce the energies of the hot electrons in the metal. Hence, semitransparent metal films are used to obtain a large photoresponse when front illumination is used.

The internal photoemission yield $Y(\hbar\omega)$ is defined as the ratio of the photoemitted electron flux I_{emi} across the barrier into the semiconductor and the flux I_{exc} of the electrons excited in the metal by the photons of energy $\hbar\omega$. *Cohen* et al. [1969] presented a most simple derivation of the spectral dependence of the yield. They assumed that in the metal

- the probability of photoexcitation is independent of the energy of the initial and the final states,
- the lifetime of the excited electrons is energy-independent, and

- the density of states of the electrons is constant in an energy interval of width $\hbar\omega$ below and above the Fermi energy W_F, and that
- the conduction band of the semiconductor is parabolic.

Internal photoemission yield spectroscopy (IPEYS) and ballistic electron emission microscopy (BEEM) are similar in that hot electrons are injected from the metal across the metal–semiconductor interface. Hence, the same arguments as in the derivation of the BEEM collector current (3.52) apply for IPEYS and one obtains the flux of the electrons photoinjected from the metal into the semiconductor as

$$I_{emi} \propto (\hbar\omega - \hbar\omega_Y)^2 \quad . \tag{3.54}$$

With the above assumptions, the flux of the excited electrons is

$$I_{exc} = \int_{W_F}^{W_F+\hbar\omega} D_m(W)\mathrm{d}W \cong \hbar\omega \quad , \tag{3.55}$$

since the density of states $D_{\mathrm{m}}(W)$ of the metal electrons is assumed to be energy-independent. Hence, this simple model gives the spectral dependence of the internal photoemission yield as

$$Y(\hbar\omega) \propto \frac{(\hbar\omega - \hbar\omega_Y)^2}{\hbar\omega} \quad , \tag{3.56}$$

where $\hbar\omega_Y$ is the threshold energy. *Fowler* [1931] calculated the temperature dependence of the photoemission yield for clean metal surfaces. His zero-temperature result $Y_{\mathrm{m}} \propto (\hbar\omega - \hbar\omega_0)^2$ differs from relation (3.55) in that the photon energy $\hbar\omega$ in the denominator is missing. Nevertheless, *Fowler*'s function was mostly used in determining the IPEYS thresholds of Schottky contacts.

As a typical example, the results of an internal photoemission yield study of Pt/*p*-Si(001) diodes [*Turan* et al. 2001] are considered. The two diagrams of Fig. 3.19 display the spectral dependence of the square root of the measured yield times the respective photon energy and the square root of the measured yield, respectively. For photon energies larger than 0.3 eV, the data are well fitted by relation (3.55) and linear least-squares fits yield photothresholds of 0.274 ± 0.003 and 0.271 ± 0.007 eV, respectively. Within the margins of experimental error both values are identical. However, the standard deviation is smaller by a factor of two for the fit of the $(Y{\cdot}\hbar\omega)^{1/2}$ versus $\hbar\omega$ plot. Table 3.5 summarizes threshold energies obtained from linear least-squares fits to experimental $(Y{\cdot}\hbar\omega)^{1/2}$ and $Y^{1/2}$ versus $\hbar\omega$ curves.

The experimental $(\mathrm{Y}{\cdot}\hbar\omega)^{1/2}$ versus $\hbar\omega$ values displayed in Fig. 3.19 deviate from the straight line towards larger yields slightly above and below the threshold. This behavior is not a special feature of the specific Pt/*p*-Si(001) contact studied but is rather a characteristic of all internal photoemission yield spectra observed with Schottky diodes. The linear spectral dependence of $(\mathrm{Y}{\cdot}\hbar\omega)^{1/2}$ versus $\hbar\omega$ is a zero-temperature approximation. The shape of the Fermi–Dirac distribution at finite temperatures easily explains the existence of such tails in the IPEY spectra as

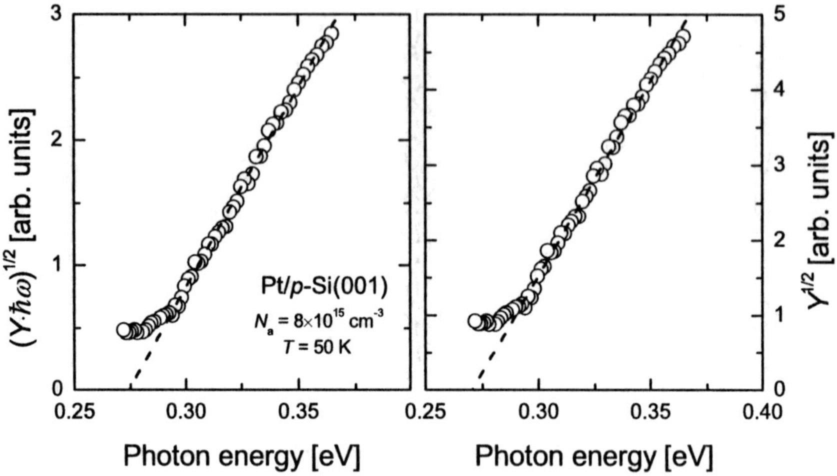

Fig. 3.19. Spectral dependence of the internal photoemission yield of a Pt/*p*-Si(001) diode: $[Y(\hbar\omega)\cdot\hbar\omega]^{1/2}$ and $Y(\hbar\omega)^{1/2}$ versus $\hbar\omega$, respectively. The dashed lines are a linear least-squares fits to the data for photon energies larger than 0.3 eV. Data from *Turan* et al. [2001]

the result of photon-assisted thermionic-emission. *Fowler*'s calculations revealed the photoemission response from metal surfaces to be nonlinear at finite temperatures. *Crowell* et al. [1972] argued that the usual linear fits to the $(Y\cdot\hbar\omega)^{1/2}$ versus $\hbar\omega$ plots represent tangents at the respective photon energies $\cdot\hbar\omega_t$ and the intercepts on the $\hbar\omega$ abscissa are *apparent* photothresholds $\hbar\omega_Y^{app}$ rather than their *true* values $\hbar\omega_Y^{true}$. Using *Fowler*'s universal $Y_F(\cdot\hbar\omega)$-relationship, they evaluated the universal thermionic-emission correction $(\hbar\omega_Y^{true}-\hbar\omega_Y^{app})/k_BT$, which is displayed as a function of $(\hbar\omega_t-\hbar\omega_Y^{app})/k_BT$ in Fig. 3.20. Provided that $(\hbar\omega_t-\hbar\omega_Y^{app})\geq 4k_BT$, the correction $(\hbar\omega_Y^{true}-\hbar\omega_Y^{app})$ is smaller than k_BT. This means that the corrections become negligible at temperatures well below room temperature but no accurate thresholds are obtained for $(\hbar\omega_t-\hbar\omega_Y^{app})<2.5\ k_BT$.

Anderson et al. [1975] further improved the evaluation of the internal photoemission yield spectra of Schottky diodes. They treated the combined effects of thermal excitation and electron tunneling through the top the barriers and applied their calculations to Au/*n*-Si(111) and Au/*n*-GaAs contacts. *Anderson* et al. [1975] compared their theoretical results with the experimental data of Au/*n*-Si(111) diodes at various temperatures [*Crowell* et al. 1964] and noted that neither the purely thermionic theory nor the combination with quantum-mechanical effects predict a large enough photoresponse at photon energies below and slightly above the threshold energy.

The theoretical studies of *Crowell* et al. [1972] and *Anderson* et al. [1975] considered no nanometer-scale lateral inhomogeneities of the barrier heights. At that time the existence of such nonuniformities was not realized. Obviously, such nm-

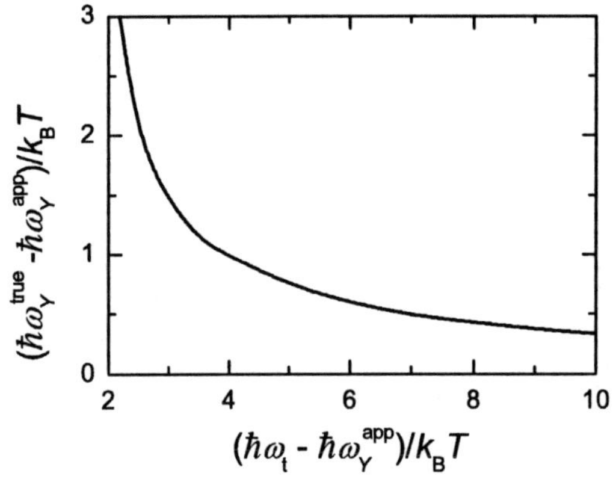

Fig. 3.20. Universal thermionic-emission correction $(\hbar\omega_Y^{true} - \hbar\omega_Y^{app})/k_B T$ to apparent IPEYS photothresholds $\hbar\omega_Y^{app}$ obtained from linear least-squares fits to $(Y(\hbar\omega)\cdot\hbar\omega)^{1/2}$ versus $\hbar\omega$ plots as a function of $(\hbar\omega_t - \hbar\omega_Y^{app})/k_B T$ above the low-energy limit $\hbar\omega_t$; $\hbar\omega_Y^{true}$ is the true photothreshold. After *Crowell* et al. [1972]

scale patches contribute to the tails observed in the internal photoemission yield spectra of real Schottky diodes. Unfortunately, no combined BEEM–IPEYS studies have been performed so far.

In combination with scanning near-field optics microscopy (SNOM), internal photoemission yield spectroscopy may be applied to study the lateral variations of Schottky barrier heights on the sub-μm-scale. *Almeida* et al. [1996] reported on a successful IPEYS–SNOM test with Pt/*n*-GaP(001) diodes. A result of their studies is shown in Table 3.5.

Table 3.5 summarizes threshold energies of diamond, Si, GaN, GaP, GaSb, and some III–VI Schottky contacts determined from internal photoemission yield spectra.

3.9 Core-Level Photoemission Spectroscopy

3.9.1 Chemical Shifts and Interface Band-Bending Determination by PES

Ultraviolet, soft X-ray, and X-ray photoemission spectroscopy or, for short, UPS, SXPS, and XPS, respectively, are versatile tools for studying properties of metal–semiconductor contacts. The continuum of the sychrotron radiation of electron storage rings is a most attractive light source for such studies. Laboratory sources utilize, for example, the He I and He II lines with photon energies of 21.2 and 40.8 eV for UPS, the Zr($M\zeta$) line with 151.2 eV for SXPS, and the Al($K\alpha$) and Mg($K\alpha$) lines with 1486.6 and 1253.6 eV, respectively, for XPS.

Table 3.5. Threshold energies of Schottky contacts determined from internal photoemission yield spectra. The photothresholds obtained from linear least-squares fits to $[Y(\hbar\omega)\cdot\hbar\omega]^{1/2}$ versus $\hbar\omega$ curves were not corrected for photo-assisted thermionic emission and the image-force lowering.

Substrate	Metal	Φ_{Bn}^{IPEYS}	Φ_{Bp}^{IPEYS}	Reference
C(111)	Al		2.2	*Mead & McGill* [1976]
(111)	Au		2.0	
Si(111)	Au	0.8		*Crowell* et al. [1962]
(111)	Au	0.78		*Crowell* et al. [1964]
(111)	Au	0.80	0.29	*Cohen* et al. [1969]
(111)	Au	0.75		*Archer & Yep* [1970]
(111)	Au	0.83		*Blauärmel* et al. [2000]
(001)	Au	0.79		*Blauärmel* et al. [2000]
(111)	Ir	0.88		*de Sousa Pires* et al. [1979]
(001)				*de Sousa Pires* et al. [1979]
(111)	Pd	0.71		*Crowell* et al. [1962]
(001)	Pt		0.22	*Turan* et al. [2001]
(001)	$CrSi_2$	0.66	0.44	*Aniltürk & Turan* [1999]
(111)	IrSi	0.88		*de Sousa Pires* et al. [1979]
(001)	IrSi	0.88		*de Sousa Pires* et al. [1979]
(111)	$ErSi_2$	0.29	0.74	*Arnaud D'Avitaya* et al.
SiC: 4*H*-(-1-210)	Mo	1.66		*Bai* et al. [2003]
4*H*-(1-100)	Pt	1.62		*Shigiltshoff* et al. [2002]
6*H*-(0001)	Au	1.42	1.45	*Aboelfotoh* et al. [2003]
	Ti	0.98	2.04	*Aboelfotoh* et al. [2003]
	Ni	1.06	1.85	*Aboelfotoh* et al. [2003]
	Cu	1.12	1.60	*Aboelfotoh* et al. [2003]
	Pt	1.38		*Shigiltshoff* et al. [2002]
$Al_{1-x}Ga_xN$(0001)	Ni			*Qiao* et al. [2000]
GaN(0001)	Ni	0.97		*Qiao* et el. [2000]
GaP(001)	Pt	1.41		*Almeida* et al. [1996]
GaSb	Au	0.6		*Mead & Spitzer* [1964]
GaSb	Au	0.6		*Rotelli* et al. [1997]
GaS(0001)	*			*Kurtin & Mead* [1969]
GaSe(0001)	*			*Kurtin & Mead* [1969]
GaTe(0001)	*			*Kurtin & Mead* [1969]

* See Fig. 6.8

Photoemission from core levels of substrate atoms may be used to study both the band bending and the chemical environment of these atoms. First of all, the core levels follow the spatial variations of the band edges and, as a consequence, the core-level energies of semiconductor atoms right at metal–semiconductor interfaces and beyond the depletion layer in the interior differ by the interface band-bending. Second, covalent bonds between unlike atoms are generally partly ionic.

Naturally, this is also true for the chemical bonds between metal and semiconductor atoms at metal–semiconductor interfaces. The respective charge transfer changes the electrostatic potential experienced by the core electrons of semiconductor atoms right at the metal–semiconductor interfaces. As a consequence, this leads to *chemical* shifts of the core-level binding energies of the interface atoms. Away from the interface the bonds in the semiconductor substrate become polarized. The respective bond-charge displacements and, therefore, the chemical shifts of the core levels decay exponentially into the semiconductor. Hence, the core levels of semiconductor atoms a few atomic layers away from the interface show no such chemical shifts.

However, their core-level energies still differ by the interface band-bending from the respective values of atoms beyond the depletion layer. With sufficiently high energy resolution, core-level photoemission signals may be decomposed into contributions from atoms right at the interface and a few atomic layers off into the depletion layer. The two lines are customarily addressed as *interface* and *bulk* components, respectively, although the label *bulk* is somewhat misleading. The energy difference between both lines then equals the chemical shift due to the partly ionic bonds between metal and semiconductor atoms while the kinetic energy of the *bulk* component reflects the interface band-bending.

The depth resolution of photoemission spectroscopy depends on the escape depth of the photoemitted electrons. As an example, Fig. 3.21 displays data for some semiconductors. The escape depths typically vary between approximately 0.4 and 20 nm for kinetic energies around the minimum at approximately 50 eV and at 10 and 1000 eV, respectively. Hence, core-level photoemission spectroscopy as a probe of metal–semiconductor interfaces is limited to metal films whose thickness does not exceed approximately 2 to 4 nm.

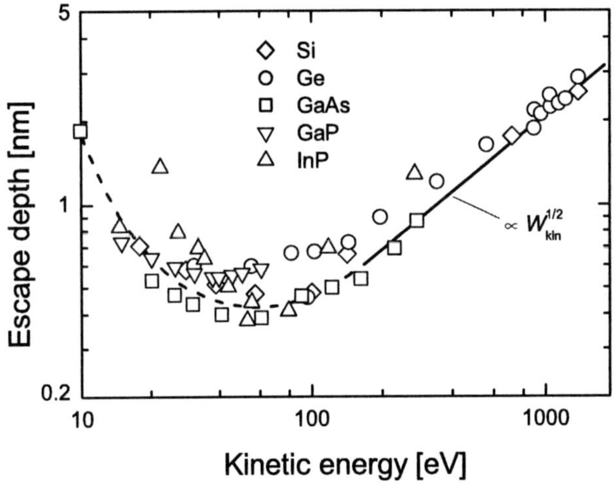

Fig. 3.21. Electron escape depth as a function of the kinetic electron energy

3.9.2 Growth Modes of Metals on Semiconductors

Metal atoms evaporated on semiconductor surfaces will eventually form continuous metal films. However, the initial growth mode depends strongly on the respective metal–semiconductor pair under study and also on the experimental conditions such as, for example, the substrate temperature and the evaporation rate. The interfaces of immiscible metal–semiconductor systems are abrupt.

Quite generally, three different growth modes are distinguished. They are commonly labeled as

- *Frank–van der Merwe* [1949] or layer-by-layer growth, which proceeds by successive nucleation and growth of two-dimensional monolayers,
- *Volmer–Weber* [1926] or island growth where the atoms evaporated aggregate in three-dimensional clumps which eventually coalesce, and
- *Stranski–Krastanov* [1939] growth where three-dimensional islands develop and grow on one or a few ad-layers initially formed.

Criteria for the occurrence of one or another of these growth modes may be based on equilibrium considerations and are obtained from the principle of minimum free energy [*Bauer* 1958]. Two-dimensional layer-by-layer growth, for example, is expected to occur if the surface free-energy is lower for deposited adatoms forming bonds with substrate atoms rather than with other adatoms. Which growth mode will be actually observed with a specific system depends on the surface free-energies σ_{ad} and σ_{sub} of the ad-layer and the substrate, respectively, the interface energy γ_i and also the strain energy in case of lattice mismatch between substrate and ad-layer. For lattice-matched materials, the strain energies may be neglected and the condition for wetting of the substrate by the ad-layer is given by the inequality [*Bauer* 1958]

$$\sigma_{sub} > \sigma_{ad} + \gamma_i \quad . \tag{3.57}$$

In such cases, *Frank–van der Merwe* or layer-by-layer growth may be expected. No wetting and *Volmer–Weber* or island growth will be observed if the inequality has the opposite sign. *Stranski–Krastanov* growth generally occurs when wetting happens but, for example, the lattice-mismatch is so large that the strain energy eventually dominates. The following examples illustrate the different growth modes.

Cesium wets semiconductor surfaces and shows *Frank–van der Merwe* or layer-by-layer growth. At room temperature, the sticking coefficient of the second layer is almost zero but at low substrate temperatures the growth of the second cesium layer is easily observed. As an example, the upper diagram of Fig. 3.22 displays XPS data for Cs deposited on a Si(111)-7×7 surface at low temperature. The increase of the Cs($3d_{5/2}$) core-level signal as a function of evaporation time may be fitted by two straight lines. The kink reflects the completion of the first continuous Cs layer and the onset of nucleation and growth of the second layer. The same growth behavior was observed with cesium on clean GaN [*Kampen* et al. 1998, Eyckeler et al. 1998], GaP [*Linz* et al. 1993], GaAs [*Spicer* et al. 1975, *Grunwald* 1987], as well as 6*H*- and 4*H*-SiC surfaces [*van Elsbergen* et al. 1996, *van Elsber-*

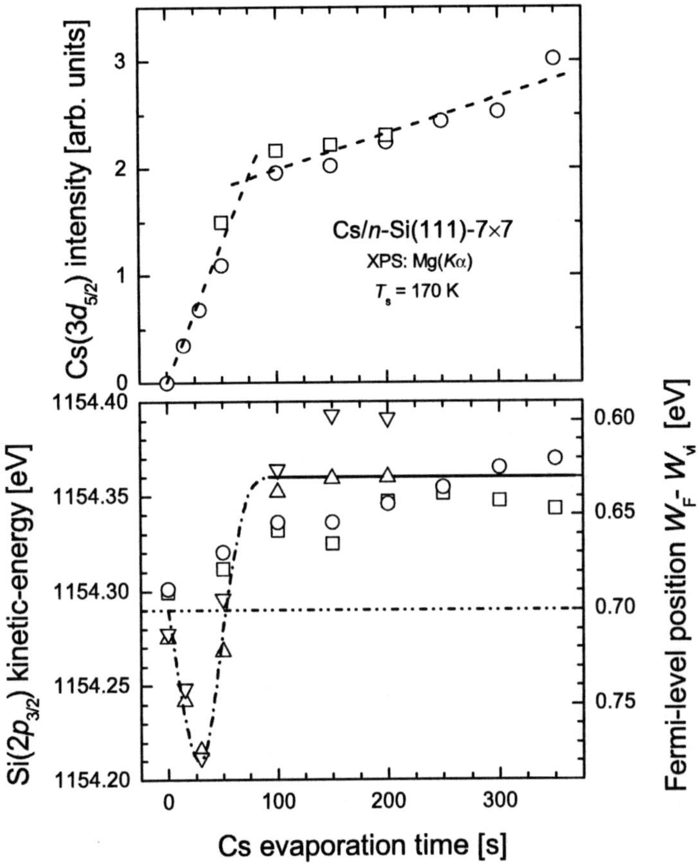

Fig. 3.22. Intensity of Cs($3d_{5/2}$) (upper diagram) and kinetic energy of Si($2p_{3/2}$) core-level signals (lower diagram) recorded with Mg($K\alpha$) radiation (1253.6 eV) versus the evaporation time of Cs evaporated on Si(111)-7×7 surfaces at 170 K. The dashed lines in the upper diagram are linear least-squares fits to the data points for Cs evaporation times below 50 s and above 100 s, respectively. The energy position $W_F - W_{vi}$ of the Fermi level relative to the valence-band maximum at the clean Si(111)-7×7 surface (dash-dot-dot line) was obtained from the energy positions of three surface states in the electron distribution curves recorded with He I radiation (21.2 eV). The dash-dotted line in the lower diagram is meant to guide the eye. After *Weyers* et al. [1999]

gen 1998]. During the formation of the second Cs layer a Fermi-edge develops as the high-energy termination of the energy distribution curve of the photoelectrons and the work function assumes the value characteristic of cesium metal [*Grunwald* 1987, *Linz* et al. 1993, *van Elsbergen* et al. 1996, *Kampen* et al. 1998]. Cesium thus grows in a layer-by-layer fashion on clean semiconductor surfaces at low temperature and the continuous Cs films already become metallic during the deposition of the second layer.

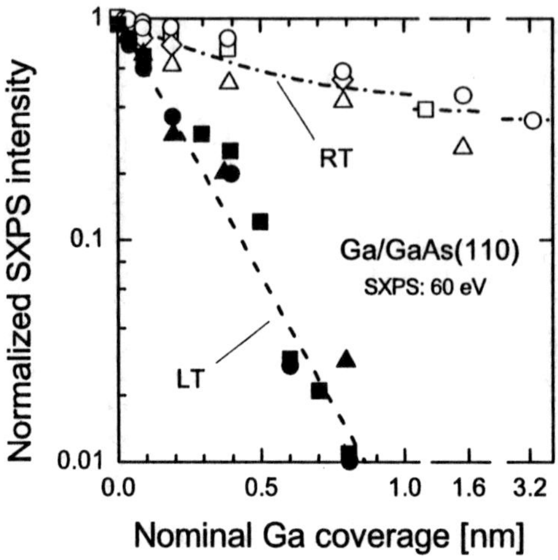

Fig. 3.23. Intensity of As(3*d*) signals recorded by using SXPS as a function of the *nominal* amount of Ga evaporated on cleaved GaAs(110) surfaces kept at room and at low temperature. The dashed line is a linear least-squares fit to the low-temperature data while the dash-dotted lines are meant to guide the eye only. After *Kahn* et al. [1989]

The strong influence of the substrate temperature on the growth mode is also illustrated by the Ga-on-GaAs(110) data displayed in Fig. 3.23. They were recorded using SXPS during the deposition of gallium on cleaved GaAs(110) surfaces at room and low temperature [*Kahn* et al. 1989]. With increasing *nominal* Ga coverage the As(3*d*) core-level signal decays exponentially during deposition at low temperature but gradually reduces to only abaout 30% of the value observed with the initially clean surface during deposition of a *nominal* equivalent of 3.2 nm or approximately 16 layers of Ga. These data indicate *Frank–van der Merwe* or layer-by-layer growth at low temperature but no wetting and *Volmer–Weber* or island growth instead at room temperature. The scanning micrographs shown in Fig. 3.24 demonstrate the presence of islands even after the *nominal* deposition of 150 layers of indium on GaAs(110) at room temperature. [*Savage* and *Lagally* 1986]. Theoretical investigations of the initial growth steps confirmed the *Volmer–Weber* growth mode for Al deposited on GaAs(110) surfaces at room temperature [*Zunger* 1981, *Ihm* and *Joannopoulos* 1982]. The Al atoms are very mobile on GaAs(110) surfaces and all possible adsorption sites turned out to be unstable against the formation of Al–Al bonds, i.e., the formation of Al clusters and three-dimensional islands.

The most typical example of *Stranski–Krastanov* growth is germanium deposited on silicon surfaces. The surface free energy is lower for Ge than for Si [*Miedema* 1978, *Mezey* and *Giber* 1982] and the lattice constant of Ge is by approximately 4% larger than that of Si. *Bauer*'s free-energy criterion (3.56) suggests

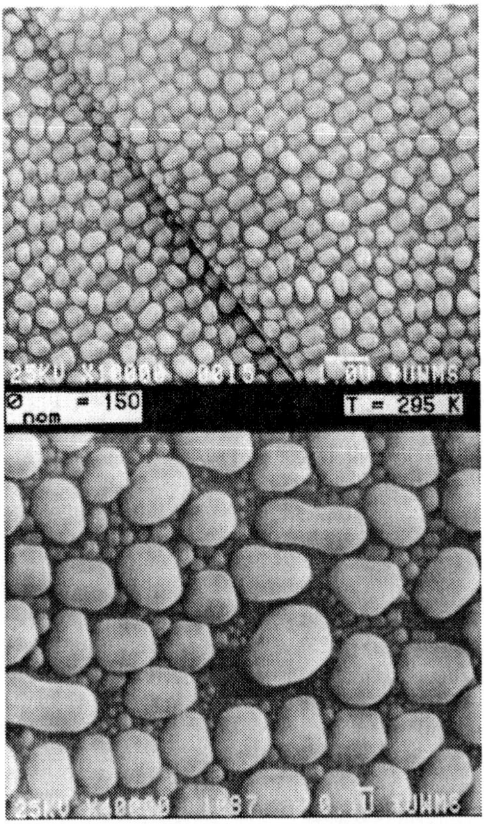

Fig. 3.24. Scanning electron micrographs of *nominally* 150 layers of In condensed on a cleaved GaAs(110) surface at room temperature. From *Savage* and *Lagally* [1986]

Volmer–Weber or island growth for Si on Ge but *Frank–van der Merwe* or layer-by-layer growth for Ge on Si. Indeed, three Ge layers were found to grow layer by layer on Si(001) at 500°C but then nucleation and growth of three-dimensional Ge islands sets in [*Tsaur* et al. 1981, *Asai* et al. 1985, *Sakamoto* et al. 1987, *Mareé* et al. 1987]. This transition from initial *Frank–van der Merwe* growth to the *Stranski–Krastanov* mode is caused by the large lattice strain which is relieved by the generation of misfit dislocations. Silicon evaporated on Ge(001) surfaces, on the other hand, grows into three-dimensional islands right from the beginning [*Mareé* et al. 1987, *Copel* et al. 1989].

3.10 Barrier Heights of Real Schottky Diodes 4: XPS

The formation of Schottky contacts has been extensively studied with core-level photoemission spectroscopy (PES). The determination of interface band-bending at metal–semiconductor interfaces by using this experimental technique requires continuous and metallic films which have a thickness of less than approximately 4 nm. However, most metals evaporated on semiconductor surfaces initially form islands which eventually coalesce at a *nominal* film thickness well beyond the value critical for the detection of the core-level PES signal from the underlying semiconductor. For such metal–semiconductor systems, the determination of interface band-bending from core-level photoemission lines is extremely difficult if not impossible.

Cesium grows layer-by-layer on semiconductors at low temperatures and, therefore, the barrier heights of such contacts can be easily determined by applying core-level PES. As an example, the lower diagram of Fig. 3.22 displays the kinetic energy of the Si($2p_{3/2}$) photoemission signal as a function of the evaporation time of cesium on a *n*-Si(111)-7×7 surface at 170 K [*Weyers* al. 1999]. Emission from the sample holder provided the kinetic energy of electrons emitted from the Fermi edge. The energy distance $W_F - W_{vi}(\Gamma)$ from the valence-band maximum to the Fermi level was determined with the clean Si(111)-7×7surface from the kinetic energy of the He I photoemission signals of three characteristic clean-surface surface states. The amount of cesium deposited results from the data depicted in the diagram above. The variation of the kinetic energy of the Si($2p_{3/2}$) core levels and, therefore, of the energy distance from the Fermi level to the band edges correlates with the Cs coverage. In the submonolayer-coverage regime, the energy separation between the Fermi level and the valence-band edge first increases to a maximum value of 0.77 eV. This change in surface band-bending is explained by the removal of the clean-surface surface states and their replacement by cesium-induced surface states which pin the Fermi level at 0.77 eV above the valence-band top [*Mönch* 1970]. With further increase of the Cs coverage to above the first continu-

Table 3.6. Barrier heights of laterally homogeneous Schottky contacts in eV as obtained by photoemission spectroscopy. The image-force lowering is not considered.

Substrate	Metal	Interface	Φ_{Bn}^{PES}	Φ_{Bp}^{PES}	References
Si(111)	Na	$(7\times7)^i$	0.54		*Paggel* et al. [1998]
(111)	Na	$(3\times1)^i$	0.61		*Paggel* et al. [1998]
(111)	Cs	$(7\times7)^i$		0.63	*Weyers* et al. [1999]
GaN(0001)	Cs			3.20	*Eyckeler* et al. [1998]
GaP(110)	Cs		0.93	1.34	*Linz* et al. [1993]
GaAs(110)	Cs		0.70		*Spicer* et al. [1975]
(110)	Cs		0.67		*Grunwald* [1987]
SiC: 4H-(0001)	Cs		0.90		*van Elsbergen* [1998]
6H-(0001)	Cs		0.57		*van Elsbergen* et al. [1996]

ous layer, the Fermi level becomes eventually pinned at 0.63 ± 0.07 eV above the valence-band maximum. Since now the Cs film is metallic this latter pinning position gives the barrier height $\Phi_{Bn}^{PES} = 0.49 \pm 0.07$ eV for Cs/*n*-Si(111)-(7×7)^{i} contacts.

Barrier heights of well-characterized metal–semiconductor interfaces which were determined by using photoemission spectroscopy are summarized in Table 3.6.

3.11 Metal-Induced Core-Level Shifts and the Intrinsic Interface Electric-Dipole Layer

Diatomic, heteropolar molecules generally exhibit electric-dipole moments. This is described as a partial ionic character of covalent bonds. Hence, the covalent bonds between metal and semiconductor atoms right at metal–semiconductor interfaces are also partly ionic and give rise to intrinsic electric-dipole layers at such

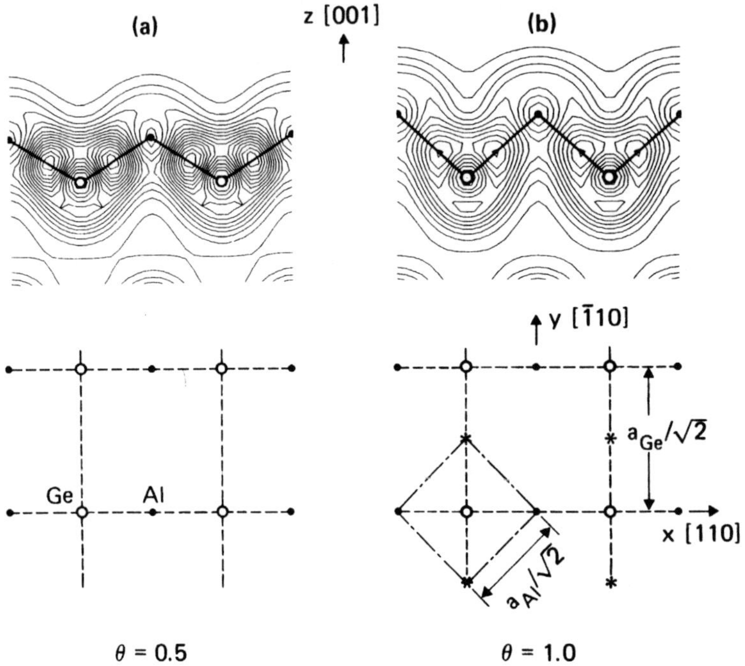

Fig. 3.25. Charge-density contours in the ($\bar{1}$10) plane of an Al/Ge(001) interface corresponding to Al in bridge-bonded sites for 0.5 and 1 monolayer. The contour spacings are 0.005 a.u. Atomic arrangements are also shown for both coverages; stars and filled circles represent Al atoms, and open circles Ge atoms. From *Batra* and *Ciraci* [1986]

contacts. The interfacial dipoles polarize the bonds in the semiconductor layers next to the interface but the corresponding displacements of the bonding charges decay exponentially into the semiconductor within a few layers. In a very simple first approximation, this polarization of bonds away from the actual interface may be neglected and the interface bonds are then described as covalently bonded *interface molecules*.

This approach is similar to the *surface-molecule* concept which is successfully used to characterize the electronic properties of adatoms on semiconductor surfaces; see *Mönch* [2001]. Strong support of such a simple approach comes from a theoretical study of the electronic properties of aluminum adatoms on Ge(001) surfaces in bridge-bonded sites. Figure 3.25 displays the charge-density contours which were obtained in a self-consistent calculation [*Batra* and *Ciraci* 1986]. The Al atoms are found to form covalent bonds with substrate Ge atoms. The bonding charges are slightly shifted away from the Al towards the Ge atoms. This behavior is easily explained when *Pauling*'s intuitive but most successful electronegativity concept is applied to interface bonds at metal–semiconductor contacts.

Pauling [1939/1960] correlated the amount of ionic character or the ionicity, Δq_1, of single bonds in diatomic molecules A–B with the difference $X_A - X_B$ of the atomic electronegativities of the atoms forming the molecule. A revised version of the relation originally proposed by *Pauling* is that of *Hanney* and *Smith* [1946]

$$\Delta q_1 = 0.16\,|X_A - X_B| + 0.035\,|X_A - X_B|^2 \ . \tag{3.58}$$

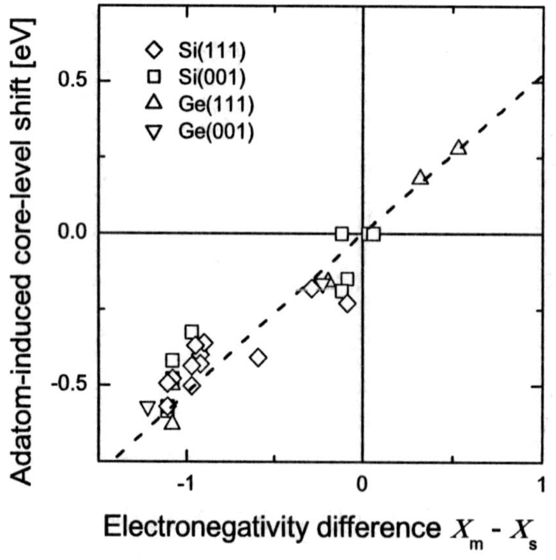

Fig. 3.26. Chemical shifts of Si(2*p*) and Ge(3*d*) core-levels induced by metal adatoms on silicon and germanium surfaces, respectively, as a function of the difference $X_m - X_s$ of the metal and the semiconductor electronegativities in Pauling units. The data are summarized in Table 3.6

In a simple point-charge model, the atoms are charged by $+\Delta q_1$ and $-\Delta q_1$, where the more electronegative atom becomes negatively charged. In a more realistic picture the bond charge is slightly displaced towards the more electronegative atom in heteropolar molecules, while it is in the middle between both atoms in homopolar diatomic molecules. The charge-contour plots of Fig. 3.25 show this be-havior. The Al–Ge bond charges are shifted towards the Ge substrate atoms, which are somewhat more electronegative than the Al adatoms.

By using the simple point-charge model, the dipole moments of diatomic molecules with $X_A - X_B \neq 0$ may be expressed as

$$p = \Delta q_1 e_0 d_{cov} \quad . \tag{3.59}$$

The bond length

$$d_{cov} = r_{cov}^A + r_{cov}^B - c_X \mid X_A - X_B \mid \tag{3.60}$$

is taken as the sum of the covalent radii of the two atoms, A and B, minus a correction term that accounts for the slight reduction of bond lengths due to the partial ionic character of the covalent bond in heteropolar molecules.

With increasing adatom coverage the distance between the adatoms reduces. Each dipole is exposed to the electrostatic field of all other dipoles, which points in the direction opposite to the alignment of the dipoles themselves. This depolarization will gradually decrease the effective dipole moment as a function of adatom coverage [*Topping* 1927] but the chemical trend of the adatom-induced electric-dipole moments will remain unaltered. The same argument also holds for the adatom-induced core-level shifts of the substrate atoms. Experimentally observed core-level shifts induced by metal adatoms on silicon and germanium surfaces are summarized in Table 3.7 and plotted in Fig. 3.26 as a function of the electronega tivity difference $X_m - X_s$ of the metal and the semiconductor atoms. All metal adatoms which are less or more electronegative than the two elemental semiconductors cause core-level shifts towards smaller and larger binding energies, respectively. Furthermore, only the electronegativity difference $X_m - X_s$ rather than the specific semiconductor matters. *Pauling*'s electronegativity concept indeed predicts such behavior. The dashed line is a linear least-squares fit to the data points ($r_1 = 0.96$). The slope parameter is 0.526 ± 0.015 eV per Pauling unit. In view of relation (3.57), which is the basic assumption of *Pauling*'s concept, the metal adatoms induce *effective* shifts of the Si(2*p*) and Ge(3*d*) core levels by approximately 3.29 eV per electron charge transferred.

Covalent bonds between metal and semiconductor atoms still exist at metal–semiconductor interfaces. This has been demonstrated by *Zhang* et al. [1986]. They applied density-functional theory in the local-density approximation to Al/GaAs(110) interfaces and used *ab initio* pseudopotentials not only for the semiconductor but also for the metal atoms. The charge-density contours displayed in Fig. 3.27 clearly show covalent bonds right at the Al/GaAs(110) interface. The bond charge is slightly shifted towards both the interfacial Ga and As atoms. Again, *Pauling*'s electronegativity concept predicts such behavior since Al atoms are more electropositive than As and, although to a lesser degree, Ga atoms.

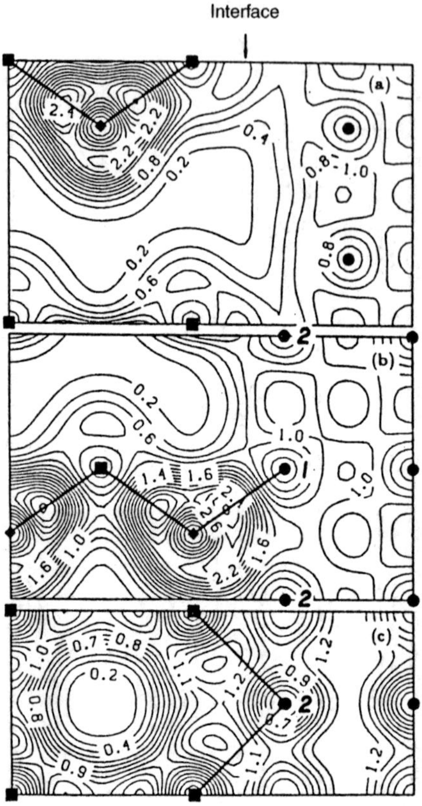

Fig. 3.27. Charge contour plots in the $(1\bar{1}0)$ (a and b) and the (100) plane (c) of Al/GaAs(110) contacts. Solid squares are gallium atoms, solid diamonds are arsenic atoms, and aluminum atoms are denoted by solid circles. From *Zhang* et al. [1986]

Naturally, the ionicity of the covalent bonds changes when the metal coverage on semiconductor surfaces is gradually increased from well below a monolayer up to a uniform metallic layer. Figure 3.28 displays core-level shifts as observed by *Prietsch* et al. [1989] during the evaporation of Na onto cleaved *p*-GaAs(110) surfaces at low temperatures. They recorded the Na(2*p*) and the As(3*d*) core-level signals using 80-eV photons and decomposed the photoemission lines into first- and further-layer components. The Na(2*p*) data represented by the gray squares were assigned to Na islands formed already at submonolayer coverages. The Na(2*p*)and the As(3*d*) first-layer core-level signals shift with opposite sign as the Na coverage increases. In the submonolayer Na coverage regime, the binding energies of the As(3*d*) and of the Na(2*p*) first-layer components become larger and smaller, respectively. This finding implies a reduction of the ionicity of the interfacial Al–As bonds as the Na coverage is increased. These observations are to be expected. As the Na coverage increases, the mutual depolarization of the partly ionic covalent surface bonds will decease their electric-dipole moment. This effect

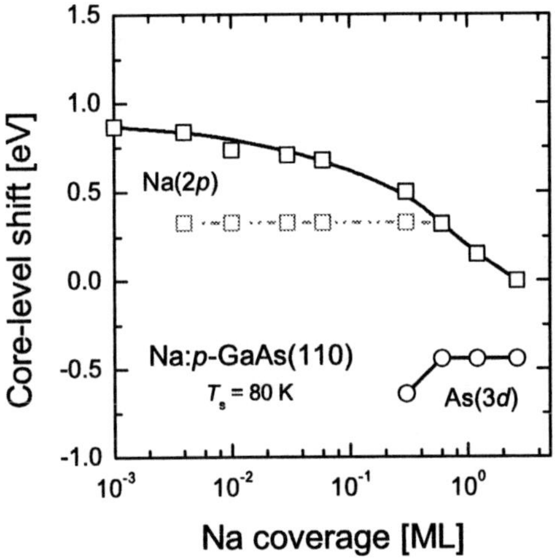

Fig. 3.28. Na(2*p*) and As(3*d*) first-layer core-level shifts as a function of Na coverage in monolayers evaporated on a cleaved GaAs(110) surface at 80 K. The Na(2*p*) data points shown in gray are assigned to 2-dimensional Na islands. Data from *Prietsch* et al. [1989]

will be further enhanced as the free electron gas forms when the Na overlayer becomes metallic. The following most important conclusion may be drawn from Figs. 3.26 through 3.28 and the underlying chemical concept. First of all, the data establish the existence of an *intrinsic* electric-dipole layer at semiconductor surfaces covered with metal adatoms and also at metal–semiconductor interfaces. The charge transfer across such interfaces varies proportionally to the difference of the metal and the semiconductor electronegativities.

3.12 Determination of Band Offsets at Heterostructures

3.12.1 Internal Photoemission Yield Spectroscopy

The band-edge discontinuities at semiconductor heterostructures were studied by applying a large variety of experimental techniques. However, the analysis of the corresponding original data requires additional simulations and the band offset searched for is a fitting parameter only. An instructive example are *C*/*V* measurements of, for example, *p*-$Al_{0.2}Ga_{0.8}As$/*p*-$Al_{0.5}Ga_{0.5}As$ heterostructures studied by *Zubkov* et al. [1999]. From their measured *C*/*V* characteristics, they determined the hole-density profiles of the heterostructures and compared them with calculated data. These simulations required not only the doping profile and the layer sequence of the samples but also the material parameters as an input and treated the valence-band offset and the built-in charge as adjustable parameters. However, there are also experimental methods which provide a more direct access to band

Table 3.7. Metal adatom-induced shifts $\Delta W_{sb} = W_b(nl) - W_s(nl)$ of Si(2*p*) and Ge(3*d*) core-levels in eV and differences $X_m - X_s$ of the Pauling electronegativities of the metal adatom and the semiconductor substrate atom

Substrate	Metal	ΔW_{sb}	$X_m - X_s$	Reference
Si(111)	Li	–0.43	–0.92	*Grehk* et al. [1993]
(111)	Li	–0.40	–0.92	*Weitering* et al. [1996]
(001)	Li	–0.37	–0.92	*Grehk* et al. [1995]
(111)	Na	–0.44	–0.97	*Okuda* et al. [1994]
(111)	Na	–0.50	–0.97	*Zhang* et al. [1996]
(001)	Na	–0.33	–0.97	*Chao* et al. [1997]
(111)	K	–0.48	–1.08	*Grupp, Taleb-Ibrahimi* [1998]
(001)	K	–0.45	–1.08	*Weitering* et al. [1993]
(001)	K	–0.42	–1.08	*Chao* et al. [1995]
(001)	Rb	–0.48	–1.08	*Chao* et al. [1998]
(111)	Cs	–0.57	–1.11	*Lin* et al. [1991a], *Magnusson* et al. [1991]
(111)	Cs	–0.49	–1.11	*Park* et al. [1995]
(001)	Cs	–0.57	–1.11	*Lin* et al. [1991a]
(001)	Cs	–0.58	–1.11	*Chao* et al. [1996]
(001)	Ag	0	+0.33	*Samsavar* et al. [1988]
(111)	Mg	–0.41	–0.59	*An* et al. [1995]
(111)	Ca	–0.36	–0.9	*Olmstead* et al. [1987]
(111)	Sr	–0.37	–0.95	*Olmstead & Bringans* [1990]
(111)	Al	–0.18	–0.29	*Andersen* et al. [1991]
(001)	Al	–0.18	–0.29	*Yeom* et al. [1996a]
(111)	Ga	–0.23	–0.09	*Higashiyama* et al. [1987]
(001)	Ga	–0.15	–0.09	*Bringans* et al. [1987]
001)	In	0	–0.12	*Rich* et al. [1987]
(001)	In	–0.19	–0.12	*Yeom* et al. [1996b]
(001)	Sn	0	+0.06	*Rich* et al. [1988]
Ge(111)	Na	–0.50	–1.08	*Soukiassian* et al. [1989]
(111)	Na	–0.63	–1.08	*Kim* et al. [1998]
(111)	Cs	–0.57	–1.22	*Lin* et al. [1991b]
(111)	Au	+0.28	+0.53	*Göthelid* et al. [1994]
(111)	Ga	–0.31	–0.20	*Bringans* et al. [1987]
(001)	In	–0.33	–0.23	*Rich* et al. [1990]
(111)	Pb	+0.18	+0.32	*Carlisle* et al. [1993]

offsets of heterostructures. Two of the widely used techniques of this kind are internal photoemission yield and core-level X-ray photoemission spectroscopy.

The application of internal photoemission yield spectroscopy (IPEYS) for the determination of band-edge discontinuities at semiconductor heterostructures is schematically explained in Fig. 3.28. Electrons may be excited from the valence-

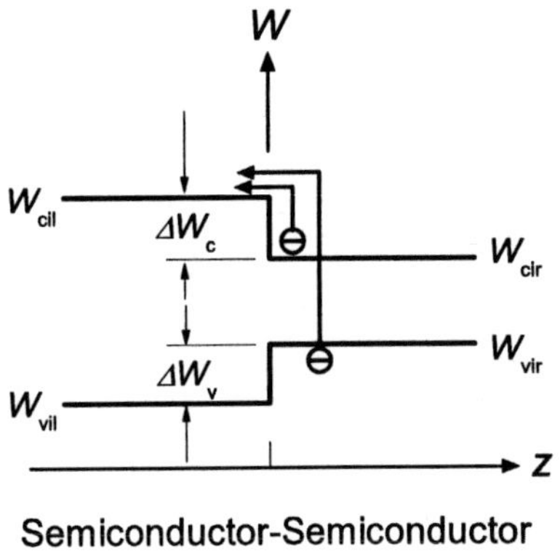

Fig. 3.29. Energy diagram of internal photoemission yield spectroscopy (IPEYS) at semiconductor heterostructures (schematic)

band of the semiconductor on the right-hand side into the conduction band of the semiconductor on the left-hand side of the junction. The spectral dependence of the internal photoemission yield is again described by relation (3.56) and the threshold photon energy $\hbar\omega_Y$ of the corresponding $[Y(\hbar\omega)\cdot\hbar\omega]^{1/2}$ versus $\hbar\omega$ plot equals the sum $\Delta W_c + W_{gr}$ of the conduction-band offset and the band-gap energy of the semiconductor on the right-hand side. Provided that the conduction band of the right-hand semiconductor is populated by electrons right at the interface then there will exist a lower IPEYS threshold energy which directly measures the conduction-band discontinuity ΔW_c. If, on the other hand, the semiconductor on the right-hand side is doped *p*-type and holes are present right at the interface, then the lowest IPEYS threshold energy equals the valence-band offset ΔW_v. The consistency of the band-offset data may thus be checked by IPEYS when the measurements are performed with heterostructures where the type of doping of one of the semiconductors is changed. *Afanas'ev* et al. [1996] and *Afanas'ev* and *Stesmans* [2000] performed such complementary studies with SiO_2/*n*-SiC and SiO_2/*p*-SiC heterostructures, respectively.

3.12.2 Core-Level X-Ray Photoemission Spectroscopy

Valence-band offsets are most reliably determined from the energy positions of core level lines in X-ray photoelectron spectra recorded with bulk samples of the two semiconductors in contact and with the respective interface itself. This technique was pioneered by *Grant* et al. [1978].[8] Since the escape depths of the respective photoelectrons are in the order of 2 nm only, one of the two semiconductors has to be sufficiently thin. This condition may be easily met when heterostructures are grown by molecular beam epitaxy (MBE) and X-ray photo-electron

[8] For details the reader is referred to an article by *Kraut* et al. [1980].

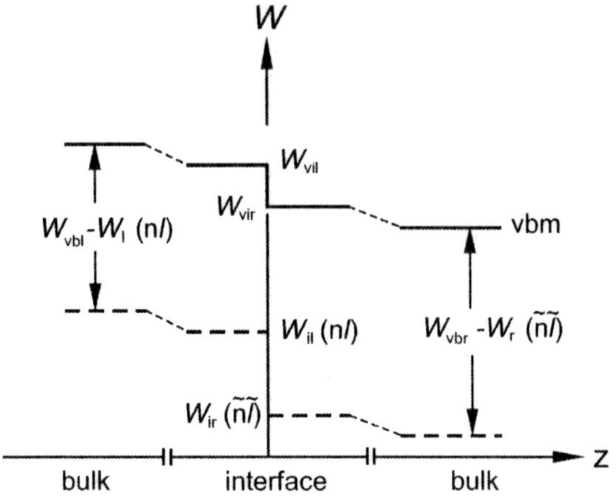

Fig. 3.30. Energy-band diagram at semiconductor heterostructures (schematic). The differing interface and bulk positions of the valence-band maxima (vbm), conduction-band minima (cbm) and core levels, (nl) on the left-hand and ($\tilde{\text{n}}\tilde{l}$) on the right-hand side of the interface, account for possible band bending due to the existence of space-charge layers. The energy differences $W_{vl} - W_l(\text{n}l)$ and $W_{vr} - W_r(\tilde{\text{n}}\tilde{l})$ from the core levels to the respective valence-band maximum do not vary locally

spectra are recorded during growth interrupts. The XPS method for determining valence-band offsets is explained by the schematic energy diagram displayed in Fig. 3.30. At abrupt interfaces, band-edge discontinuities are established within a few atomic layers [*Okabayashi* et al. 2002]. The regions of band bending depend on the doping profiles on both sides of the contact and will extend over a few Debye lengths. In Fig. 3.30 this is indicated by the z-axis breaks and the dashed lines connecting the core-level positions at the interface and in the bulk of the semiconductors. Here, only the band-edge offsets will be considered. The valence-band discontinuity results as

$$\begin{aligned}\Delta W_v &= W_{vir} - W_{vil} \\ &= W_{ir}(\text{n}l) - W_{il}(\tilde{\text{n}}\tilde{l}) + [W_{vbr} - W_{br}(\text{n}l)] - [W_{vbl} - W_{bl}(\tilde{\text{n}}\tilde{l})] \ ,\end{aligned} \tag{3.61}$$

where (nl) and ($\tilde{\text{n}}\tilde{l}$) denote the core levels considered with the semiconductors on the left, l, and the right side, r, of the interface, respectively. The subscripts i and b characterize interface and bulk properties, respectively. The energy difference $W_{ir}(\text{n}l) - W_{il}(\tilde{\text{n}}\tilde{l})$ between core levels of the two semiconductors forming the heterostructure are determined from energy-distribution curves of photoelectrons excited by X-rays during MBE growth of the heterostructure. The energy positions $W_{vbr} - W_{br}(\text{n}l)$ and $W_{vbl} - W_{bl}(\tilde{\text{n}}\tilde{l})$ of the core levels relative to the valence-band maxima in each of the two semiconductors are evaluated separately.

4. Laterally Inhomogeneous Schottky Contacts

4.1 Potential Distribution

The evaluation of current–voltage curves characterizes each individual Schottky contact by an effective barrier height and an ideality factor. The ideality factors n are generally larger than n_{if}, the value determined by the image-force effect only. In other words, the barrier heights depend more strongly on the applied voltage than because of the Schottky effect. Obviously, *real* metal–semiconductor interfaces are in one way or another not ideal. Ballistic-electron-emission spectroscopy, on the other hand, revealed spatial variations of the local BEEM thresholds or barrier heights on the nm-length scale. The corresponding histograms are well described by Gaussian functions, as verified by the experimental PtSi/n-Si(001) data shown in Fig. 3.16. Hence, regions of reduced barrier height with dimensions smaller than the depletion-layer width embedded in areas of larger but constant barrier height might be a plausible description of *real* Schottky contacts. Historically, *Freeouf* et al. [1982a, b] simulated the current transport in such *patchy* metal–semiconductor interfaces many years before *Kaiser* and *Bell* [1988] developed the BEEM technique and *Fowell* et al. [1990] published the first BEEM studies with lateral nm-resolution.

The current transport through Schottky contacts is determined by their barrier height. In contrast to ideal, i.e., laterally uniform contacts, the barrier heights of patchy Schottky diodes vary locally at their interfaces. The spatial variations of the potential in the semiconductor are solutions of the three-dimensional Poisson equation

$$\nabla^2 V(x.y.z) = -\rho(x,y,z)/\varepsilon_b\varepsilon_0 \tag{4.1}$$

with the boundary condition

$$\Phi_{Bn}(x,y) = \Phi_{Bn}^{\text{hom}} + e_0\varphi_{Bn}(x,y) = e_0 V(x,y,0) + W_n \quad . \tag{4.2}$$

Here, ∇^2 is the Laplace operator. The semiconductor is assumed to be homogeneously and non-degenerately doped n-type so that the Fermi-level position $W_n = W_{cb} - W_F = -k_B T \ln(N_d/N_c)$ is constant throughout the bulk. In order to illustrate the influence of patches with reduced barrier height and lateral dimensions smaller than the depletion-layer width on the current transport the potential distribution at the edge of metal–semiconductor contacts will be considered first.

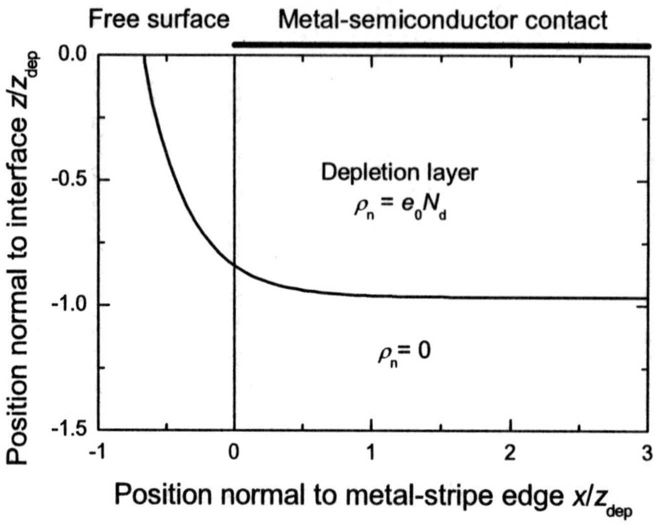

Fig. 4.1. Equipotential curve or constant charge-density contour at the edge of a stripe-shape metal–semiconductor contact. After *Wasserstrom* and *McKenna* [1970]

At the edge of a metal film on a semiconductor the space-charge layer will laterally spread along the uncovered surface. *Wasserstrom* and *McKenna* [1970] modeled finite-size Schottky contacts by stripes of constant interface potential on homogeneously doped semiconductors and solved Poisson's equation (4.1). They assumed the stripes to be much wider than the Debye length of the semiconductor. Figure 4.1 displays one typical example of their numerical results. Along the uncovered semiconductor surface the depletion layer extends approximately three-quarters of its width z_{dep} directly below the metal–semiconductor interface.

These findings of *Wasserstrom* and *McKenna* have direct implications for *patchy* metal–semiconductor contacts. At the border between regions of different barrier heights, the potential distribution will not change abruptly but the depletion layer of the region with the larger barrier height will laterally reach into the region of lower barrier height. If the distance between opposite edges of such lateral barrier-height inhomogeneities becomes smaller than the depletion-layer width outside the patch the corresponding potential tails will overlap in front of the region with reduced barrier height. At the actual interface the Fermi level is pinned or, in other words, the lower as well as the larger barrier heights are fixed. Therefore, the potential distribution will show a saddle point in front of the region of reduced barrier height [*Freeouf* et al. 1982a, b; *Ohdomari* and *Aochi* 1987; *Bastys* et al. 1988; *Tung* 1992]. Figure 4.2 explains this behavior. For the model parameters used in the figure, the patch is said to be *pinched off*. This terminology means that majority carriers originating from outside the patch must go over a potential barrier that is larger than the barrier of the patch itself in order to reach the interface. As a consequence, the current–voltage characteristics of such patchy Schottky contacts are described by effective barrier heights that are lower than the values of

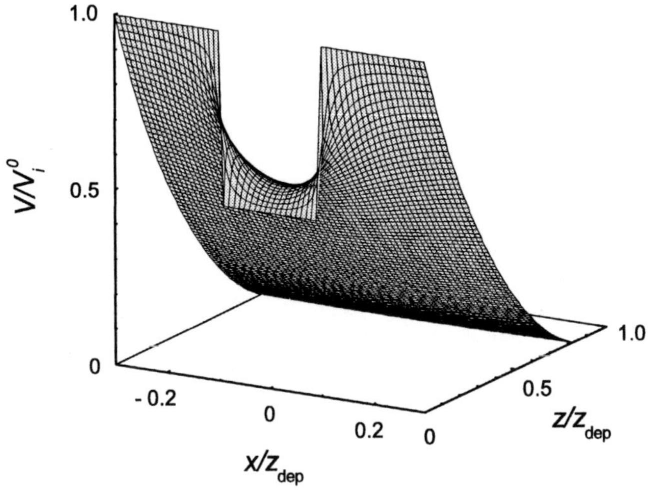

Fig. 4.2. Calculated potential distribution underneath and around a patch of reduced interface potential embedded in a region of larger interface band-bending height. The lateral dimension and the interface-potential reduction of the patch are set to two tenths of the depletion-layer width and one half of the interface band-bending of the surrounding region

the laterally uniform regions and, what is even more important, by ideality factors that are larger than the values determined by the image-force effect only.

The potential distribution within the depletion layers of patchy Schottky contacts has been determined by solving Poisson's equation (4.1) either numerically [*Freeouf* et al. 1982a, b; *Ohdomari* and *Aochi* 1987; *Bastys* et al. 1988; *Sullivan* et al. 1991] or analytically [*Tung* 1992] but then with additional simplifying assumptions. *Tung* [1992] treated the variation of the interface potential due to the inhomogeneity $e_0\varphi_{Bn}(x,y)$ of the barrier height as a perturbation. This term is described as the potential of a dipole layer with a locally varying dipole moment $2\varepsilon_b\varepsilon_0\varphi_{Bn}(x,y)$ per unit area. Consequently, the potential in the depletion layer is approximated by

$$V(x,y,z) = \\ = (V_i^0 - V_c)(1 - z/z_{dep})^2 + \int \frac{\varphi_{Bn}(x',y')}{2\pi} \frac{z}{[z^2 + (x'-x)^2 + (y'-y)^2]^{3/2}} \mathrm{d}x'\mathrm{d}y', \quad (4.3)$$

where z_{dep} is the width of the depletion layer outside the patch, i.e., in the region of uniform barrier height, and V_c is the applied-voltage drop across the depletion layer. The first term on the right-hand side of (4.3) is the solution of the unperturbed case where the zero-bias interface potential V_i^0 or, what is equivalent, the barrier height $\Phi_{Bn} = \Phi_{Bn}^{hom} = e_0 \,|\, V_i^0 \,| + W_n$ is uniform across the whole interface.

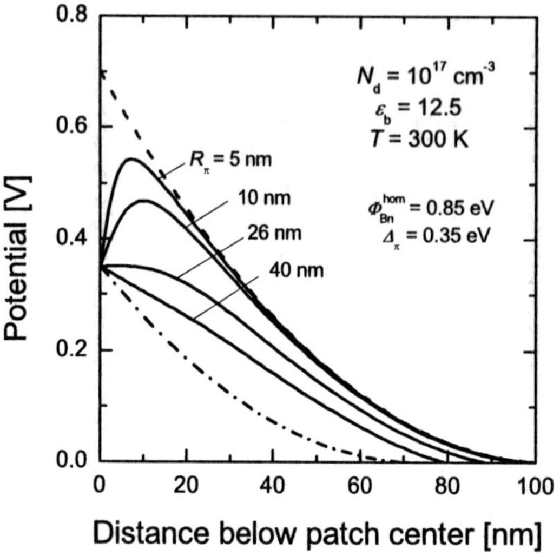

Fig. 4.3. Potential distribution below the center of circular patches with a barrier height lowered by 0.35 eV and with radii varying between 5 and 40 nm embedded in a region of constant barrier height of 0.85 eV and a depletion layer width of 103 nm. The dashed and the dotted lines represent the potential distribution below the large-barrier area and very wide patches (R_π>> 40 nm). Pinch-off sets in for patch radii smaller than 26 nm

A most simple and illustrative example of a Schottky barrier-height inhomogeneity is a circular patch of radius R_π and reduced barrier height $\Phi_{Bn}^{\pi} = \Phi_{Bn}^{hom} - \Delta_\pi$ embedded in a region of uniform barrier height Φ_{Bn}^{hom}. The approximate solution (4.3) of Poisson's equation then gives the potential distribution in the z-direction below the center of the patch ($r = 0$) as

$$\begin{aligned} V(0,z) - V_c &= (V_i^0 - V_c)\cdot\left(1 - \frac{z}{z_{dep}}\right)^2 - \int_0^{R_\pi} \frac{\Delta_\pi}{2\pi e_0} \frac{z}{(z^2 + r^2)^{3/2}} 2\pi r \, \mathrm{d}r \\ &= (V_i^0 - V_c)\cdot\left(1 - \frac{z}{z_{dep}}\right)^2 - \frac{\Delta_\pi}{e_0}\left[1 - \frac{z}{\left(z^2 + R_\pi^2\right)^{1/2}}\right]. \end{aligned} \quad (4.4)$$

Figures 4.3 to 4.5 illustrate the potential distribution (4.4) below the center of circular patches.

As Figs. 4.3 and 4.4 indicate, no pinch-off will occur for large patch radii R_π and/or small differences $\Delta_\pi = \Phi_B^{hom} - \Phi_B^{\pi}$ between the barrier heights of the embedding region and of the patch itself. Pinch-off means that the initial slope of the potential (4.4) is positive. The first derivative of (4.4) thus sets the condition for pinch-off as

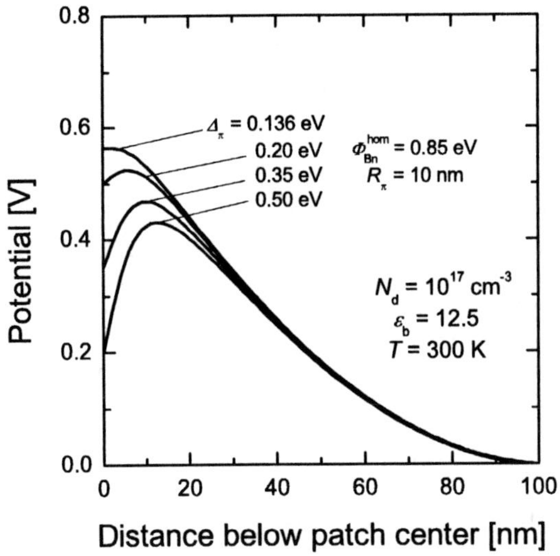

Fig. 4.4. Potential distribution below the center of circular patches with a barrier height lowered between 0.136 and 0.5 eV and with a radius of 10 nm embedded in a region of constant barrier height of 0.85 eV and a depletion layer width of 103 nm

$$\frac{\Delta_\pi}{e_0(V_i - V_c)} > \frac{2R_\pi}{z_{dep}} \quad . \tag{4.5}$$

For the parameters used in Figs. 4.3 and 4.4 condition (4.5) gives a critical patch radius of 26 nm and a critical barrier-height lowering of 0.136 eV, respectively, for pinch-off to occur. The corresponding potential distributions are displayed in both figures.

The area integral on the right-hand side of relation (4.3) cannot be expressed analytically for all cases. *Tung* [1992] simplified the calculation of the potential distribution around circular patches with lowered barrier height by replacing the spatially extended dipole layer by a point dipole. This approximation results in a singularity at the core of the dipole, but it nevertheless provides an adequate description of the current transport in patchy Schottky contacts since it is the saddle-point potential, which is some distance away from the interface, that determines the current transport.

Tung's point-dipole approximation gives the potential distribution around a circular patch with radius R_π and a barrier height lowered by Δ_π, so that the total moment of the point dipole is $2(\varepsilon_b \varepsilon_0 / e_0)\Delta_\pi \pi R_\pi^2$, as

$$V(r,z) \approx (V_i^0 - V_c)\left(1 - \frac{z}{z_{dep}}\right)^2 + V_c - \frac{\Delta_\pi R_\pi^2}{2} \frac{z}{(z^2 + r^2)^{3/2}} \quad , \tag{4.6}$$

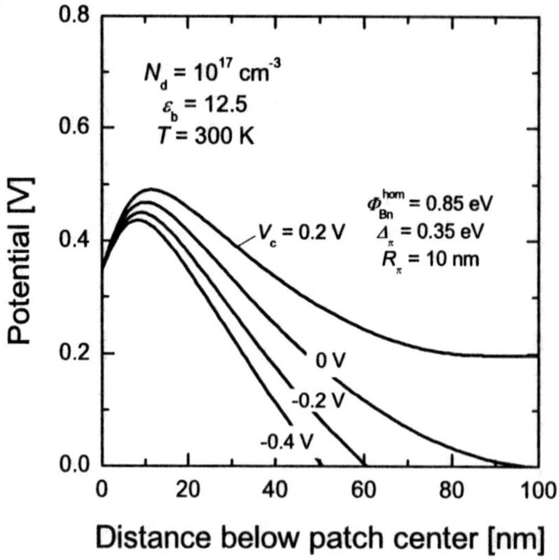

Fig. 4.5. Potential distribution below the center of a circular patch with a barrier height lowered by 0.35 eV and with a radius of 10 nm embedded in a region of constant barrier height of 0.85 eV and a depletion layer width of 103 nm as a function of the applied-voltage drop across the depletion layer

where r is the radial distance from the patch center or, what is the same, from the position of the point dipole.

By setting the first derivative of relation (4.6) to zero one obtains the location z_{sad} and the barrier height $\Phi_{Bn}^{sad} = e_0 V(0, z_{sad}) + W_n$ of the saddle point in front of circular pinched-off patches. With the assumption $\Delta_\pi R_\pi^2 << 2e_0 V_i^0 z_{dep}^2$ one finds the position of the saddle point as

$$z_{sad} = z_{dep}\left(\frac{\Delta_\pi R_\pi^2}{2e_o(V_i^0 - V_c)z_{dep}^2}\right)^{1/3} = z_{dep}\Gamma_\pi \tag{4.7}$$

and the potential at the saddle point as

$$V_\pi^{sad} = \left[1 - 3\left(\frac{1}{2}\frac{\Delta_\pi}{e_0\left(V_i^0 - V_c\right)}\frac{R_\pi^2}{z_{dep}^2}\right)^{1/3}\right]\left(V_i^0 - V_c\right) = (1 - 3\Gamma_\pi)(V_i^0 - V_c) \ . \tag{4.8}$$

The dimensionless space-charge parameter Γ_π is defined as

$$\Gamma_\pi = \left(\frac{1}{2}\frac{\Delta_\pi}{e_0\left(V_i^0 - V_c\right)}\frac{R_\pi^2}{z_{dep}^2}\right)^{1/3} \ . \tag{4.9}$$

Lateral inhomogeneities of the barrier height are thus characterized by their patch parameters

$$\gamma_\pi = 3\left(\Delta_\pi R_\pi^2 / 4\right)^{1/3} \tag{4.10}$$

or, what is the same and was the basic assumption, by their dipole moments $2(\varepsilon_b \varepsilon_0 / e_0)\Delta_\pi \pi R_\pi^2$.[1]

Olbrich et al. [1997, 1998] deliberately fabricated Schottky contacts with laterally inhomogeneous barrier heights by growing Co clusters on n-$GaAs_{0.67}P_{0.33}(001)$ surfaces prior to the deposition of thick Au films. They then studied these patchy interfaces using scanning ballistic-electron-emission spectroscopy. Figure 4.6 displays BEEM profiles through such Au-embedded Co islands of different lateral dimensions. The analysis of the BEEM current–voltage

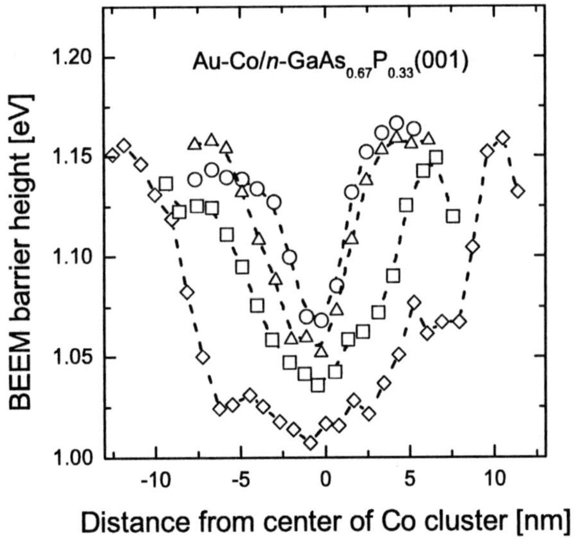

Fig. 4.6. BEEM profiles underneath Co clusters embedded in a Au film on an n-$GaAs_{0.67}P_{0.33}$ substrate. Diameter of Co clusters: (O) 5 nm, (Δ) 7 nm, (□) 10 nm, (◇) 18 nm. Data from *Olbrich* et al. [1997]

[1] *Rau* et al. [1992] proposed a somewhat different but very interesting approach. They decomposed the variations $\varphi_{Bn}(x,y)$ of the barrier height right at the metal–semiconductor interface into a Fourier series and solved Poisson's equation. Considering one Fourier component with a wavelength smaller than the depletion layer width only, they verified the saddle point of the potential in front of the barrier-height minima. To simulate real Schottky contacts they assumed that, first, only a small portion of the interface exhibits potential fluctuations with lateral dimensions smaller than the depletion layer width and, second, both the lateral size and the amplitude of the potential may be modeled by Gaussian distributions. Unfortunately, *Rau* et al. only announced but never published a complete derivation of the analytical expressions given in their publication mentioned above.

curves yielded an average BEEM barrier height of 1.18 eV at the Au/n-$GaAs_{0.67}P_{0.33}$ interface outside of the patches. The broad and trough-shaped potential in front of the 18-nm Co island indicates the absence of pinch-off and a BEEM barrier height of approximately 1.03 eV at the Co/n-$GaAs_{0.67}P_{0.33}$ interface. The n-$GaAs_{0.67}P_{0.33}$ substrates had a doping level of 5×10^{16} cm^{-3}, so that the depletion layers are approximately 160 nm wide. Relation (4.5) then predicts that for the barrier height lowering $\Delta_\pi = 0.15$ eV underneath the Co clusters pinch-off will set in for patch diameters $2R_\pi < 16$ nm, in perfect agreement with the experimental observations. The barriers in front of the smaller Co islands studied increase by up to 50 meV when the diameters of the clusters decrease from 18 to 5 nm. Figure 4.7 displays the averages of the BEEM barrier heights and the zero-bias saddle-point barrier heights $\Phi_{Bn}^{sad0} = e_0 V_\pi^{sad0} + W_n$, which were computed using relation (4.8), both as a function of the diameter of the Co clusters. The calculated curve excellently fits the experimental data. These findings verify that BEEM probes the saddle point barrier heights Φ_{Bn}^{sad} in front of patches of lowered barrier height embedded in an area of larger barrier height rather than the *true* barrier heights $\Phi_{Bn}^{hom} - \Delta_\pi$ of the patches.

The lateral variations of the BEEM barrier heights of real Schottky contacts are well described by Gaussian distributions. This is demonstrated by the data of PtSi/n-Si(001) contacts shown in Fig. 3.16 as an example. Therefore, the saddle-point barrier heights are also Gaussian-distributed. The saddle-point potential (4.8) together with definitions (4.9) and (4.10) of the space-charge Γ_π and the patch pa-

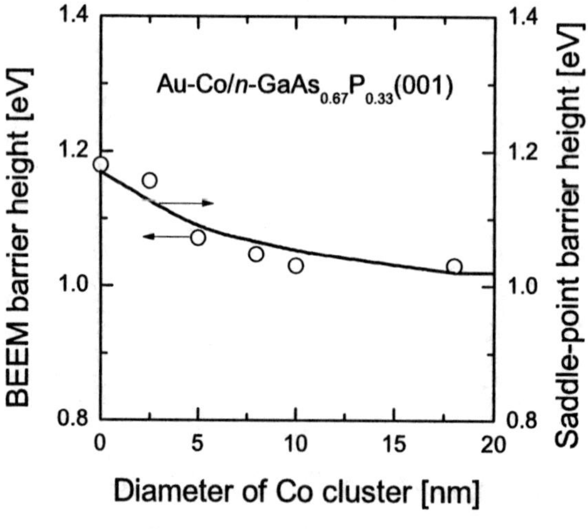

Fig. 4.7. BEEM barrier heights and calculated saddle-point barrier heights underneath Co clusters embedded in a Au film on an n-$GaAs_{0.67}P_{0.33}$ substrate as a function of the diameter of the Co islands. Data from *Olbrich* et al. [1997, 1998]

rameter γ_π, respectively, and the depletion-layer width (2.18) gives the lowering

$$\delta\Phi_\pi^{sad} = \Phi_{Bn}^{hom} - (e_0 V_\pi^{sad} + W_n) = 3\Gamma_\pi e_0 (V_i^0 - V_c) = \gamma_\pi \left(\frac{e_0 (V_i^0 - V_c) k_B T}{L_D^2} \right)^{1/3} \quad (4.11)$$

of the barrier height right at the saddle point relative to the larger value Φ_{Bn}^{hom} of the surrounding region. Hence, the patch parameters γ_π are also Gaussian-distributed in real Schottky diodes.

4.2 Current Transport in Schottky Contacts with One Circular Patch

Assuming thermal emission over the barrier, the current transport across a metal–semiconductor contact containing pinched-off patches is determined by the potential distributions around the saddle points in front of the patches and their variations with the applied voltage. For that purpose the potential (4.6) around the saddle-point location (4.7) may be expanded in a power series with respect to the radial coordinate r. At the saddle point the first derivative of (4.6) is zero *per definitionem* so that the second derivative provides the first non-vanishing and, therefore, quadratic term. Hence, one obtains

$$V(r, z_{sad}) = V_\pi^{sad} + \frac{3(V_i^0 - V_c)}{2\Gamma_\pi z_{dep}^2} r^2 + \ldots \quad . \quad (4.12)$$

The thermal-emission current in the forward direction through a Schottky contact containing one circular pinched-off patch only is then given by

$$\begin{aligned} I_{te}^\pi (V_c > 0) &\approx \int_0^\infty A_R^* T^2 \exp\left(-\frac{e_0 [V(r, z_{sad}) - V_c]}{k_B T} \right) 2\pi r^2 \mathrm{d}r \\ &= A_R^* T^2 \frac{2\pi\Gamma_\pi z_{dep}^2 k_B T}{3e_0 (V_i^0 - V_c)} \exp\left(-\frac{\Phi_{Bn}^{hom} - 3\Gamma_\pi e_0 (V_i^0 - V_c)}{k_B T} \right) \exp\left(\frac{e_0 V_c}{k_B T} \right) \quad . \end{aligned} \quad (4.13)$$

The current in the reverse direction compensates the forward current at zero bias so that the total current–voltage characteristics results as

$$I_{te}^\pi = A_R^* T^2 \frac{2\pi\Gamma_\pi z_{dep}^2 k_B T}{3e_0 (V_i^0 - V_c)} \exp\left(-\frac{\Phi_{Bn}^{hom} - 3\Gamma_\pi e_0 (V_i^0 - V_c)}{k_B T} \right) [\exp\left(\frac{e_0 V_c}{k_B T} \right) - 1] . \quad (4.14)$$

Relation (4.13) may be further simplified. The prefactor

$$A_\pi^{eff} = \frac{2\pi\Gamma_\pi z_{dep}^2 k_B T}{3e_0 (V_i^0 - V_c)} \quad (4.15)$$

has the dimension of an area and may be described as the effective area of the Schottky contact containing one singular circular patch of lowered barrier height.

According to relation (4.8), the second term in the exponent of the first exponential equals the lowering

$$\delta\Phi_{\pi}^{sad} = \Phi_{Bn}^{hom} - \Phi_{Bn}^{sad} =$$

$$= 3\left(\frac{1}{2}\frac{\Delta_{\pi}}{e_0(V_i^0 - V_c)}\frac{R_{\pi}^2}{z_{dep}^2}\right)^{1/3} e_o(V_i^0 - V_c) = 3\Gamma_{\pi} e_o(V_i^0 - V_c) \tag{4.16}$$

of the barrier height Φ_{Bn}^{sad} at the saddle point with respect to the value Φ_{Bn}^{hom} of the region surrounding the patch.

Like the image-force lowering (2.29), the patch-induced reduction (4.16) of the barrier height also depends on the applied voltage. Expansion of (4.16) in a power series gives

$$\delta\Phi_{\pi}^{sad} = \delta\Phi_{\pi}^{sad0} + \Gamma_{\pi}^{0} e_0 V_c + \ldots \; . \tag{4.17}$$

Consequently, the *effective* barrier height

$$\Phi_{Bn}^{eff} = \Phi_{Bn}^{hom} - \delta\Phi_{Bn}^{sad0} = \Phi_{Bn}^{hom} - 3\Gamma_{\pi}^{0} e_0 V_i^0 \tag{4.18}$$

and the *patch-induced* ideality factor

$$n_{\pi} \approx 1 + \Gamma_{\pi}^{0} = 1 + \frac{\delta\Phi_{\pi}^{sad0}}{3e_0 V_i^0} \tag{4.19}$$

characterize the current transport in a Schottky contact containing one pinched-off circular patch of reduced barrier height [*Tung* 1992].

The Schottky contact considered in Fig. 4.5 contains one circular patch of lowered barrier height which is characterized by the zero-bias space-charge parameter $\Gamma_{\pi}^{0} = 0.133$. Hence, its *effective* barrier height is lower by 0.28 eV than without the patch and its *patch-induced* ideality factor has the value 1.133. Both values are larger than the barrier-height reduction $\delta\Phi_{if}^{0} = 40$ meV and the ideality factor $n_{if} = 1.02$ which are caused by the image-force effect.

The reduction of the barrier height and the ideality factor are both determined by the space-charge factor Γ_{π}^{0}. The combination of the corresponding relations (4.16) and (4.18) yields the linear correlation

$$\Phi_{Bn}^{eff} = \Phi_{Bn}^{hom} - 3\Gamma_{\pi}^{0}(\Phi_{Bn}^{hom} - W_n)\cdot(n_{\pi} - 1) \tag{4.20}$$

between the effective barrier heights and the patch-induced ideality factors of Schottky contacts containing one circular region of reduced barrier height with lateral dimensions smaller than the depletion layer width embedded in an area with larger and constant barrier height. This simple model for small space-charge parameters Γ_{π}^{0} already provides a *phenomenological* explanation of the linear correlation (3.41) experimentally observed between the effective barrier heights and the ideality factors of real Schottky contacts.

Tung [1992] performed a similar analysis for Schottky contacts containing stripes of reduced barrier heights with lateral dimensions smaller than the depletion layer width embedded in areas of larger and constant barrier height. For such configurations his results give a factor of 4 instead of 3 in relation (4.20). The dashed line in Fig. 1.19 demonstrates that this result of *Tung*'s small-signal approach confirms the outcome of the earlier numerical simulations done by *Freeouf* et al. [1982,a, b].

4.3 Current Transport in Schottky Contacts with Many Circular Patches

The experimental results presented in Sect. 3.5 demonstrate that the effective barrier heights and the ideality factors of real Schottky contacts differ from one diode to another even if they were prepared under identical experimental conditions. However, both quantities were found to vary in a correlated way in that the effective barrier heights decrease as the ideality factors become larger. Such behavior may be modeled by patches of reduced barrier height with lateral dimensions smaller than the depletion layer embedded in an area of larger and constant barrier height. Using ballistic-electron emission spectroscopy, on the other hand, nanometer-scale variation of the respective BEEM threshold voltages were detected in real Schottky contacts. The histograms of such data are well fitted by Gaussian distributions as the data displayed in Fig. 3.16 demonstrate.

The reduction of the effective barrier height and the increase of the ideality factor induced by an individual circular patch are both determined by the patch parameter $\gamma_\pi = 3(\Delta_\pi R_\pi^2/4)^{1/3}$. This is shown in Sects. 4.1 and 4.2. Considering the BEEM results, a Gaussian distribution of the patch parameters γ_π, i.e.,

$$N(\gamma_\pi) = \frac{A\rho_\pi}{\sqrt{2\pi}\sigma_\pi}\exp\left(-\frac{\gamma_\pi^2}{2\sigma_\pi^2}\right) \tag{4.21}$$

is a well-founded assumption for real Schottky contacts. The area density of the patches is denoted as ρ_π. The standard deviation σ_π of the distribution may be interpreted as an average patch parameter

$$\sigma_\pi = <\gamma_\pi> = 3<(\Delta_\pi R_\pi^2/4)^{1/3}> \ . \tag{4.22}$$

By integration over the total area of such "patchy" Schottky contacts *Tung* [1992] obtained the total thermionic current as

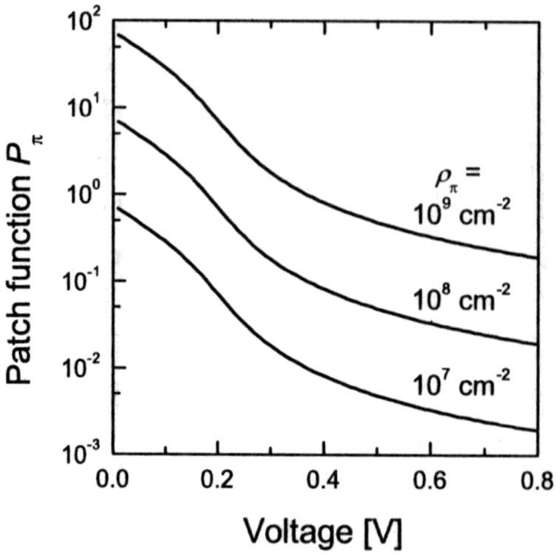

Fig. 4.8. Patch function P_π versus the applied voltage. The model parameters assumed were $\Phi_{Bn}^{hom} = 0.7$ eV, $\sigma_\pi = 2\times10^{-4}$ eV$^{1/3}$cm$^{2/3}$, $R_s = 80\ \Omega$, and $N_d = 1.4\times10^{15}$ cm^{-3}. The area densities ρ_π are given in the Figure (see also Fig. 4.9)

$$I_{te}^{tot} = AA_R^* T^2 \exp\left(-\frac{\Phi_{Bn}^{hom}}{k_B T}\right)\left[\exp\left(\frac{e_0 V_c}{k_B T}\right)-1\right] \times\left\{1+\frac{4\pi\sigma_\pi^3 \rho_\pi}{<\delta\Phi_\pi^{sad}>}\exp\left[\frac{1}{2}\left(\frac{<\delta\Phi_\pi^{sad}>}{k_B T}\right)^2\right]\right\} = I_{te}\{1+P_\pi\}\ . \tag{4.23}$$

The terms on the right-hand side in front of the braces are the thermionic-emission current (3.17) over a homogeneous Schottky contact with barrier height Φ_{Bn}^{hom}. The patch function P_π in (4.23) describes the lateral inhomogeneities of the barrier height that are assumed to have a Gaussian distribution of their patch paramcters γ_π as one patch with an effective area

$$A_\pi^{eff} = \frac{4\pi\sigma_\pi^3 \rho_\pi}{<\delta\Phi_\pi^{sad}>} A \tag{4.24}$$

and an effective barrier height

$$\Phi_{Bn}^{eff} = \Phi_{Bn}^{hom} - <\delta\Phi_\pi^{sad}>^2 / 2k_B T\ . \tag{4.25}$$

Figure 4.8 shows the dependence of the patch function P_π on the applied voltage. The patch function drastically decreases with increasing external bias so that, as intuitively expected, the influence of the patches on the current transport is largest at low applied voltage. This is further illustrated in Fig. 4.9.

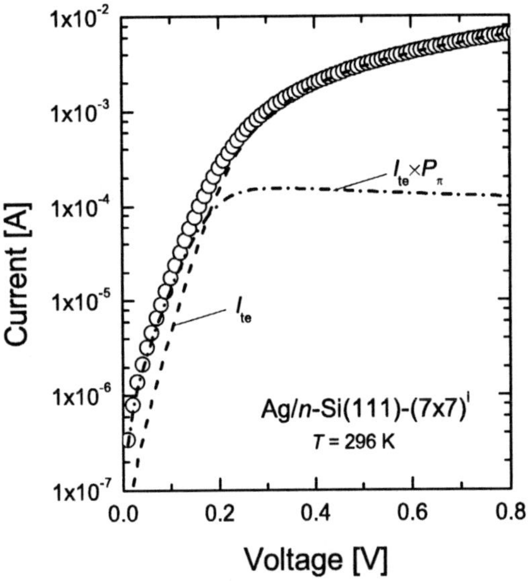

Fig. 4.9. Decomposition of a current–voltage curve measured with a Ag/*n*-Si(111)-(7×7)^{i} contact. A least-squares fit of relation (4.20) to the experimental *I*/*V* data yields Φ_{Bn}^{hom} = 0.7 eV, ρ_{π} = 1×10^{8} cm^{-2}, σ_{π} = 2×10^{-4} eV$^{1/3}$cm$^{2/3}$, R_s = 80 Ω, and N_d = 1.4×10^{15} cm^{-3}

Figure 4.9 explains the decomposition of the forward *I*/*V* characteristic of an Ag/*n*-Si(111)-(7×7)^{i} contact into the two contributions I_{te} and $I_{te}\times P_{\pi}$ of the laterally uniform contact and of the patches, respectively. Up to a bias of approximately 0.2 V the latter patch-determined term dominates, while for larger applied voltages the first term takes over. The two dashed curves are the result of least-squares fitsof relation (4.23) to the experimental data with the homogeneous barrier height Φ_{Bn}^{hom} = 0.7 eV, the average patch parameter σ_{π} = 2×10^{-4} eV$^{1/3}$cm$^{2/3}$, and the area density ρ_{π} = 1×10^{8} cm^{-2} of patches. The application of the standard *I*/*V* relation (3.38) to the *I*/*V* curve of this Ag/*n*-Si(111)-(7×7)^{i} contact, on the other hand, yielded an effective barrier height Φ_{Bn}^{eff} of 0.646 eV and an ideality factor n = 1.26.

Schmitsdorf and *Mönch* [1999] applied both the phenomenological "standard" *I*/*V* relation (3.38) as well as relation (4.23) derived for "patchy" *Schottky* contacts to a total of 68 *real* Pb/*n*-Si(111)-(1×1)^{i} contacts. Figure 4.10 displays the effective barrier heights and the ideality factors that resulted from least-squares fits of relation (3.38) to the *I*/*V* curves of all these diodes. The dashed line is a linear least-squares fit to the data for n < 1.4. It again corroborates the linear Φ_{Bn}^{eff} versus

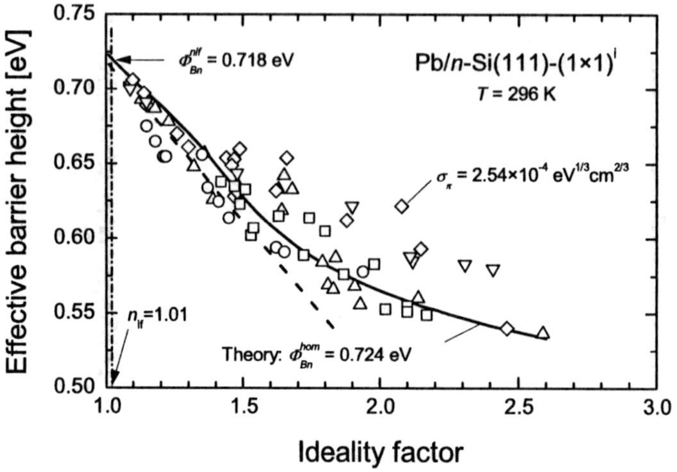

Fig. 4.10. Effective barrier heights and ideality factors of Pb/*n*-Si(111)-(1×1)[i] contacts at room temperature. Each of the symbols indicates a different Si(111) substrate. The dashed line is a linear least-squares fit to the data for $n < 1.4$. The solid line is the result of numerical simulations (for details see text). From *Schmitsdorf* and *Mönch* [1999]

n correlation (3.41). The extrapolation to $n_{if} = 1.01$, the ideality factor determined by the image-force effect only, yields a barrier height Φ_{Bn}^{nif} of 0.718 ± 0.02 eV. For ideality factors larger than approximately 1.4, the effective barrier heights level off. *Freeouf* et al. [1982,a, b] found such behavior in their simulations of the current transport in Schottky contacts containing one stripe of reduced barrier height and a width smaller than the depletion layer embedded in an area of larger barrier height. Their data are shown in Fig. 1.11.

On the other hand, *Schmitsdorf* and *Mönch* also fitted the *I/V* curves of all 68 diodes to relation (4.22) derived for patchy contacts assuming circular patches and a Gaussian distribution (4.20) of the patch parameter γ_π. This analysis yielded a mean homogeneous barrier height $<\Phi_{Bn}^{hom}> = 0.724 \pm 0.02$ eV, a mean patch parameter $<\sigma_\pi> = (1.92 \pm 0.3)\times10^{-4}$ eV$^{1/3}$cm$^{2/3}$, and a mean series resistance $<R_s> = 103 \pm 32\ \Omega$, while the area density ρ_π of the patches varied between 1×10^4 and 5×10^8 per cm^2. Within the margins of experimental error, the average homogeneous barrier height $<\Phi_{Bn}^{hom}>$ matches the extrapolated barrier height Φ_{Bn}^{nif}. This finding gives numerical evidence that homogeneous barrier heights of real Schottky contacts may be indeed extracted from Φ_{Bn}^{eff} versus n plots by extrapolation to n_{if}, the ideality factor determined by the image-force effect only. Furthermore, lateral inhomogeneities of the barrier height obviously explain the experi mentally observed correlation between effective barrier heights and ideality factors.

The current–voltage characteristics of patchy Schottky contacts may be calculated using relation (4.23). Application of the "standard" relation (3.38) to such simulated *I/V* curves then provides effective barrier heights and ideality factors of such model contacts. The solid line in Fig. 4.10 is the result of such simulations where the above average values were used and the area density ρ_π of the patches was increased from zero up to 1.2×10^9 per cm^2. Obviously, this theoretical curve excellently describes the experimental effective barrier heights over the whole range of ideality factors. Specifically, the $\Phi_{Bn}^{eff}(n)$ dependence of patchy Schottky contacts turns out to be linear for ideality factors smaller than 1.4. The results of these simulations thus justify the determination of the barrier heights of laterally homogeneous contacts by linear extrapolation of the Φ_B^{eff} versus n plots to n_{if}, the ideality factor controlled by the image-force effect only, or, in other words, from data obtained with patchy metal–semiconductor contacts.

In the range of large ideality factors, the solid curve levels off and tends to approach a constant value. The scatter of the experimental data around the results of the simulations is certainly due to the corresponding input parameters. As an example, the marked data point in Fig. 4.10 belongs to a diode that is individually characterized by the average patch parameter $\sigma_\pi = 2.54\times10^{-4}$ $eV^{1/3}cm^{2/3}$. This value deviates by more than 25% from the mean value $\langle\sigma_\pi\rangle = 1.92\times10^{-4}$ $eV^{1/3}cm^{2/3}$ that is assumed in the simulations.

4.4 Test of the Patch Concept with Modified Contacts

A most direct test of the concept that the correlated variations of the effective barrier heights and the ideality factors are caused by nm-sized patches of lowered barrier height has been performed by *Lee* et al. [2002]. They studied Ti/4*H*-SiC and Ti/6*H*-SiC Schottky contacts without and with artificial patches of lowered barrier height and lateral dimensions smaller than the depletion layer width. For this purpose they deliberately deposited size-selected nm-sized Au aerosol particles with a diameter of 20 nm on clean SiC substrates and covered them with thick Ti layers. The thickness of the depletion layers, on the other hand, ranged between 240 and 670 nm. Gold has a larger electronegativity than titanium. Irrespective of the semiconductor considered, the Schottky barrier heights of nm-sized gold particles embedded in titanium are thus smaller than the surrounding regions on semiconductors doped *p*-type but they are larger if the substrates are doped *n*-type, see Eqs. (5.22) and (5.24.). Hence, nm-sized Au particles embedded in Ti contacts are expected to act as *additional* patches of lowered barrier height on substrates doped *p*-type and, as a consequence, to decrease the barrier heights and to increase the ideality factors in addition to what is observed with Ti contacts free of such particles. On *n*-type substrates, on the other hand, no such changes should occur since the nm-sized Au patches have larger barrier heights than the surrounding regions.

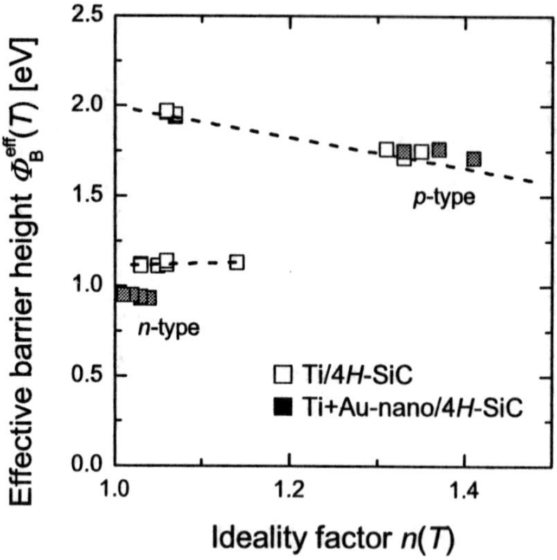

Fig. 4.11. Effective barrier heights and ideality factors of Ti contacts on *p*- and *n*-Si4H-SiC. The filled symbols distinguish contacts containing embedded nm-sized Au particles. Data from *Lee* et al. [2001, 2002]

Lee et al. [2001, 2002] prepared Ti contacts on both *p*- and *n*-4*H*-SiC substrates with and without embedded nm-sized Au particles. For one modified and unmodified contact each, they reported effective barrier heights and ideality factors which they determined from *I*/*V* curves measured at different temperatures. Figure 4.11 displays their data. The effective barrier heights of the *p*-type contacts decrease with increasing ideality factor. Most importantly, the contacts containing embedded nm-sized Au particles exhibit large ideality factors. The *n*-type contacts, on the other hand, are characterized by ideality factors ranging close to $n_{if} \approx 1.01$, the value determined by the image-force effect only. These findings directly confirm the predictions of the concept that patches with reduced barrier heights and lateral dimensions smaller than the depletion-layer width explain the reduction of the effective barrier heights of real Schottky contacts with increasing ideality factors.

4.5 *I*/*V* and BEEM Test of the Patch Concept

A detailed test of the concept of laterally inhomogeneous metal–semiconductor interfaces has been performed by *Im* et al. [2001]. They prepared Pd/*n*-6*H*-SiC(0001) contacts and *simultaneously* measured their *I*/*V* characteristics and the nm-scale distribution of their local BBEM barrier heights. Figure 4.12 displays the effective barrier heights as a function of the ideality factors which they evaluated from the measured *I*/*V* characteristics of eight of their diodes. The Φ_{Bn}^{eff} versus *n*

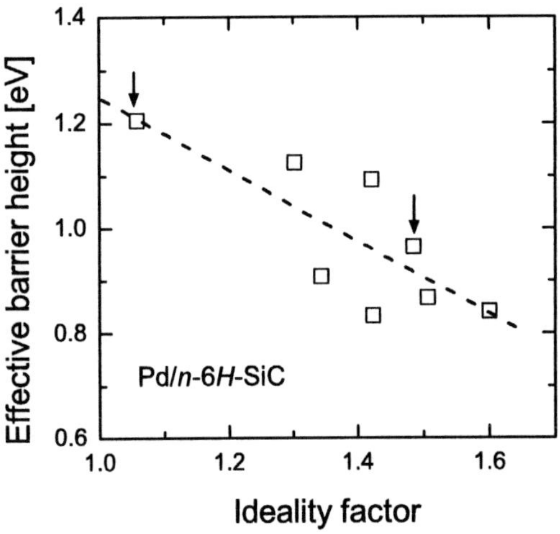

Fig. 4.12. Effective barrier heights versus ideality factors determined from *I*/*V* curves of Pd/*n*-6*H*-SiC(0001) contacts at room temperature. Data from *Im* et al. [2001]

plot again reveals a linear correlation between both quantities; the extrapolation to the ideality factor $n_{if} = 1.01$, which is determined by the image-force effect only, gives a barrier height Φ_{Bn}^{nif} of 1.24 ± 0.09 eV. Figure 4.13 shows the BEEM barrier-height distributions of a "good" and a "bad" diode. The arrows in Fig. 4.12 mark the corresponding samples. The "good" diode is characterized by $n = 1.06$, whose value is closest to $n_{if} = 1.01$, while the "bad" diode selected has the much larger ideality factor $n = 1.49$. The local BEEM barrier heights were obtained by fits of the *Bell–Kaiser* square-law (3.51) to the I_{coll}/V_{tip} characteristics measured at 800 different locations over the sample surfaces, each separated by at least 30 nm. *Im* et al. estimated the standard deviation of each of the BEEM barrier heights as $\sigma_{BEEM} \approx 20$ meV. Most remarkably, the two BEEM barrier-height distributions are essentially identical. They are again well described by Gaussian distributions which are centered at 1.271 eV and 1.268 eV, respectively, and have standard deviations of 29 meV and 33 meV, respectively. The latter two values are very close to the margins of experimental error σ_{BEEM}.

Their BEEM data led *Im* et al. to a slight refinement of *Tung*'s description of laterally patchy Schottky contacts. Figure 4.13 explains their model. The patches are assumed to occupy a fractional area $C_{i\pi} = \Sigma_{\gamma\pi} A_{i\pi}(\gamma_{i\pi})/A$ where $A_{i\pi}(\gamma_\pi)$ is the effective area of an individual patch characterized by the patch parameter γ_π . The distribution of the patch parameters is again assumed to be Gaussian and, consequently, the distribution of the saddle-point or, what is the same, the local BEEM barrier heights Φ_{Bn}^{BEEM} of the patches will be a Gaussian function. It is centered at the barrier height Φ_{Bn}^{hom} of the embedding area $(1 - C_{i\pi})A$ and has a standard deviation

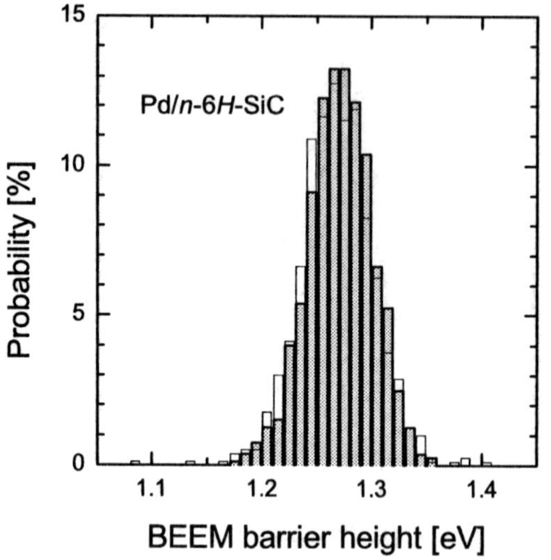

Fig. 4.13. Histograms of BEEM barrier heights of a "good" (gray solid bars) and a "bad" (empty bars) Pd/n-6H-SiC(0001) diode as determined by fitting 800 BEEM I_{coll}/I_{tip} spectra, each using the *Bell–Kaiser* square-law (3.51). Data from *Im* et al. [2001]

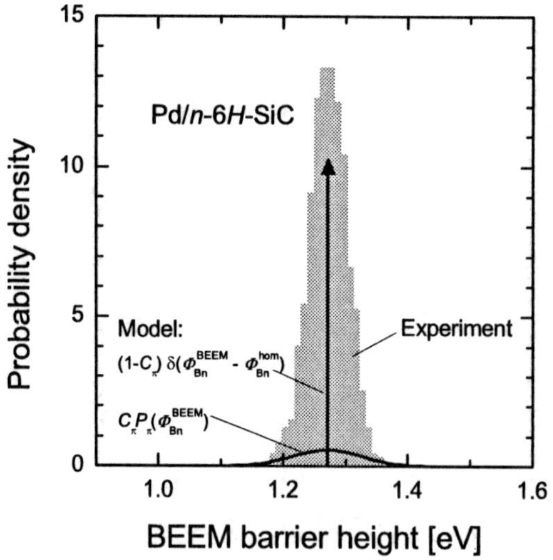

Fig. 4.14. Gaussian and Dirac delta function distributions of barrier heights of intrinsic patches and of the embedding area, respectively, of a model diode, and histogram of experimental BEEM barrier heights of a real Pd/n-6H-SiC(0001) diode (same data as in Fig. 4.13). The model parameters are $\Phi_{Bn}^{hom} = 1.27$ eV, $\sigma_{\pi} = 1.03\times10^{-4}$ eV$^{1/3}$cm$^{2/3}$, $\rho_{\pi} = 6\times10^{9}$ cm^{-2}, and $C_{\pi} = 0.107$

$$\sigma_{i\pi} = e_0 (V_i^0 N_d / \varepsilon_b \varepsilon_0)^{1/3} \sigma_{i\gamma} \quad . \tag{4.26}$$

The model barrier-height distribution of the entire diode may then be written as

$$P_{\mathrm{mod}}^{i\pi}(\Phi_{Bn}^{BEEM}) = \frac{C_{i\pi}}{\sqrt{2\pi}\sigma_{i\pi}} \exp\left(-\frac{(\Phi_{Bn}^{BEEM} - \Phi_{Bn}^{\mathrm{hom}})^2}{2\sigma_{i\pi}^2} \right) + (1 - C_{i\pi})\delta_{\mathrm{D}}(\Phi_{Bn}^{BEEM} - \Phi_{Bn}^{\mathrm{hom}}) \ , \tag{4.27}$$

where the Dirac δ function $\delta_{\mathrm{D}}(\Phi_{Bn}^{BEEM} - \Phi_{Bn}^{\mathrm{hom}})$ describes the laterally homogeneous area of the diode. The overall standard deviation of this model barrier-height distribution is then given by

$$\sigma_{\mathrm{model}} = \sqrt{C_{i\pi}}\sigma_{i\pi} = \sqrt{C_{i\pi}}\, e_0 (V_i^0 N_d / \varepsilon_b \varepsilon_0)^{1/3} \sigma_{i\gamma} \quad . \tag{4.28}$$

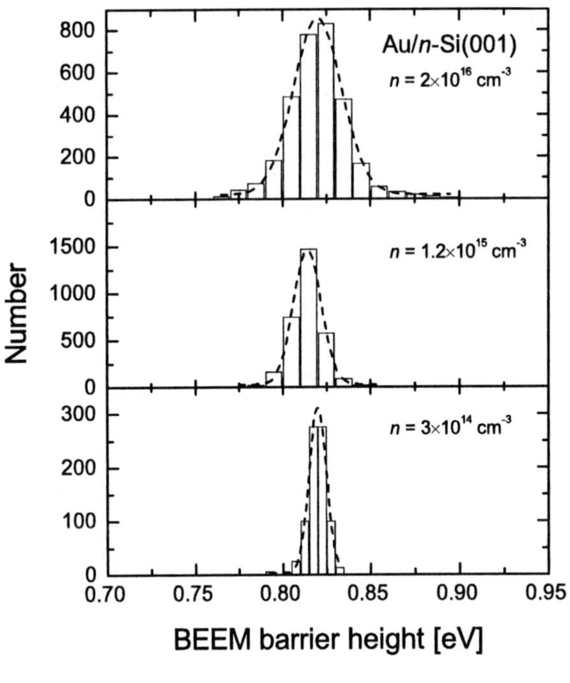

Fig. 4.15. Histograms of BEEM barrier heights on Au/*n*-Si(001) contacts with different donor densities of the silicon substrates. The dashed lines are Gaussian least-squares fits to the data. The centers and the standard deviations of the three Gaussians are 0.820 ± 0.0003 eV and 0.014 ± 0.0002 eV, 0.814 ± 0.0002 eV and 0.008 ± 0.0001 eV, and 0.820 ± 0.0001 eV and 0.005 ± 0.0001 eV, respectively. Data from *Morgan* et al. [1996]

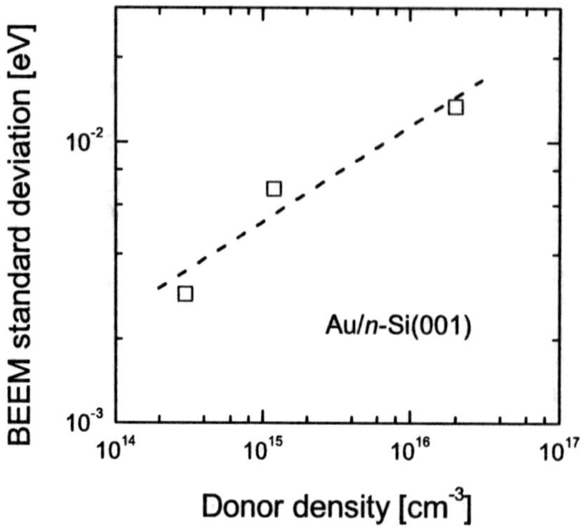

Fig. 4.16. Standard deviations of the Gaussian fits to the BEEM barrier-height histograms displayed in Fig. 4.14 as a function of the donor density. The dashed line is a linear least-squares fit

Morgan et al. [1996a] measured BEEM barrier-height distributions of Au Schottky diodes prepared on differently doped *n*-Si(001) substrates. Their results are displayed in Fig. 4.15. The experimental data are again excellently described by Gaussian functions and their standard deviations increase as the doping level becomes larger. Figure 4.15 shows the standard deviations obtained from the barrier height distributions given in Fig. 4.15 as a function of the donor density of the Si substrates. The dashed line is a linear least-squares fit to the data points and it has a slope parameter of 0.34 ± 0.09[2], i.e., these experimental data excellently confirm the variation of the model standard deviation predicted by relation (4.28).

In summary, *Im* et al. arrived at the following view. Both "good" as well as "bad" Pd/*n*-6*H*-SiC diodes contain intrinsic nm-size patches of lowered barrier height. They occupy a fraction $C_{i\pi}$ of approximately 5% of the total diode area and their area density and average patch parameters are $\rho_{i\pi} \approx 5\times10^{9}\ \mathrm{cm}^{-2}$ and $\sigma_{i\pi} = \langle\gamma_{i\pi}\rangle \approx 1\times10^{-4}\ \mathrm{eV}^{1/3}\mathrm{cm}^{2/3}$, respectively. These patches cover approximately 5% of the total contact area. "Bad" diodes with ideality factors $n > n_{if}$ additionally contain a few gross defects of extrinsic nature. *Skromme* et al. [2000] also concluded the existence of such extrinsic defects from their *I*/*V* studies of metal/*n*-4*H*-SiC diodes.

[2] A more rigorous treatment has to consider the experimental uncertainty σ_{error} of the measured BEEM barrier heights. The experimental standard deviations σ_{exp} of the barrier height distributions should then be replaced by the actual variations, which may be approximated by $\sigma_{\mathrm{actual}} \approx (\sigma^2_{\mathrm{exps}} - \sigma^2_{\mathrm{error}})^{1/2}$. With the assumption $\sigma_{\mathrm{error}} \approx 4$ meV, one obtains a slope parameter of 0.346 ± 0.094, which value agrees with that simply obtained from the experimental standard variations.

4.6 Origin of Lateral Barrier-Height Inhomogeneities

4.6.1 Natural or Intrinsic Nonuniformities

The conventional model of band bending at Schottky contacts assumes a uniform space charge in the depletion layer. More rigorous approaches, however, have to consider that the space charge is made up rather of discrete donors or acceptors. These ionized impurities are randomly distributed. The density of the charged dopant atoms and, as a consequence of this, also the potential of the depletion layer will exhibit spatial fluctuations. Calculations indeed showed potential fluctuations superimposed on the quadratic potential distribution that is characteristic of the usual jellium model of the space charge [*Boudville* and *McGill* 1985]. Potential fluctuations at the interface or, to be more precise, at the maximum of the barrier that results from the image-force lowering represent lateral inhomogeneities of the barrier height. These nonuniformities are of natural or intrinsic origin and they thus also exist at otherwise ideal Schottky contacts. *Schottky* [1942] and *Bethe* [1942] had already postulated the most likely existence of patches with reduced barrier height due to the unavoidable fluctuations in the spatial distribution of dopants.

Figure 4.17 displays the calculated widths δV_i of the distribution of these natural potential fluctuations at the interface or, what is the same, of the barrier height as a function of the doping density. The two data points are for random distributions of dopants [*Bondarenko* et al. 1998]. The solid line [*van Schilfgaarde* 1990],

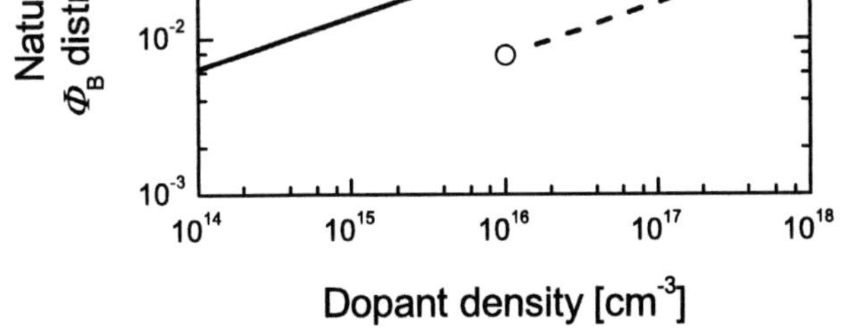

Fig. 4.17. Width of barrier-height distributions due to spatial dopant fluctuations as function of dopant density. Solid line after *van Schilfgaarde* [1990] and O data from *Bondarenko* et al. [1998]

$$\delta V_i = \frac{3e_0}{5\varepsilon_b\varepsilon_0}\left(\frac{4\pi}{3}N_{d,a}\right)^{1/3} , \tag{4.29}$$

is the result for the average potential perturbation in an ordered array of model impurities. Obviously, this approximation overestimates the fluctuations. The width of the barrier height distributions scale with the inverse of the average distance $<d_{d,a}>= N_{d,a}^{-1/3}$ between the ionized donors or acceptors.

From their simultaneous *I/V* and BEEM studies of Pd/*n*-6*H*-SiC contacts *Im* et al. [2001] concluded the existence of intrinsic patches of reduced barrier height which cover approximately 5% of the total diode area with an area density $\rho_{i\pi} \approx 5\times10^9$ cm^{-2}. These values correspond to an average patch radius $<R_{i\pi}>$ of approximately 18 nm. For a donor density of 3×10^{15} cm^{-3}, the doping level of the *n*-6*H*-SiC substrates used by *Im* et al., Fig. 4.16 gives a potential fluctuation of approximately 50 meV caused by the random distribution of the impurities. Insertion of these values in relation (4.10) then yields an average patch parameter $<\gamma_{i\pi}> = 3(e_0\delta V_i<R_\pi>^2/4 \approx 1.02\times10^{-4}$ eV$^{1/3}$cm$^{2/3}$. This estimate agrees with the value of 1×10^{-4} eV$^{1/3}$cm$^{2/3}$ evaluated by *Im* et al. for their Pd/*n*-6*H*-SiC diodes. Hence, the potential fluctuations caused by the random distribution of the dopants obviously explain the *intrinsic* lateral inhomogeneities of the barrier heights of Schottky contacts which were suggested by *Im* et al. [2001] from the detailed analysis of their simultaneous *I/V* and BEEM measurements of Pd/*n*-6*H*-SiC(0001) contacts.

4.6.2 Extrinsic Nonuniformities

Defects other than the natural fluctuations of the ionized dopant atoms may also cause lateral nm-size inhomogeneities of the barrier heights in *Schottky* contacts. It is very possible that, for example, anion microclusters [*Freeouf* and *Woodall* 1981] or antisite defects [*Spicer* et al. 1988] are responsible for barrier height fluctuations at metal interfaces with compound semiconductors. Local intermixing or alloy formation may cause patchy interfaces even of elemental semiconductors. However, none of these very possible mechanisms has been verified so far. Hitherto, dislocations at $CoSi_2$/Si(001) interfaces are the only example of defects that were definitely identified the cause of lateral inhomogeneities of the barrier height.

High-resolution transmission electron microscopy showed that epitaxial $CoSi_2$/Si(001) interfaces contain interfacial dislocations of edge-type with Burgers vectors $\vec{b} = a_0/4<111>$ [*Bulle-Lieuwma* et al. 1991]. BEEM images recorded with $CoSi_2$/*n*-Si(001) interfaces also displayed such dislocations [*Sirringhaus* et al. 1996]. Figure 3.14 shows BEEM I_{coll}/V_{tip} curves recorded on top of such a dislocation and in the defect-free region nearby. Least-squares fits of the quadratic I_{coll}/V_{tip} relationship yield barrier heights of 0.68 ± 0.04 eV on top of this dislocation and of 0.76 ± 0.05 eV at a spot in the surrounding, non-defected region. In ad-

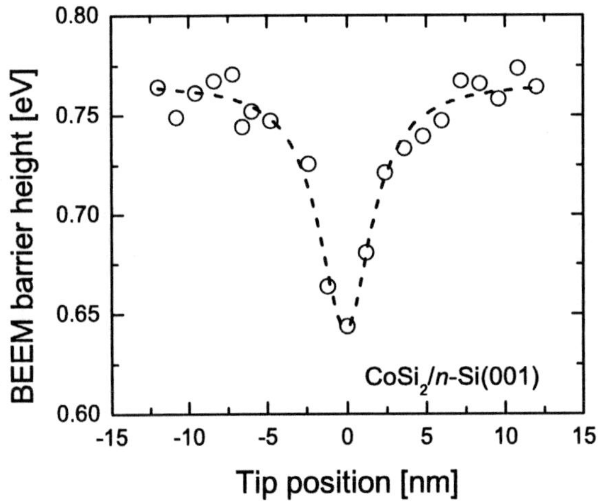

Fig. 4.18. Distribution of BEEM barrier heights along a line-scan perpendicular to a "linear defect" at a $CoSi_2$/*n-Si*(001) interface. The dashed line is a least-squares fit of a Lorentzian to the experimental data. The "linear defect" and the adjacent defect-free regions have BEEM barrier heights of 0.65 and 0.76 eV, respectively. The FWHM of the Lorentzian fit measures 4 nm. Data from *Sirringhaus* et al. [1996]

dition to these dislocations, the BEEM images of $CoSi_2$/*n*-Si(001) interfaces show other "linear defects". However, their atomic structure has not been resolved. Nevertheless, Fig. 4.18 displays the distribution of BEEM barrier heights along a line scan perpendicular to such a "linear defect". These data may be used to estimate the barrier-height lowering at these "linear defects".

The profile shown in Fig. 4.18 clearly reveals that the saddle-point barrier in front of the "linear defect" is 0.1 eV lower than the barrier height in the defect-free regions on both sides. As for circular patches (see Sect. 4.1), the potential barriers in front of semi-infinite strips with lowered barrier heights exhibit saddle points. At the saddle points the barrier heights are lowered by [*Tung* 1992]

$$\delta\Phi_{\sigma}^{sad} = 4\left(\frac{1}{2\pi}\frac{\Delta_{\sigma}L_{\sigma}}{e_0 V_i z_{dep}}\right)^{1/2} e_0 V_i \ , \qquad (4.30)$$

where L_σ and Δ_σ denote the width of the strip and the barrier-height reduction with regard to the value of the surrounding homogeneous regions, respectively. The dashed Lorentzian in Fig. 4.18 is a least-squares fit to the lateral profile of the BEEM barrier heights. As an upper limit, the lateral extension L_σ of the "linear defect" may be approximated by the width of the fitting curve, i.e., $L_\sigma = 40$ nm. Considering the donor density $N_d \approx 1\times10^{15}$ cm^{-3} of the silicon substrates and the observed saddle-point reductions $\delta\Phi_{\sigma}^{sad}$ of 0.08 and 0.1 eV at the dislocations and the "linear defects", respectively, relation (4.30) yields a barrier-height lowering Δ_σ that is close to or even larger than the homogeneous barrier height of 0.76 eV in the adjacent defect-free regions. In other words, the barrier heights of the interfacial dislocations and the "linear defects" at $CoSi_2$/*n*-Si(001) interfaces are small or even zero. For more precise estimates, silicon substrates with higher doping levels are required. The depletion layers are then narrower so that the dislocations and the "linear defects" will be less effectively pinched off.

5. The IFIGS-and-Electronegativity Theory

5.1 Band-Structure Lineup and Electronegativity: A Chemical Approach

5.1.1 Barrier Heights of Schottky Contacts

Studies of core-level photoemission revealed the partial ionic character of the covalent bonds between metal and substrate atoms on semiconductor surfaces covered with metal adatoms. Hence, *intrinsic* electric-dipole layers exist at metal-semiconductor interfaces because of the ionicity of the covalent interface bonds. The chemical shifts of the core levels and by this the charge transfer were found to vary proportionally to the difference $X_m - X_s$ of the metal and the semiconductor electronegativities; see Fig. 3.25. In analogy to the *surface-molecule* concept of ad-atoms at surfaces, the partially ionic interface bonds may be described as *interface molecules*. In a point-charge model, the dipole layer made up by these *interface molecules* may be viewed as a parallel-plate capacitor. The voltage drop across the dipole layer is then obtained as

$$\Delta_{ms} = \frac{e_0^2}{\varepsilon_i \varepsilon_0} \Delta q_{ms}^{eff} d_{ms}^{eff} N_{ms} \tag{5.1}$$

where N_{ms} is the density of *interface molecules* per unit area and ε_i is an interface dielectric constant. The *effective* charge transfer Δq_{ms}^{eff} and the *effective* bond length d_{ms}^{eff} of the *interface molecules* will differ from the values (3.57) and (3.58), respectively, which are characteristic of single bonds between the same metal and semiconductor atoms, i.e., of corresponding *surface molecules*. First, the mutual electrostatic depolarization of the interface dipoles will lead to a charge transfer lower than Δq_1 even at surfaces [*Topping* 1927]. Second, a possible interaction with the electron gas of the neutral metal on the metal side of the interface and, third, the bond polarization on the semiconductor side have to be considered in addition. These effects are difficult to estimate. Nevertheless, the electrostatic-energy term

$$\Delta_{ms} = S_X (X_m - X_s) \tag{5.2}$$

will vary proportionally to the difference $X_m - X_s$ of the metal and the semiconductor electronegativities.

For metals and the elemental semiconductors, their atomic electronegativities are used. Compound semiconductors are treated in analogy to molecules. The average electronegativity of diatomic molecules is taken as the geometric mean of the two corresponding atomic values. In generalizing this concept, the average electronegativity of binary compound semiconductors AB is defined as

$$X_{AB} = (X_A X_B)^{1/2} \quad . \tag{5.3}$$

The elements that constitute semiconductors are placed in the middle of the periodic table and their electronegativities scatter around 2 Pauling-units. The mean of the electronegativities of the group–IV, the III–V, and the II–VI semiconductors is

$$\langle X_s \rangle \approx 2 \pm 0.1 \text{ Pauling-units} \quad .$$

The electronegativities of the metals, on the other hand, vary between 0.79 and 2.54 Pauling-units for cesium and gold, respectively.

Provided that the electronegativity difference is zero, then the dipole term (5.2) vanishes. Silicon and silver have approximately the same electronegativities. However, the barrier height of Ag/*n*-Si(111)-$(1\times1)^{i}$ contacts is by no means near to zero but rather measures 0.7 eV; see Table 3.5. This value will thus be close to the zero-charge-transfer barrier height Φ_{Bn}^{zct} of *n*-type silicon Schottky contacts. Hence, the chemical trend of the barrier heights of ideal Schottky contacts may be quite generally expressed as

$$\Phi_{Bn}^{hom} = \Phi_{Bn}^{zct} + \Delta_{ms} = \Phi_{Bn}^{zct} + S_X (X_m - X_s) \quad . \tag{5.4}$$

This *chemical* electronegativity concept of the partial ionic character of covalent bonds at metal–semiconductor interfaces thus predicts the barrier heights to vary linearly as a function of the electronegativity difference $X_m - X_s$.[1] As an example, Fig. 5.1 displays the barrier heights of *n*-Si(111)-$(1\times1)^{i}$ Schottky contacts;, see Table 3.3. The dashed line, which is a linear least-squares fit to the data, clearly demonstrates that the experimental data indeed follow the linear relationship (5.4). Calculations of both the zero-charge-transfer barrier height Φ_{Bn}^{zct} and the slope parameter S_X are not possible in the framework of this simple chemical concept and require further physical input.

[1] This approach differs conceptually from an earlier one where *Kurtin* et al. [1969] replaced the metal work function and the semiconductor electronaffinity in the *Schottky–Mott* rule (1.2) by the respective electronegativities. Both quantities are surface rather than interface properties, while the electronegativities are characteristic of the materials themselves. This early attempt failed to explain the chemical trends of the then known Schottky barrier heights. The reason was that experimental data of laterally inhomogeneous Schottky contacts were considered. The importance of lateral inhomogeneities of the barrier heights with dimensions less than the depletion layer width was realized much later.

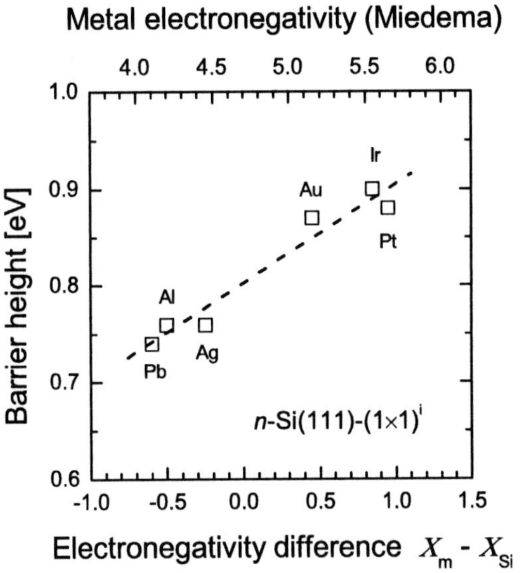

Fig. 5.1. Barrier heights of laterally homogeneous n-Si(111)-(1×1)^{i} Schottky contacts as a function of the electronegativity difference. Data from Table 3.3

5.1.2 Interface Dipoles at Semiconductor Heterostructures

The electronic properties of contacts between two semiconductors or, as they as commonly called, semiconductor heterostructures are determined by the lineup of their band structures across the interface and the existence of space-charge layers. At abrupt interfaces band-edge discontinuities are established within a few atomic layers, while the regions of band bending depend on the doping profiles on both sides of the heterostructure and will extend over a few Debye lengths. Here, only the band-edge offsets will be considered. Figure 5.2 schematically illustrates the band lineup at the interface between two semiconductors. The valence-band and the conduction-band offsets, $\Delta W_v = W_{vi}^l - W_{vi}^r$ and $\Delta W_c = W_{ci}^r - W_{ci}^l$, add up to the difference $\Delta W_g = W_g^l - W_g^r$ of the band gaps of the two semiconductors in contact, i.e.,

$$\Delta W_v + \Delta W_c = \Delta W_g \quad . \tag{5.5}$$

The valence-band rather than the conduction-band offsets are the more fundamental quantities.

There are two particularly simple cases of semiconductor heterostructures at which intrinsic effects determine the band lineup. A first example is epitaxial and

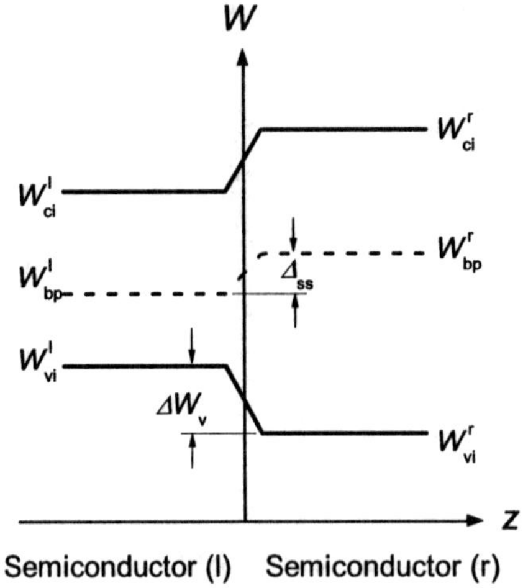

Fig.5.2. Band lineup at semiconductor heterostructures exhibiting interface dipoles (schematic)

nonpolar interfaces between group-IV, III–V, and II–VI semiconductors with perfect lattice matching and no extrinsic defects of any kind. The interface bonds will be covalent, as in the bulk of semiconductors, and no unsaturated dangling bonds will exist at such interfaces. The second example is interfaces between layered semiconductors as in, for example, III–VI compounds. While the bonds within the layers are also covalent, the interlayer forces are of the van der Waals type. Therefore, neither covalent nor dangling bonds are present at interfaces between layered semiconductors. Such heterostructures are considered in Sect. 6.2.5.

As at metal–semiconductor contacts, the covalent bonds at interfaces between two semiconductors are partly ionic too. Hence, intrinsic interface dipoles will also exist in semiconductor heterostructures and the same arguments which apply to Schottky contacts are valid for semiconductor heterostructures too. In analogy to (5.2), the voltage drop across the dipole layer caused by the interfacial bonds may now be expressed as

$$\Delta_{ss}^{\mathrm{int}} = D_X (X_{\mathrm{sr}} - X_{\mathrm{sl}}) \ , \tag{5.6}$$

where X_{sr} and X_{sl} are the electronegativities of the semiconductors on the right-hand and the left-hand side, respectively, of the heterostructure.

The orientation of the intrinsic interface dipoles in semiconductor heterostructures is easily derived from the electronegativities of the semiconductors. The signs resulting from this simple approach agree with theoretical findings that were obtained by using a linear-muffin-tin-orbital method and potentials constructed with the atomic-sphere approximation [*Christensen* 1988, *Lambrecht* et al. 1990].

The electronegativities of the elemental group–IV and the III–V and II–VI compound semiconductors are almost equal; see the preceding Sect. 5.1.1. Therefore, the dipole term $D_X(X_{\mathrm{sr}} - X_{\mathrm{sl}})$ will be small and may be neglected for practical purposes. In this respect, semiconductor heterostructures drastically differ from metal–semiconductor contacts since the dipole term $S_X(X_m - X_s)$ determines the chemical trends of their barrier heights.

5.2 Barrier Heights and Interface States: A Phenomenological Approach

The valence electrons in semiconductors may be viewed in terms of covalent bonds and of energy bands. Likewise, electronic interface states are the equivalent *physical* description of *chemical* interface bonds at semiconductor interfaces. The charge transferred from the metal into the semiconductor has negative or positive sign depending on whether the metal has a smaller or larger electronegativity than the semiconductor. Hence, the electronic interface states should form a continuum within the band gap. The character of the interface states has to be donor-like below and acceptor-like above a certain energy level, which is called their branch point (bp) or, more intuitively, their charge neutrality level (CNL).

The precise energy position of the Fermi level within the band gap and thus the value of the barrier height depends on the occupation of the interface states. This is explained schematically in Fig. 5.3, where the interface states are assumed to have a constant density of states throughout the band gap.[2] The net charge density in the continuum of interface states

$$Q_{is} = e_0 D_{is} \cdot (W_{bp} - W_F) = e_0 D_{is} \cdot (\Phi_{Bn} - \Phi_{bp}^n) \tag{5.7}$$

is negative, vanishes, or becomes positive when the Fermi level W_F is above, coincides with, or is below, respectively, the branch point W_{bp} of the interface states. Hence, the zero-charge-transfer barrier height Φ_{Bn}^{zct} in relation (5.4) and the energy distance from the branch point to the conduction-band minimum at the interface are identical, i.e.,

$$\Phi_{Bn}^{zct} \equiv \Phi_{bp}^n = W_{ci} - W_{bp} \quad . \tag{5.8}$$

The condition of charge neutrality at the interface

$$Q_m + Q_s = Q_m + (Q_{sc} + Q_{is}) = 0 \tag{5.9}$$

requires that the charge densities on the metal and the semiconductor side of the contact, Q_m and Q_s, are equal in size but opposite in sign. The combination of the chemical electronegativity concept and the physical interface-states model demands that

$$-Q_m = Q_s \approx Q_{is} \propto X_m - X_s \quad , \tag{5.10}$$

so that relation (5.2) is retrieved. The assumption that the space-charge density Q_{sc} is small and may be neglected is generally justified.

[2] *Cowley* and *Sze* [1965] were the first to consider such a model.

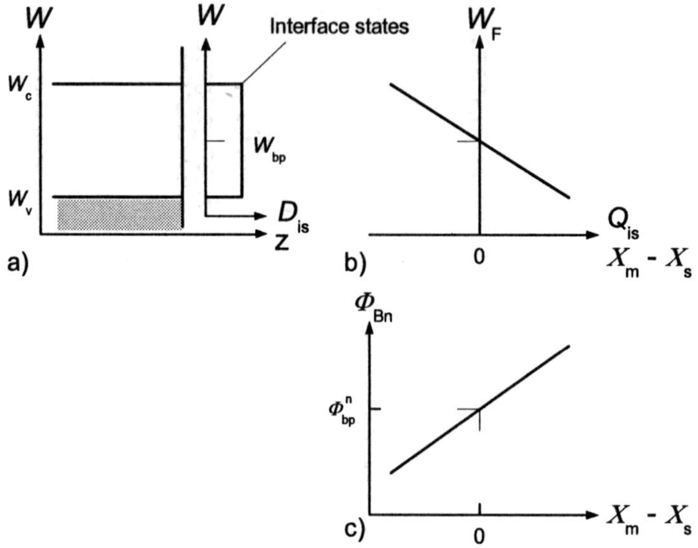

Fig. 5.3. Band diagram (a), charge transfer (b), and barrier height (c) at metal–semiconductor contacts containing a continuum of electronic interface states with constant density of states across the band gap (schematic)

The interface charge densities Q_m and Q_s reside on either side of the interface and form an electric double-layer. Its width δ_{is} is of atomic dimensions. In thermal equilibrium, the energy barrier Δ_{is} of such a double layer is related to the interface charge-density by

$$\Delta_{is} = -\frac{e_0}{\varepsilon_i \varepsilon} Q_m \delta_{is} \ , \tag{5.11}$$

where ε_i is the dielectric constant of the interface. Combining (5.7) to (5.11), and (2.16), one obtains

$$\Delta_{is} = \left[s_b(\Phi_{Bn} - W_n)\right]^{1/2} + s_i(\Phi_{Bn} - \Phi^n_{bp}) \ , \tag{5.12}$$

with

$$s_b = (2e_0^2 \varepsilon_b / \varepsilon_i \varepsilon_0) N_d \delta_{is} \tag{5.13}$$

and

$$s_i = (e_0^2 / \varepsilon_i \varepsilon_0) D_{is} \delta_{is} \ . \tag{5.14}$$

The parameter s_b may be estimated when reasonable values of δ_{is} and ε_i are assumed. For intimate and abrupt metal–semiconductor interfaces, the thickness of the interfacial double layer should be of atomic dimensions, i.e., typically $\delta_{is} \approx 0.2$ nm, and the interface dielectric constant may be approximated by the vacuum value, i.e., $\varepsilon_i \approx 1$. For a bulk dielectric constant $\varepsilon_b = 12$ and a donor den sity $N_d = 10^{17}$ cm^{-3}, one obtains $s_b \approx 0.002$ eV. Since ε_i should range between 1 and the dielectric constant of the semiconductor bulk the preceding value of s_b is overestimated. Therefore, the first term on the right-hand side of relation (5.12) may be neglected. This conclusion is identical with the condition $Q_{sc} << Q_{is}$ as assumed above.

In analogy to a metal and a semiconductor being separated by a vacuum layer of a thickness much larger than atomic dimensions, the energy barrier Δ_{is} may be expressed as the work-function difference of the metal and the semiconductor in contact. The work functions of solids generally differ. In thermal equilibrium, a contact potential

$$\Delta_{cp} = \phi_l - \phi_r \tag{5.15}$$

exists between two solids facing each other in vacuum. It equals the difference of their work functions. For $\phi_l > \phi_r$, the solid on the left-hand (l) and on the right-hand side (r) will be charged negatively and positively, respectively. As in small molecules, such charge transfer suggests that the work functions of solids correlate with their electronegativities [*Gordy* and *Thomas* 1956]. Figure 5.4 displays the work functions of polycrystalline metals [*Michaelson* 1977] as a function of *Miedema*'s electronegativities [1973, 1980].[3] The data display a linear correlation and a linear least-squares fit gives

$$\phi_m = 0.86 X_{\text{Mied}} + 0.59 \text{ eV} \quad , \tag{5.16}$$

while

$$\phi_m = 1.79 X_{\text{Paul}} + 1.11 \text{ eV} \tag{5.17}$$

is obtained when *Pauling*'s electronegativities are used instead. The contact potential between two metals facing each other in vacuum may thus be written as

$$\Delta_{cp} = \mathrm{A}_X (X_l - X_r) \quad , \tag{5.18}$$

where A_X is the respective proportionality coefficient in relations (5.16) or (5.17) depending on which of the two electronegativity scales is considered.

Applying the electronegativity concept to metal–semiconductor contacts, the energy barrier Δ_{is} may be written as [*Mönch* 1996a]

---

[3] *Pauling* derived his electronegativities from properties of small molecules. *Miedema* and co-workers [1973, 1980], on the other hand, calculated the heats of formation of metallic alloys and intermetallic compounds by including a charge transfer between the constituents. In their studies, they established an alternative electronegativity scale for metals and a few nonmetals which are used for alloying. Their values relate to those given by *Pauling* as

$$X_{\text{Mied}} = 1.93\, X_{\text{Paul}} + 0.87 \quad . \tag{5.19}$$

In dealing with interfaces between solids, *Miedema*'s electronegativities will be preferred. *Pauling*'s scale, on the other hand, will be applied when a local and molecule-like bond picture is more adequate.

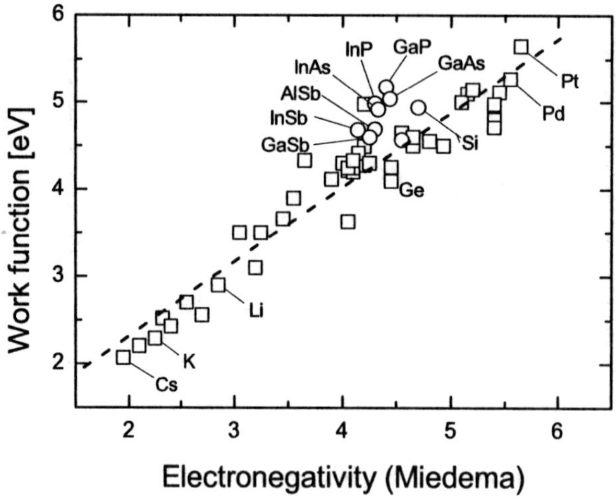

Fig. 5.4. Work functions of metals and dielectric work functions of semiconductors as a function of *Miedema*'s electronegativities. □ data of metals from *Michaelson* [1977]; ○ data of semiconductors (see text)

$$\Delta_{is} = \mathrm{A}_X (X_m - X_s) - (\Phi_{Bn} - \Phi^n_{bp}) \quad . \tag{5.20}$$

Here, the *dielectric work function* of the semiconductor

$$\phi_{sd} = W_{vac} - W_{bp} = \mathrm{A}_X X_s + \mathrm{B}_X \tag{5.21}$$

is assumed to scale with the semiconductor electronegativity according to relations (5.16) or (5.17), which are derived from the metal data plotted in Fig. 5.4.

The dielectric work function is a characteristic property of semiconductors. The electronegativity concept predicts no charge transfer at interfaces between, for example, metals and semiconductors having identical electronegativities, i.e., if $X_m - X_s$. Provided that the semiconductor bulk is doped such that its Fermi–level position $W_{cb} - W_F$ equals the branch-point energy $W_c - W_{bp}$, the net charge in the continuum of interface states as well as the space charge will be zero since no interface band-bending will occur on the semiconductor side of such contacts. As a consequence, the energy barrier Δ_{is} vanishes, in agreement with relation (5.20).

The dielectric work functions $\phi_{sd} = W_{vac} - W_{bp} = I - (W_{bp} - W_{vs})$ of the elemental group-IV and some of the II–V compound semiconductors are also plotted in Fig. 5.4. The ionization energy $I = W_{vac} - W_{vs}$ is defined as the energy distance from the valence-band maximum to the vacuum level at the semiconductor surfaces. These data fit the chemical trend observed with the metals, although on the average the dielectric work functions of semiconductors are slightly larger than the work functions of the metals having the same electronegativities. The deviations seem to correlate with the ionicity of the semiconductors. Germanium has the largest metallicity of the semiconductors considered and its dielectric work function perfectly fits the chemical trend observed for the metals.

Combining relations (5.12) and (5.20) where the first term on the right-hand side is neglected, one obtains

$$\Phi_{Bn} = \Phi^{n}_{bp} + [\mathrm{A}_X/(1-s_i)](X_\mathrm{m} - X_\mathrm{s}) = \Phi^{n}_{bp} + S_X(X_\mathrm{m} - X_\mathrm{s}) \ . \qquad (5.22)$$

Most importantly, this *phenomenological* interface-states model reproduces relation (5.4), which was derived by applying *Pauling*'s electronegativity concept to metal–semiconductor interfaces. In addition, it also explains the slope parameter

$$S_X = \frac{\mathrm{A}_X}{1 + (e_0^2/\varepsilon_i\varepsilon_0)D_{is}\delta_{is}} \qquad (5.23)$$

in terms of an interface dielectric-constant ε_i, the density of states D_{is}, and the extension δ_{is} of the electronic interface states. However, it is left to a microscopic model to determine the branch-point energies, the densities of states, and the extensions of the electronic interface states as well as the interface dielectric-constant.

With the assumption that the semiconductor substrate is doped *p*-type one obtains

$$\Phi_{Bp} = \Phi^{p}_{bp} - S_X(X_\mathrm{m} - X_\mathrm{s}) \ , \qquad (5.24)$$

where $\Phi^{p}_{Bp} = W_{bp} - W_{vi}$ is the energy distance from the valence-band maximum to the branch point of the interface states. For a continuum of interface states, the barrier heights Φ_{Bn} and Φ_{Bp} of one and the same metal contact on a semiconductor doped *n*- and *p*.type, respectively, thus add up to the width of the band gap.

The slope parameter assumes the two limiting values $S_X = 0$ for extremely large and $S_X = \mathrm{A}_X$ for vanishing density of states D_{is} of the electronic interface states. The first or *Bardeen* [1947] case means that the barrier height is the same irrespective of the metal considered or, in other words, the Fermi level is *pinned* at the branch point of the interface states. The second condition gives

$$\Phi_{Bn} = \Phi^{n}_{bp} + \mathrm{A}_X(X_\mathrm{m} - X_\mathrm{s}) = \Phi^{n}_{bp} + (\phi_\mathrm{m} - \phi_{sd}) = \phi_\mathrm{m} - \chi_s \ ,$$

i.e., the *Schottky–Mott* relation (1.2) is retrieved.

5.3 Interface-Induced Gap States

5.3.1 Metal-Induced Gap States

At clean metal surfaces the wave-functions of the electrons decay exponentially into vacuum. Figure 1.5a explains this behavior. When the vacuum is replaced by a semiconductor or, more generally speaking, by a dielectric the propagation of the wave-functions across the interface is somewhat more complicated. *Heine* [1965] had already pointed out that in the energy range from the top of the valence

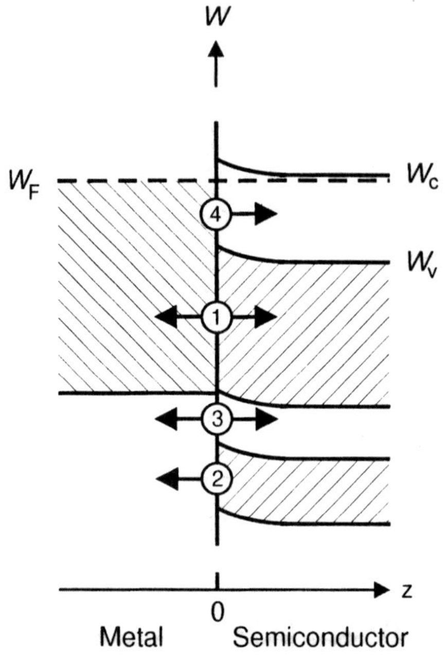

Fig. 5.5. Types of states at metal–semiconductor interfaces (schematic): ①: Bulk-like states on both sides of the interface; ②: semiconductor states exponentially decaying into the metal; ③: localized interface states; ④: metal states exponentially decaying into the semiconductor or metal-induced gap states (MIGS)

band up to the Fermi level, where the conduction band of the metal overlaps the semiconductor band-gap, the wave-functions of the metal electrons tunnel into the semiconductor as they do into vacuum at clean surfaces. This behavior is schematically shown in Fig. 1.5b. These tails of the metal wave-functions are described by complex wave-vectors and they thus derive from the complex band structure of the semiconductor. Most importantly, these metal-induced gap states or, for short, MIGS, as *Louie* and *Cohen* [1976] named them later, are an intrinsic property of the semiconductor.

Louie and *Cohen* [1976] performed model calculations for a jellium–silicon interface. They adjusted the jellium electron density such as to correspond to aluminum and described the silicon in the pseudopotential formalism. Their computations revealed four different types of electronic states to exist at such interfaces. They are explained schematically in Fig. 5.5. First, there are states which are bulk-like on either side of the boundary. Second, below the bottom of the conduction band of the metal, bulk-like semiconductor states penetrate into the metal. Third, truly localized interface states may be present in low-lying semiconductor energy gaps. Fourth, in the energy range where the metal conduction-band overlaps the energy gap of the semiconductor, a continuum of gap states exists. These are the metal-induced gap states. Their wave-functions are bulk-like in the metal and de-

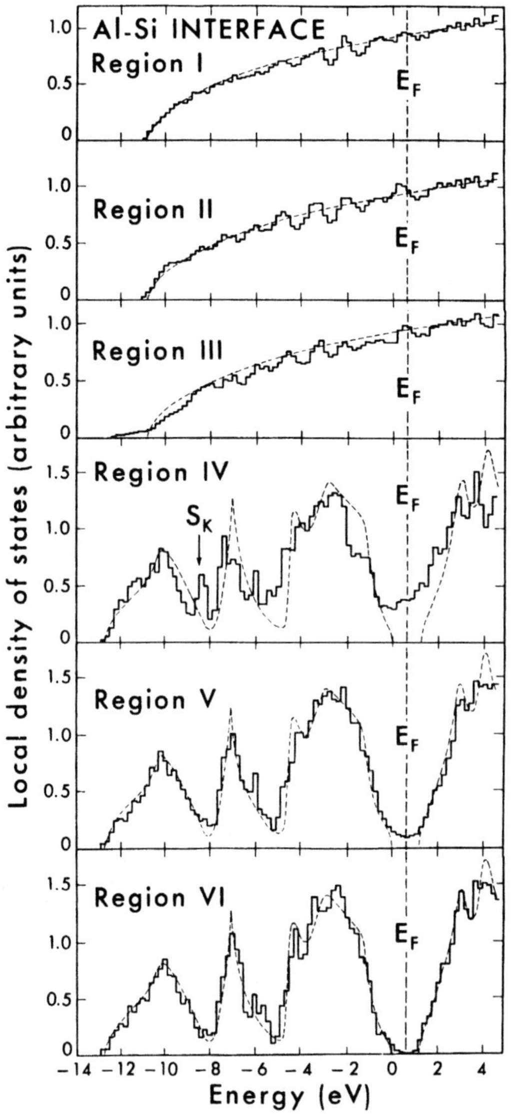

Fig. 5.6. Local density of states for a Al-jellium/Si(111) interface. Each of the six regions has a width of 0.314 nm, which is the distance between Si-bilayers in the {111} direction. From *Louie* and *Cohen* [1976]

cay exponentially into the semiconductor, i.e., they have complex wave-vectors. Figure 5.6 displays the computed local densities of states in three regions on both sides of the Al-jellium/Si(111) interface. Each of these regions is 0.314 nm wide, which is the distance between two adjacent silicon bilayers in the {111} direction. The jellium–silicon interface is at the border between regions III and IV. Already approximately 1 nm away from the interface on either side, the local densities of states are bulk-like. At the actual interface, the density of states is only slightly changed, while in the semiconductor the band gap is filled by a continuum of states. They exhibit an U-shaped energy distribution and decay exponentially into the semiconductor. The states close to mid-gap show the shortest decay length.

The results of the early and simple model calculations for Al-jellium/silicon interfaces were excellently confirmed by later studies based on density-functional theory and *ab-initio* pseudopotentials for both metal and semiconductor. These calculations demonstrated the existence of metal-induced gap states at Ag/MgO [*Blöchl* et al. 1990], Al/GaAs [*Berthod* et al. 1998], Al/3*C*-SiC [*Hoekstra* and *Kohyama* 1998], and Al-, Ag-, and Au/3*C*-GaN interfaces [*Picozzi* et al. 1998, 2000].

The computational results excellently confirmed *Heine*'s [1965] earlier concept that in the energy range between the valence-band top of the semiconductor and the Fermi level the metal wave-functions decay exponentially into the semiconductor. These interface states have complex wave-vectors and thus derive from the complex band structure of the semiconductor. This concept is explained in Sect. 5.4 by treating a simple one-dimensional lattice in a nearly free electron approximation [*Maue* 1935, *Goodwin* 1939a].

5.3.2 Semiconductor Heterostructures

Heine's simple MIGS concept [1965] is easily applied to semiconductor heterostructures. As an example, heterostructures with band-lineups similar to the one depicted in Fig. 5.2 will be considered. Within the valence-band discontinuity the wave-functions of the valence electrons of the semiconductor on the left-hand side are expected to tail into the semiconductor on the right-hand side. These semiconductor-induced gap states are a property of the semiconductor on the right-hand side of the heterostructure and again derive from the continuum of its virtual gap states. The phrase "semiconductor-induced gap states" was not adopted but was rather replaced by the more general term "interface-induced gap states" or, for short, IFIGS which also subsumes metal-induced gap states.

Pickett et al. [1978] and *Pickett* and *Cohen* [1978] calculated the electronic structure of nonpolar (110) interfaces of Ge–GaAs, AlAs–GaAs, and Ge–ZnSe heterostructures. The lattice mismatch is smaller than 0.2% and was neglected. The calculations used self-consistent pseudopotentials and the exchange–correlation potential was derived in the local-density approximation. These studies discovered both true interface states and interface-induced gap states. The true interface states were only found at the Ge–GaAs and Ge–ZnSe interfaces, i.e., in heterostructures between a non-polar and a polar semiconductor. These localized interface states derive from Ge–cation (Ga, Zn), Ge–anion (As, Se), Ge–Ge and GaAs or ZnSe bonds parallel to the interface. The band-structure lineup at these three heterostructures is of the type depicted in Fig. 5.2 and the interface-induced gap states appeared within the valence-band offset of the Ge–ZnSe heterostructure and within the heteropolar ZnSe and GaAs gaps. These IFIGS derive from bulk Ge valence states and decay quickly into the heteropolar gaps, i.e., they have all the characteristics of the interface-induced gap states that follow from *Heine*'s fundamental concept. These calculations detected no interface-induced gap states within the valence-band discontinuities of the Ge–GaAs and AlAs–GaAs heterostructures. However, the corresponding valence-band offsets, 0.25 and 0.35 eV, are almost equal to the margins of computational error, ± 0.2 eV.

5.3.3 Valence-Band Offsets at Semiconductor Heterostructures

The continuum of interface-induced gap states is the fundamental mechanism that determines the band-structure lineup at all types of semiconductor interfaces. Hence, the valence-band offsets of semiconductor heterostructures contain both a zero-charge-transfer and an electric-dipole term. The charge neutrality level of the interface-induced gap states is identical with their branch point and, hence, *Tejedor* and *Flores* [1978] expressed the valence-band offsets at semiconductor heterostructures as

$$\Delta W_v = W_{vi}^l - W_{vi}^r = \Phi_{bp}^{pl} - \Phi_{bp}^{pr} + \Delta_{ss} \quad . \tag{5.25}$$

At ideal lattice-matched and non-polar semiconductor heterostructures, the interfacial electric-dipole term Δ_{ss} is determined by the net charge in the interface-induced gap states only or, what is equivalent, by the ionicity of the covalent interface bonds, i.e., relation (5.25) may be written as

$$\Delta W_v = W_{vi}^l - W_{vi}^r = \Phi_{bp}^{pl} - \Phi_{bp}^{pr} + D_X(X_{\mathrm{sr}} - X_{\mathrm{sl}}) \quad . \tag{5.26}$$

To a first approximation the slope parameter D_X of heterostructures may be equated with the slope parameter S_X of Schottky barriers since the IFIGS determine the dipole contribution to both valence-band offsets and barrier heights.

The electronegativities of the elemental group-IV and the III–V and II–VI compound semiconductors are almost equal. Hence, the intrinsic dipole term $\Delta_{ss}^{\mathrm{int}} = D_X(X_{\mathrm{sr}} - X_{\mathrm{sl}})$ will be small and relation (5.26) may be approximated by

$$\Delta W_v = W_{vi}^l - W_{vi}^r \approx \Phi_{bp}^{pl} - \Phi_{bp}^{pr} \tag{5.27}$$

for practical purposes.

5.4 Virtual Gap States: One-Dimensional Model[4]

The periodic potential of a one-dimensional solid may be approximated by retaining the first Fourier coefficient only, i.e., by

$$V(z) = V_0 - V_1 \cos(g_1 z) \quad , \tag{5.28}$$

[4] See *Maue* [1935] and *Goodwin* [1939a]

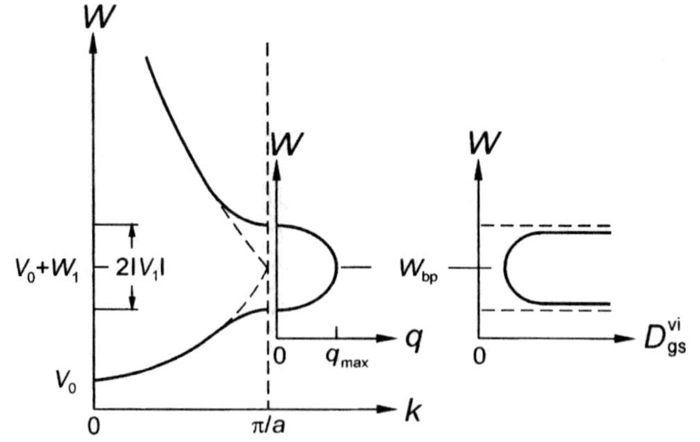

Fig. 5.7. Band structure of a one-dimensional solid for real and complex wave-vectors and density of virtual gap states (schematic)

where $g_1 = \pi/a$ is the shortest vector of the reciprocal lattice and a is the lattice parameter. Provided tat the Fourier coefficient V_1 of the potential is small compared to the kinetic energy of the electrons, their wave-functions may be approximated by the first two terms of a Fourier expansion, i.e., one obtains

$$\psi_k(z) = \mathrm{A}\exp(ikz) + \mathrm{B}\exp[i(k - g_1)z] \ . \tag{5.29}$$

Substituting (5.28) and (5.29) into Schrödinger's equation gives the secular equation

$$\begin{bmatrix} (\hbar^2/2m_0)k^2 + V_0 - W & V_1 \\ V_1 & (\hbar^2/2m_0)(k - g_1)^2 + V_0 - W \end{bmatrix} \begin{bmatrix} \mathrm{A} \\ \mathrm{B} \end{bmatrix} = 0 \ . \tag{5.30}$$

For the following, it is convenient to measure the wave-vectors from the edge of the Brillouin zone, i.e., $k = \pi/a - \kappa$. Solving (5.30) yields the energy dispersion

$$W(\kappa) - V_0 = W_1 + (\hbar^2/2m_0)\kappa^2 \pm [V_1^2 + 4W_1(\hbar^2/2m_0)\kappa^2]^{1/2} \ , \tag{5.31}$$

where the definition

$$W_1 = (\hbar^2/2m_0)(g_1/a)^2 \tag{5.32}$$

is used. The left-hand diagram of Fig. 5.7 schematically displays the resulting well-known band structure for real wave-vectors k.[5] It exhibits a single energy gap of width $|2V_1|$ at the boundary of the first Brillouin zone. For small deviations κ of the wave-vector from the zone boundary, the bands disperse parabolically.

The energy $W(\kappa)$ is a continuous function of κ^2. Positive as well as negative values of κ^2, i.e., real and complex wave-vectors are permitted likewise. In the

[5] See textbooks on solid state physics, e.g. by *Ibach* and *Lüth* [1991].

bulk, only real wave-vectors are *physically* meaningful since for complex wave-vectors the Bloch waves (5.29) would grow exponentially in the limit of $z \to \infty$. Hence, they could not be normalized. At surfaces and interfaces, on the other hand, complex wave-vectors become physically relevant. The periodic wave-functions of surface states, for example, decay exponentially from the surface into the metal or the semiconductor and are fitted by a pure exponential tail in the vacuum. Figure 1.5c schematically illustrates the wave-functions of surface-states. Such wave-functions are readily normalized and thus represent electrons bound to surfaces.

Surface states, and the same holds for interface states, at the end of finite one-dimensional solids will thus have complex wave vectors

$$k = \pi / a + \mathrm{i}q \tag{5.33}$$

and energy levels within the band gap at the boundary of the first Brillouin zone. According to the Fourier expansion (5.29), the corresponding wave-functions may be written as

$$\begin{aligned} \psi_q(z) &= \exp(-qz)[\mathrm{A}\exp(\mathrm{i}\pi z / a) + \mathrm{B}\exp(-\mathrm{i}\pi z / a)] \\ &= \mathrm{A}'\exp(-qz)\cos(\pi z / a + \varphi) \ , \end{aligned} \tag{5.34}$$

where A′ is a constant. The ratio

$$\mathrm{A/B} = \exp(\mathrm{i}2\varphi) \tag{5.35}$$

of the two Fourier coefficients defines the phase factor φ. It is obtained by the substitution of complex wave-vectors (5.33) into the secular equation (5.24). Specifically, the imaginary part yields

$$\sin 2\varphi = -W_1 q / V_1 g_1 \ . \tag{5.36}$$

Across the band gap, the phase factor φ changes from $-\pi/2$ to 0 for a positive value and from 0 to $\pi/2$ for a negative value of the first Fourier coefficient V_1 of the potential.

For complex k-values (5.33) the solutions of the secular equation (5.29) have the form

$$W(q) - V_0 = W_1 - (\hbar^2 / 2m_0)q^2 \pm [V_1^2 - 4W_1(\hbar^2 / 2m_0)q^2]^{1/2} \tag{5.37}$$

or

$$(\hbar^2 / 2m_0)q^2 = -[W(q) - V_0] - W_1 \pm \{V_1^2 + 4W_1[W(q) - V_0]\}^{1/2} \ . \tag{5.38}$$

The diagram in the middle of Fig. 5.7 illustrates the variation of the imaginary part q of the complex wave-vector (5.33) across the gap. The complex band structure of a one-dimensional solid thus contains a loop at the Brillouin-zone boundary with complex wave-vectors in the gap between the two bulk bands. At the edges of the bulk energy bands, the imaginary part q of the wave-vector vanishes. It passes through a maximum at $V_1^2 / 4W_1$ below mid-gap and there it assumes its maximum value which is given by

$$(\hbar^2/2m_0)q_{\max}^2 = V_1^2/4W_1 \ . \tag{5.39}$$

The inverse of q is the decay length of the respective wave-function in the linear chain. The maximum value q_{max} slightly below the mid-gap position thus gives a minimum decay length. At the edges of the bulk energy bands, $1/q$ approaches infinity so that the wave-functions become delocalized and behave as Bloch waves in the bulk.

The density of states of these solutions of Schrödinger's equation results as [*Garcia–Moliner* and *Flores* 1979]

$$D_{gs}^{vi}(W)\mathrm{d}W = \frac{1}{2\pi}\frac{\mathrm{d}\varphi}{\mathrm{d}W}\mathrm{d}W \approx \frac{1}{2\pi[V_1^2 - (W - V_0 - W_1)^2]^{1/2}}\mathrm{d}W \ . \tag{5.40}$$

The density of states of the solutions of Schrödinger's equation with complex wave-vectors for a one-dimensional solid thus varies as a U-shape across the band gap. This is schematically shown in the diagram on the right-hand side of Fig. 5.7.

The energy dispersion curves (5.37) or (5.38) are mathematical solutions of Schrödinger's equation with complex wave-vectors (5.33) for a one-dimensional solid with the simple cosine potential (5.28) only and contain no physical boundary conditions. Therefore, *Heine* [1965] called them the continuum of *virtual gap states* or, for short, ViGS. Nevertheless, the virtual gap states of the above simple, one-dimensional model resemble all the characteristics of the metal-induced gap states which resulted from much more realistic and involved calculations. The local density of states which *Louie* and *Cohen* [1976] computed for Al-jellium/Si(111) contacts and which are displayed in Fig. 5.6 are a convincing example. These and later calculations confirmed *Heine*'s more basic reasoning that the continuum of the virtual gap states is indeed the source of the metal-induced gap states at real metal–semiconductor interfaces.

5.5 Virtual Gap States: Three-Dimensional Model

5.5.1 Mean-Value $\underline{k}$-Points and the Dielectric Band Gap

The preceding one-dimensional model cannot be directly applied to real and therefore three-dimensional semiconductors. Their band gaps vary throughout the Brillouin zone. However, *Baldereschi*'s [1973] concept of mean-value $\underline{k}$-points in the Brillouin zone facilitates the calculation of branch-point energies for real semiconductors. Mean-value points $\underline{k}_{mv}$ in the Brillouin zone are defined such that the value which any given periodic function $f(k)$ of the wave-vector assumes at these special $\underline{k}$-points is an excellent approximation to the average value of the same function throughout the Brillouin zone, i.e.,

$$<f>=1/\Omega_k \int_{B-z} f(\underline{k})d^3k = f(\underline{k}_{mv}) \quad , \tag{5.41}$$

where Ω_k is the primitive-cell volume in k-space. For the face-centered cubic lattice, there is an infinite number of such mean-value $\underline{k}$-points. If only the first few nearest-neighbor shells are significant then the coordinates of the *best* mean-value point are $\underline{k}_{mv} = (2\pi/a_0)(0.6223, 0.2953, 0)$ where a_0 is the lattice parameter [*Baldereschi* 1973, *Chadi* and *Cohen* 1973]. From a generalization of the one-dimensional model to three dimensions, one is inclined to conclude that the branch points in the continuum of the virtual gap states might be close to the middle of the average band gap at the mean-value point in the first Brillouin zone. [6]

The width of the average band gap of a semiconductor may be obtained using a simple, classical approach.[7] The valence electrons of semiconductors may be considered as nearly free electrons. Their response to externally applied electric-fields $\underline{E}$ is then described as damped harmonic oscillations and the respective equation of motion may be written as

$$m_0 d^2\underline{r}/dt^2 + m_0\gamma d\underline{r}/dt + m_0\omega_0^2\underline{r} = -e_0\underline{E} \quad , \tag{5.42}$$

where ω_0 is a natural oscillator frequency and γ characterizes the usual velocity-dependent damping. For harmonic excitations $\underline{E} = \underline{E}_0\exp(-i\omega t)$, one obtains the forced displacements of the electrons as

$$\underline{r} = -(e_0/m_0)[(\omega_0^2 - \omega^2) - i\gamma\omega]^{-1}\underline{E} \quad . \tag{5.43}$$

The valence electrons oscillate against the positively charged cores and, therefore, the response (5.43) may also be described in terms of the polarization

$$\underline{P} = -e_0 n_v \underline{r} = (e_0^2 n_v/m_0)[(\omega_0^2 - \omega^2) - i\gamma\omega]^{-1}\underline{E} = [\varepsilon(\omega) - 1]\varepsilon_0\underline{E} \tag{5.44}$$

or, what is the same, by the frequency-dependent dielectric function

$$\varepsilon(\omega) - 1 = (e_0^2 n_v/\varepsilon_0 m_0)/[(\omega_0^2 - \omega_2) - i\gamma\omega] = \omega_p^2/[(\omega_0^2 - \omega^2) - i\gamma\omega] \quad , \tag{5.45}$$

where n_v is the density of the bulk valence electrons. For $\omega \to 0$, i.e., for the static case and by setting $\hbar\omega_0 = W_{dg}$, this classical approach gives

$$\varepsilon_\infty - 1 = (\hbar\omega_p/W_{dg})^2 \quad , \tag{5.46}$$

---

[6] *Cardona* and *Christensen* [1987] were the first to use *Baldereschi*'s concept in calculations of branch-point energies. They computed the energy bands within the linear muffin-tin orbital method and treated exchange and correlation in the local-density approximation. They avoided the scissors operation for enlarging the band gaps to the experimental values but increased the band gaps at the mean-value $\underline{k}$-point by taking the energy distance between the average of the top two valence and of the lowest two conduction bands. These averaged gaps are close to the dielectric band gaps of the III–V and II–VI compounds but for diamond the deviation is as large as 25%. *Cardona* and *Christensen*'s so called dielectric mid-gap energies are close to *Tersoff*'s [1986c] branch-point energies.

[7] See, for example, *Burns* [1985].

where ε_∞ is the electronic contribution to the static dielectric constant or, what is equivalent, the optical dielectric constant and

$$\omega_p = (e_0^2 n_v / \varepsilon_0 m_0)^{1/2} \tag{5.47}$$

is the plasmon frequency of the bulk valence electrons.[8] Intuitively, W_{dg} is called the dielectric band gap of the semiconductor. The dielectric band gaps of the group–IV elemental and many of the compound semiconductors are compiled in the fourth column of Table 5.1.

Penn [1962] also obtained relation (5.46) when he described semiconductors by an isotropic nearly free electron model in three dimensions. In his approach the nearly free electrons occupy a valence band of width W_1 which is separated from the conduction band by an energy gap of constant width W_{dg}. His quantum-theoretical model gives an electronic contribution to the static dielectric constant

$$\varepsilon_\infty - 1 = (\hbar\omega_p / W_{dg})^2 [1 - W_{dg} / 4W_1 + (W_{dg} / 4W_1)^2 / 3 - \ldots] \ . \tag{5.48}$$

For the group–IV elemental and the III–V and zinc-chalcogenide compound semiconductors $W_{dg}/4W_1$ varies between 0.07 and 0.15. Thus, the second and the third term in the square brackets of (5.48) may be neglected and the quantum-theoretical expression reduces to the classical relation (5.46). This approach then equates the average band gaps of semiconductors with their dielectric band gaps.

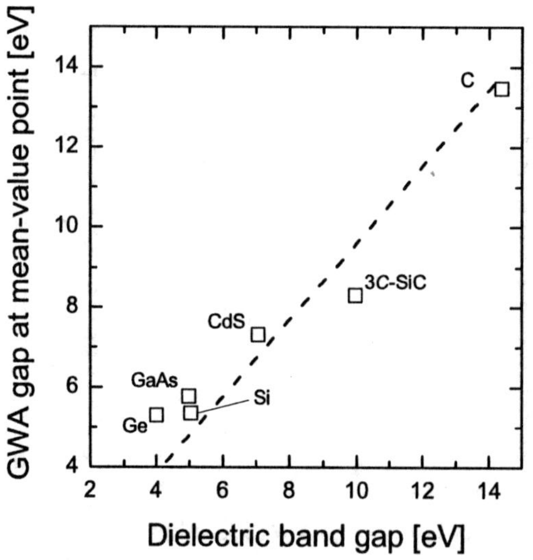

Fig.5.8. Width of the band gap at the mean-value point of diamond, Si, Ge, 3*C*-SiC, GaAs, and CdS as calculated in the GW approximation versus the width of the dielectric band gap calculated using (5.44) and (5.45). The dashed line is a linear least-squares fit and has a slope parameter of 0.96 ± 0.05. From *Mönch* [1996b]

---

[8] See, for example, *Raether* [1980].

Standard first-principles calculations of electronic band structures of semiconductors are based on density-functional theory (DFT) and the local-density approximation (LDA) for the exchange-correlation functional. This approach describes ground-state properties quite well but notoriously underestimates the band gaps. This problem was first treated by rigid shifts of the conduction bands to higher energies but lately *Hybertson* and *Louie* [1986] resolved it by the GW approximation (GWA) of the electron self-energy operator. Figure 5.8 compares the DFT-GWA-computed band gaps of diamond, Si, Ge, SiC, GaAs, and CdS at the-mean-value point $\underline{k}_{mv} = (2\pi/a_0)(0.6223, 0.2953, 0)$ in the first Brillouin zone [*Rohlfing* et al. 1993, 1995, 1996] and the dielectric band gaps of these semiconductors that were calculated using Eq. (5.47). The dashed line is a linear least-squares fit to the data points. Its slope parameter of 0.96 ± 0.05 confirms that the band gap at the mean-value $\underline{k}$-point in the first Brillouin zone indeed equals the dielectric band gap [*Mönch* 1996b].

5.5.2 Branch-Point Energy in the ViGS Continuum

Interface and surface states alike derive from the bulk bands. Correspondingly, the character of interface states changes across the band gap from predominantly donor-like to predominantly acceptor-like closer to the valence-band maximum and the conduction-band minimum, respectively. The energy W_{bp}, at which the contributions from both bands are equal in magnitude, is called the branch point. At the branch point the decay length of the respective ViGS wave-functions assumes its minimum value $1/q_{max}$. The branch point in the ViGS continuum of a one-dimensional lattice is thus below the mid-gap position by $V_1^2/4W_1$. A generalization of the one-dimensional model to three dimensions then suggests that the branch point in the ViGS continuum is found slightly below the middle of the dielectric gap at the mean-value point $\underline{k}_{mv}$ in the first Brillouin zone, i.e., to a first approximation one expects $W_{bp} \approx W_v(\underline{k}_{mv}) + W_{dg}/2$ and $W_{bp} - W_v(\Gamma) \approx W_{dg}/2 - [W_v(\Gamma) - W_v(\underline{k}_{mv})]$, respectively, where $W_v(\Gamma)$ is the energy of the valence-band maximum in the middle of the first Brillouin zone.

Tersoff [1984a, 1986a, 1986c] calculated the branch-point energies with regard to the valence-band maximum for 15 semiconductors. He used a linearized augmented plane-wave method (APW) and the local-density approximation (LDA) and he enlarged the LDA band gaps, which turned out to be too narrow, to the experimental values by rigid shifts of the conduction bands or, as this procedure is called, by applying the *scissors operator*. The third column of Table 5.1 displays *Tersoff*'s branch-point energies $[W_{bp} - W_v(\Gamma)]_{APW}$.

To obtain the energy position of the branch point within the gap at the mean-value $\underline{k}$-point the energy difference $W_v(\Gamma) - W_v(\underline{k}_{mv})$ from the valence band at the mean-value $\underline{k}$-point to its maximum at Γ in the middle of the Brillouin zone is required. *Mönch* [1996b] demonstrated that the dispersion of the DFT-GWA valence bands, and consequently the energy differences $W_v(\Gamma) - W_v(\underline{k}_{mv})$, are well represented by the empirical tight-binding approximation (ETB). This method will be briefly illustrated by treating the simple one-dimensional model considered above.

In the isolated atoms, the electrons are assumed to move in a spherically symmetric potential $U(r)$, and their s states of energy W_s with wave-functions $\phi_s(r)$ shall be considered. The wave-functions $\psi(z)$ of the electrons in the periodic potential $V(z)$ of the one-dimensional solid with lattice parameter a are taken as linear combinations of the unperturbed atomic orbitals, i.e.,

$$\psi(z) = \sum_{\mathrm{n}} \mathrm{c_n} \phi_s(r_\mathrm{n}) \quad . \tag{5.49}$$

Substituting (5.49) into Schrödinger's equation gives

$$\sum_{\mathrm{n}} \mathrm{c_n} \{(W - W_s) - [V(z) - U(r_\mathrm{n})]\} \phi_s(r_\mathrm{n}) = 0 \quad . \tag{5.50}$$

The energies W are obtained by multiplying (5.49) with $\phi_s^*(r_\mathrm{m})$ and integrating over all space. By considering nearest-neighbor interactions only, i.e., for $(\mathrm{m} - \mathrm{n}) = \pm 1$, and using the definitions

$$\int \phi_s^*(r_\mathrm{n})[V(z) - U(r_\mathrm{n})]\phi_s(r_\mathrm{n}) \mathrm{d}\tau = -\alpha \tag{5.51a}$$

and

$$\int \phi_s^*(r_\mathrm{m})[V(z) - U(r_\mathrm{n})]\phi_s(r_\mathrm{n}) \mathrm{d}\tau = -\gamma \quad , \tag{5.51b}$$

one obtains

$$\gamma \mathrm{c_{n-1}} + (W - W_s + \alpha)\mathrm{c_n} + \gamma \mathrm{c_{n+1}} = 0 \quad . \tag{5.52}$$

Since again wave-vectors $\kappa = \pi/a - k$ at the boundary of the first Brillouin zone are of interest only, it is sufficient to consider only two terms of the Fourier series expansion of the LCAO coefficients, which may then be written as

$$\mathrm{c_n} = \mathrm{A} \exp(\mathrm{in}\kappa a) + \mathrm{B} \exp(-\mathrm{in}\kappa a) \quad . \tag{5.53}$$

Substitution of (5.49) into (5.48) gives the well-known energy dispersion in the tight-binding approximation

$$W = W_s - \alpha - 2\gamma \cos(\kappa a) \tag{5.54}$$

for one-dimensional solids.[9] Near to the zone boundary, the energy bands disperse parabolically, as they do in the nearly free electron approximation (5.31). A similar energy dispersion curve but with opposite curvature is obtained for p electrons.

The elemental group-IV as well as the II–VI and III–V compound semiconductors are tetrahedrally coordinated in the bulk. In a simple picture, each atom forms four covalent or electron-pair bonds with its four nearest neighbors. *Harrison*'s [1973, 1980] bond orbital approximation starts from sp^3 hybrids constructed for isolated atoms which finally form into semiconductors. The four sp^3 hybrids are thought to point at the corners of a tetrahedron with the atom at its

[9] See textbooks on solid state physics as, e.g by *Ibach* and *Lüth* [1991].

center. These hybrids h_i are taken as linear combinations of the atomic s, p_x, p_y, and p_z orbitals, i.e.,

$$|h^i> = \frac{1}{2}\left[|s^i> + |p_x^i> + |p_y^i> + |p_z^i>\right] . \tag{5.55}$$

In compounds, the superscript i refers to the corresponding cations (c) or anions (a). The expectation value of the energy of such hybrids, the so called hybrid energy, results as

$$W_h^i = \frac{W_s^i + 3W_p^i}{4} , \tag{5.56}$$

where W_s^i and W_p^i are the atomic s- and p-term values, respectively, of the atoms considered. Consequently, four inter-orbital matrix elements, $V_{ss\sigma}$, $V_{sp\sigma}$, $V_{pp\sigma}$, and $V_{pp\pi}$, enter tight-binding studies of semiconductor band-structures. These matrix elements scale with the reciprocal square of the nearest-neighbor distances d_{nn} as

$$V_{ll'\lambda} = \eta_{ll'\lambda}(\hbar^2/m_0)d_{nn}^{-2} \tag{5.57}$$

($\hbar^2/m_0 = 0.0762$ eVnm^2). The coefficients $\eta_{ll'\lambda}$ are not specific for any special semiconductor.

The bulk energy bands of semiconductors may be calculated by using a three-dimensional tight-binding approximation which is an analog to the one-dimensional approach briefly outlined above. To mention an example, this approach gives the energy of the valence-band maximum of binary compound semiconductors as [*Chadi* and *Cohen* 1975]

$$W_v(\Gamma) = (W_p^c + W_p^a)/2 - [(W_p^c - W_p^a)^2/4 + (4V_{xx})^2]^{1/2} \tag{5.58}$$

with the inter-atomic matrix element

$$V_{xx} = (V_{pp\sigma} + 2V_{pp\pi})3 = \eta_{xx}(\hbar^2/m_0)d_{nn}^{-2} . \tag{5.59}$$

By fitting tight-binding valence bands of a number of semiconductors to band structures which were computed by using a pseudopotential scheme, *Harrison* [1981] obtained the four universal coefficients

$$\eta_{ss\sigma} = -1.32, \ \eta_{sp\sigma} = 1.42, \ \eta_{pp\sigma} = 2.22, \text{ and } \eta_{pp\pi} = -0.63 . \tag{5.60}$$

In his empirical tight-binding or, for short, ETB calculations, *Harrison* used *Fischer*'s [1972] atomic Hartree–Fock term values.

The energy position $W_{bp} - W_v(\underline{k}_{mv})$ the branch point within the energy gap of width W_{dg} at the mean-value $\underline{k}$-point may be divided up into the dispersion $W_v(\Gamma) - W(\underline{k}_{mv})$ of the top valence band from the middle of the Brillouin zone, Γ, up to the mean-value $\underline{k}$-point and the branch-point energy $\Phi_{bp}^p \equiv W_{bp} - W_v(\Gamma)$; see Fig. 5.9. The energy term $W_v(\Gamma) - W(\underline{k}_{mv})$ may be calculated by solving Schrödinger's equation of the three-dimensional ETB approach and using *Fischer*'s atomic term

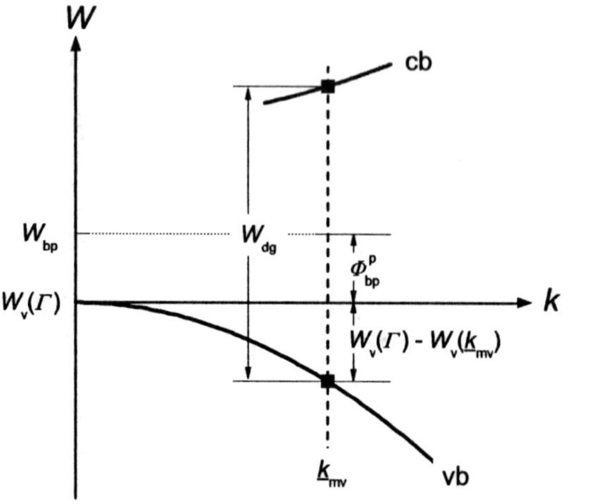

Fig. 5.9. Dispersion of the top valence (vb) and the lowest conduction band (cb) around the mean-value $\underline{k}$-point (schematic). The width of the energy gap at the mean-value point equals the dielectric band gap W_{dg}

values and *Harrison*'s universal inter-orbital matrix elements. Branch-point energies Φ^{p}_{bp} are available for the elemental semiconductors Si and Ge and for 13 of the II–VI and III–V compound semiconductors from *Tersoff*'s APW calculations. The branch-point energies $W_{bp} - W_v(\underline{k}_{mv}) = [W_v(\Gamma) - W_v(\underline{k}_{mv})]_{\mathrm{ETB}} + [\Phi^{p}_{bp}]_{\mathrm{APW}}$ of these 15 semiconductors are plotted versus the widths W_{dg} of their dielectric band gaps in Fig. 5.10. The dashed line is a linear least-squares fit to the data points of the compound semiconductors. Its slope parameter is 0.449 ± 0.007. The data points of the elemental semiconductors Si and Ge slightly deviate from this best-fit straight line. For the compound semiconductors, the energy positions of the branch points with respect to the valence-band maxima thus result as [*Mönch* 1996b]

$$W_{bp} - W_v(\Gamma) = 0.449 \cdot W_{dg} - [W_v(\Gamma) - W_v(\underline{k}_{mv})] \quad . \tag{5.61}$$

The branch points in the continua of the virtual gap states are thus below the mid-gap at the mean-value $\underline{k}$-point in the Brillouin zone by 5%. This finding corresponds with the result found above for the one-dimensional model. Relation (5.61) now makes it possible to calculate the branch-point energies of semiconductors for which no APW data are available.

The second column of Table 5.1 shows the branch-point energies $[W_{bp} - W_v(\Gamma)]_{\mathrm{ETB}}$ which were calculated using (5.61) and ETB values $[W_v(\Gamma) - W_v(\underline{k}_{mv})]_{\mathrm{ETB}}$ of the valence-band dispersion from the middle of the Brillouin zone to the mean-value $\underline{k}$-point. On the average, the ETB values of the compound semiconductors differ from *Tersoff*'s APW data by 0.12 ± 0.08 eV. Somewhat larger deviations

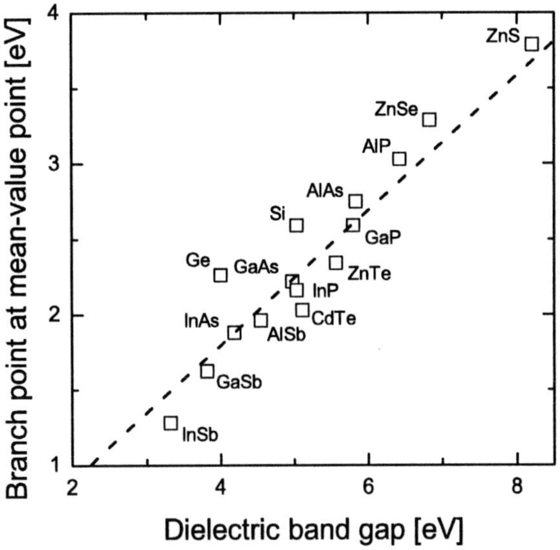

Fig. 5.10. Position of the ViGS branch point above the top valence band at the mean-value point versus the width of the dielectric band gap. The dashed line is a linear least-squares fit to the data of the binary compounds and has a slope parameter of 0.449 ± 0.007. From *Mönch* [1996b]

occur for silicon and germanium. This effect may be attributed to equating ε_∞ with ε_b for the elemental semiconductors. Multiphonon absorption adds to the electronic dielectric constant ε_∞ , so that one expects $\varepsilon_b > \varepsilon_\infty$ not only for the compound but also for the elemental semiconductors. Considering this effect would enlarge the dielectric band gap. and, as a consequence of this, the Ge and Si data points would move closer to the best-fit straight line in Fig. 5.10.

5.5.3 Decay Length of the ViGS at their Branch Point

For an estimate of the decay length $1/q_{ss}^{mg}$ of interface states at the branch point, the widths W_1 and $2|V_1|$ of the valence band and the band gap, respectively, in the one-dimensional model are now replaced by the widths ΔW_v of the bulk valence bands and of the dielectric band gaps of three-dimensional semiconductors, respectively. Typical values are $\Delta W_v \approx 12$ eV and $<W_{dg}> \approx 4.7$ eV. These data give the decay lengths $1/q_{is}^{bp} \approx 0.575$ nm for the wave-functions and $1/2q_{is}^{bp} \approx 0.288$ nm for the respective charge density of interface states at their branch point. For some semiconductors the decay lengths of the charge distribution in the top-most occupied interface states were calculated by using detailed theoretical approaches. These values are shown in the fifth column of Table 5.1.

Table 5.1 Branch-point energies $[W_{bp} - W_v(\Gamma)]_{ETB}$ and $[W_{bp} - W_v(\Gamma)]_{APW}$ calculated using the empirical tight-binding [*Mönch* 1996b] and an augmented plane-wave method [*Tersoff* 1984a, 1986a, 1986c], respectively. Also listed are the dielectric gap W_{dg}, the charge-density tailing length $1/2q_{gs}^{mi}$, and the density of states D_{gs}^{mi} of metal-induced gap states. The dielectric band gaps were calculated using (5.44) and (5.45).

Semi-conductor	$[W_{bp} - W_v(\Gamma)]_{ETB}$ [eV]	$[W_{bp} - W_v(\Gamma)]_{APW}$ [eV]	W_{dg} [eV]	$1/2q_{gs}^{mi}$ [nm]	$D_{gs}^{mi} \times 10^{14}$ [cm^{-2}eV^{-1}]
C	1.77		14.4	0.137[d]	2.00[d]
Si	0.03	0.36	5.04	0.30[e]	4.50[e]
					3.96[b]
Ge	–0.28	0.18	4.02	0.40[e]	
Sn	–0.46		2.52		
SiC	1.44		9.84		
BN	3.25		16.33		
AlN	2.97		11.92		
AlP	1.13	1.27	6.45		
AlAs	0.92	1.05	5.81	0.66[b]	
AlSb	0.53	0.45	4.51		
GaN	2.37		10.80	0.186[g]	
GaP	0.83	0.81	5.81	0.294[d]	3.12[a]
GaAs	0.52	0.50	4.97	0.30[e]	5.00[c]
				0.28[c]	3.72[a]
				0.32[f]	2.70[f]
GaSb	0.16	0.07	3.80		
InN	1.51		6.48		
InP	0.86	0.76	5.04	0.424[b]	3.45[a]
InAs	0.50	0.50	4.20		
InSb	0.22	0.01	3.33		5.05[a]
ZnS	2.05	2.22	8.12	0.09[c]	1.40[c]
				0.16[a]	1.96[a]
ZnSe	1.48	1.70	7.06	0.19[c]	2.00[c]
				0.20[b]	2.28[a]
ZnTe	1.00	0.84	5.55		
CdS	1.93		7.06		
CdSe	1.53		6.16		
CdTe	1.12	0.85	5.11		2.34[a]
HgTe		0.34			
BeTe	1.34		6.42		
MgS	1.97		6.60		
MgTe	1.19		4.54		
MnTe		1.6			
$CuGaS_2$	1.43		7.46		
$CuInS_2$	1.47		7.0		
$CuAlSe_2$	1.25		6.85		

Table 5.1. (continued)

Semi-coductor	$[W_{bp} - W_v(\Gamma)]_{ETB}$ [eV]	$[W_{bp} - W_v(\Gamma)]_{APW}$ [eV]	W_{dg} [eV]	$1/2q_{gs}^{mi}$ [nm]	$D_{gs}^{mi} \times 10^{14}$ [cm^{-2}eV^{-1}]
$CuGaSe_2$	0.93		6.29		
$CuInSe_2$	0.75		5.43		
$CuGaTe_2$	0.61		5.39		
$CuInTe_2$	0.55		4.78		
$AgGaSe_2$	1.09		5.96		
$AgInSe_2$	1.11		5.60		

[a] *Louis* et al. [1976]
[b] *Tejedor* et al. [1977]
[c] *Louie* et al. [1977]
[d] *Ihm* et al. [1978]
[e] *Tersoff* [1984a, 1986c]
[f] *Berthod* et al. [1996]
[g] *Picozzi* et al. [1998]

Remarkably, they are very close to the above estimate; this estimate followed from a simple adoption of relation (5.39), which was derived in the framework of the one-dimensional model.

5.6 Slope Parameters of Barrier Heights

The continuum of the metal-induced gap states constitutes the intrinsic interface states at metal–semiconductor contacts. Equation (5.23), which is simply derived from a phenomenological model, may then be rewritten as

$$\frac{A_X}{S_X} - 1 = \frac{e_0^2}{\varepsilon_i \varepsilon_0} \frac{D_{gs}^{mi}(W_{bp})}{2q_{gs}^{mi}(W_{bp})} , \tag{5.62}$$

where $D_{gs}^{mi}(W_{bp})$ is the density of states of the MIGS at their branch point and the thickness δ_{is} of the corresponding interfacial dipole layer is approximated by the charge decay length $1/2q_{gs}^{mi}(W_{bp})$ of the MIGS at their branch point.

The MIGS continuum derives from the virtual gap states or, in other words, from the complex band structure of the semiconductor. The simple one-dimensional ViGS model of Sect. 5.4 will be again considered. Relations (5.39) and (5.40) give the charge decay length $1/2q_{gs}^{vi}(W_{bp}) \cong 4W_1/2g_1|V_1|$ and the density of states $D_{gs}^{vi}(W_{bp}) \cong 1/2\pi|V_1|$, respectively, of the ViGS at their branch point. Hence, one obtains $D_{gs}^{vi}(W_{bp})/2q_{gs}^{vi}(W_{bp}) \cong 2W_1 a/\pi^2 4V_1^2$, i.e., the product of the ViGS density-of-states and the thickness of the corresponding interfacial dipole layer are proportional to the inverse square of the width $2|V1|$ of the band gap.

The conclusions reached from the one-dimensional model may again be transferred to three dimensions, i.e., to real metal–semiconductor contacts. Columns five and six of Table 5.1 display the densities of states D_{gs}^{mi} and the tailing lengths $1/2q_{gs}^{mi}$ that have been calculated for metal contacts on diamond, Si, GaP, GaAs, InP, ZnS, and ZnSe by applying the DFT-LDA scheme. The relevant band gaps are the energy gaps at the mean-value $\underline{k}$-point, i.e., the dielectric band gaps W_{dg}. Consequently, the respective values of $(e_0^2/\varepsilon_0)(D_{gs}^{mi}/2q_{gs}^{mi})$ are plotted versus the dielectric band gaps of the semiconductors in Fig. 5.11a. The dashed line is the linear least-squares fit

$$D_{gs}^{mi}/2q_{gs}^{mi} \propto 1/W_{dg}^{1.98\pm0.46} \tag{5.63}$$

to the data. This is exactly the same dependence that is predicted by the simple one-dimensional model as outlined above. Only the data point of diamond deviates from the general trend. This is easily explained. The widths of the valence bands of Si, Ge, and the III–V and II–VI compounds exhibit only small variations around their average value $\langle \Delta W_v \rangle = 12.75 \pm 1.07$ eV. The valence-band width of diamond, on the other hand, is measured as 24 eV and thus is almost twice as large as that observed in the other semiconductors.

The optical dielectric constant ε_∞ defines the width of the dielectric band gap through the relation (5.46). Again, the valence-electron plasmon energies of Si,

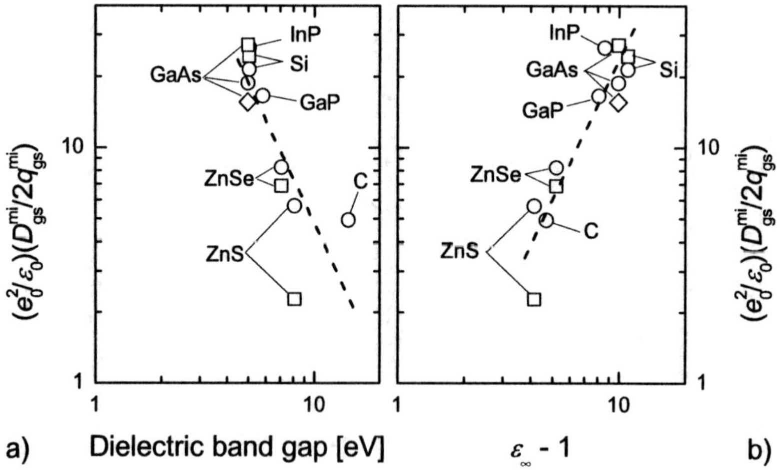

Fig. 5.11. Theoretical "slope parameters" $(e_0^2/\varepsilon_0)(D_{gs}^{mi}/2q_{gs}^{mi})$ as a function of the width of the dielectric band gap (a) and of the optical susceptibility $\varepsilon_\infty - 1$ (b). The calculated values of the densities of states D_{gs}^{mi} and of the decay lengths $1/2q_{gs}^{mi}$ of the MIGS at their branch point are listed in Table 5.1. The dashed lines are linear least-squares fits to the data points

Ge, and the III–V compound semiconductors do not differ much from their average value $<\hbar\omega_p> = 15 \pm 1.4$ eV. Hence, the same $(e_0^2/\varepsilon_0)(D_{gs}^{mi}/2q_{gs}^{mi})$ data used in Fig. 5.11a are replotted in Fig. 5.11b but now as a function of the optical susceptibility $\varepsilon_\infty - 1$. The dashed line is the linear least-squares fit

$$(e_0^2/\varepsilon_0)(D_{gs}^{mi}/2q_{gs}^{mi}) = 0.28(\varepsilon_\infty - 1)^{1.9\pm0.24} \tag{5.64}$$

to the data. Using relation (5.58), one then obtains [*Mönch* 1996a]

$$\mathrm{A}_x/S_x - 1 = 0.28(\varepsilon_\infty - 1)^2/\varepsilon_i \ . \tag{5.65}$$

The slope parameters S_X are thus determined by the optical dielectric constant ε_∞ of the semiconductor and an interface dielectric constant ε_i. The interface dielectric constant may be estimated by considering a simple electrostatic model.

The charge transferred at Schottky contacts may be modeled by a parallel-plate capacitor with an effective thickness t_{eff}. On the metal side of the interface, the charge resides within the Thomas–Fermi screening length $L_{\text{Th-F}}$ and on the semiconductor side in the metal-induced gap states that have a decay length $1/2q_{gs}^{mi}$. The effective distance between the center of mass of these charges is given by the true distances t_{m} in the metal and t_{s} in the semiconductor divided by the appropriate dielectric screening functions ε_{m} and ε_b, respectively. *Louie* et al. [1977] defined the interface dielectric constant ε_i by the relation

$$t_{eff} = t_{\text{m}}/\varepsilon_{\text{m}} + t_{\text{s}}/\varepsilon_b \equiv t_{\text{s}}/\varepsilon_i \ . \tag{5.66}$$

A typical value of the Thomas–Fermi screening length is $t_{\text{m}}/\varepsilon_{\text{m}} = 0.05$ nm, and theoretical calculations yielded decay lengths $1/2q_{gs}^{mi}$ of 0.137 nm for diamond [*Ihm* et al. 1978] ($\varepsilon_b = 5.7$) and 0.3 nm [*Louie* et al. 1977] for both GaAs ($\varepsilon_b = 12.8$) and Si ($\varepsilon_b = 11.9$). With these values relation (5.66) yields interface dielectric constants ε_i of 2 for diamond, 4 for GaAs, and 4.6 for silicon. It has to be mentioned that relation (5.66) is an extremely crude approximation in that it combines microscopic lengths and macroscopic dielectric constants.

With the approximate values $t_{\text{m}}/\varepsilon_{\text{m}} = L_{\text{TH-F}} \approx 0.05$ nm, $t_{\text{s}} = 1/2q_{gs}^{mi} \approx 0.245$ nm, and $\varepsilon_b \approx 12.5$, the relation (5.66) gives $\varepsilon_i \approx 3.5$ as an estimate of a mean interface dielectric constant. One then obtains

$$\mathrm{A}_x/S_x - 1 \approx 0.1\cdot(\varepsilon_\infty - 1)^2 \tag{5.67}$$

as an approximation for binary semiconductors. Remarkably, a bulk property of the semiconductor, its optical dielectric constant ε_∞, determines the slope parameter $S_X = \mathrm{d}\Phi_B/\mathrm{d}X_{\text{m}}$ of its Schottky contacts.

Already in 1986, *Mönch* obtained relation (5.67) in an empirical approach. Starting out from the ViGS of the one-dimensional model and *Penn*'s [1962] definition of the dielectric band gap, he argued that the expression $\mathrm{A}_X/S_X - 1$ might correlate with the optical susceptibility $\varepsilon_\infty - 1$. Figure 1.9 displays his plot of the then available slope parameters originally evaluated by *Kurtin* et al. [1969] and revised by *Schlüter* [1978, 1982] versus the optical susceptibility of 20 different semiconductors and insulators. The least-squares fit to the data, which is represented by the dashed straight line, exactly produced relation (5.63).[10] Even though the data had large margins of experimental error and, in many cases, strongly deviated from the best-fit straight line this finding was thought to be the first strong experimental indication that the MIGS are the intrinsic mechanism which determines the barrier heights of metal–semiconductor contacts.

[10] Different but unsuccessful attempts to correlate the slope parameters S_X and the dielectric constants ε_b and ε_∞ had been made earlier. *Phillips* [1963] plotted $(1 - S_X)$ versus $(1 - \varepsilon_\infty)$ and found a diagram similar to that of Fig. 1.9 but obtained no correlation. *Schlüter* [1978] looked for a correlation between the slope parameter and the static rather than the optical susceptibility. *Tersoff* [1985], finally, assumed different power laws between the slope parameters and the optical dielectric constant but found a poor correlation only.

6. The IFIGS-and-Electronegativity Concept: Experiment and Theory

6.1 Schottky Contacts

6.1.1 Introductory Remarks

Real metal–semiconductors even if they are intimate, abrupt, and free of any interfacial impurities are not *ideal*. Rather, they contain regions of lowered barrier height which have lateral dimensions smaller than the depletion-layer width and which are embedded in areas of larger but constant barrier height. The current–voltage characteristics of such *patchy* Schottky are characterized by *effective* barrier heights and ideality factors that are larger than the value n_{if} determined by the image-force effect only. Both quantities differ from one diode to another even if they are prepared under identical experimental conditions. They do not vary at random but are correlated in that the effective barrier height decreases with increasing ideality factor. The extrapolation of such Φ_{Bn}^{eff} versus n curves to the ideality factor n_{if} then gives the barrier height Φ_{Bn}^{hom} of the laterally homogeneous Schottky contact minus the image-force lowering $\delta\Phi_{if}^{0}$. Only the homogeneous barrier heights Φ_{Bn}^{hom} are a reliable basis for a comparison with theoretically valuesLocal variations of the barrier heights of real Schottky contacts on the nm-scale were directly detected by scanning ballistic-electron emission spectroscopy. The respective local BEEM barrier heights are well described by Gaussian distributions. The center values of the Gaussians again equal the barrier heights Φ_{Bn}^{hom} of the laterally homogeneous Schottky contacts minus the image-force lowering $\delta\Phi_{if}^{0}$.

Barrier heights determined by internal photoemission yield spectroscopy were always found to be larger than the effective barrier heights evaluated from the I/V characteristics of the same diodes. This is easily explained since the thresholds of the IPEYS curves again measure the barrier heights Φ_{Bn}^{hom} of the laterally homogeneous Schottky contacts minus the image-force lowering $\delta\Phi_{if}^{0}$. The tails to lower photon energies are caused by thermal effects and again by the lateral inhomogeneities of the barrier heights.

Tables 3.3 to 3.5 contain homogeneous barrier heights which were determined from Φ_{Bn}^{eff} versus n curves, the center values of Gaussian-distributed local BEEM barrier heights, and the thresholds of IPEYS curves, respectively. These data are compared in the sections to follow with the zero-charge-transfer barrier heights Φ_{bp}^{n} and the slope parameters S_X predicted by the MIGS-and-electronegativity concept. As an example, the 13 data for Au/n-Si contacts will be briefly considered. The average barrier height of intimate, abrupt, clean, and laterally homogeneous Au/n-Si contacts results as 0.831 ± 0.025 eV. The standard deviation of this mean value is smaller than the margins of error of the individual original data.

6.1.2 Silicon Schottky Contacts

The experimental barrier heights of intimate, abrupt, clean, and laterally homogeneous metal–silicon and silicide–silicon contacts are plotted versus the difference $X_m - X_{Si}$ of the *Miedema* electronegativities of the metals and silicon in Figs. 6.1 and 6.2, respectively. The experimental values of the barrier heights are taken from Tables 3.3 to 3.6. They are differentiated by the symbols ○ and □, ◇, Δ, and ∇ for data from *I/V*, BEEM, IPEYS, and PES measurements, respectively. The same set of symbols is also used in Figs. 6.3 to 6.6.

The barrier heights of the metal–silicon barrier heights which are displayed in Fig. 6.1 are obviously grouped according to their interface structures. The dashed and the dash-dotted lines are the linear least-squares fits

$$\Phi_{Bn}^{(1\times1)^i} = 0.803 + 0.098 \cdot (X_m - X_{Si}) \text{ [eV]} \tag{6.1}$$

and

$$\Phi_{Bn}^{(7\times7)^i} = 0.727 + 0.089 \cdot (X_m - X_{Si}) \text{ [eV]} \tag{6.2}$$

to the data of Schottky contacts with $(1\times1)^i$ and $(7\times7)^i$ interface structure, respectively. The slope parameters of both fits are identical within the margins of experimental error. However, the zero-charge-transfer barrier heights differ by 76 ± 2 meV. The Schottky contacts with $(7\times7)^i$ -reconstructed interfaces have the smaller barrier heights compared with the $(1\times1)^i$-unreconstructed contacts. The atomic positions of the interfacial $(7\times7)^i$ reconstruction differ from the bulk structure. On the whole, the corresponding rearrangements of the valence-charge distribution are balanced and, hence, the $(7\times7)^i$ reconstruction will contain an *extrinsic* electric-dipole layer in addition to the *intrinsic* MIGS electric-dipole layer. The effect of this extrinsic interfacial electric-dipole layer on the zero-charge-transfer barrier height is explained in Sect. 10.2.1.

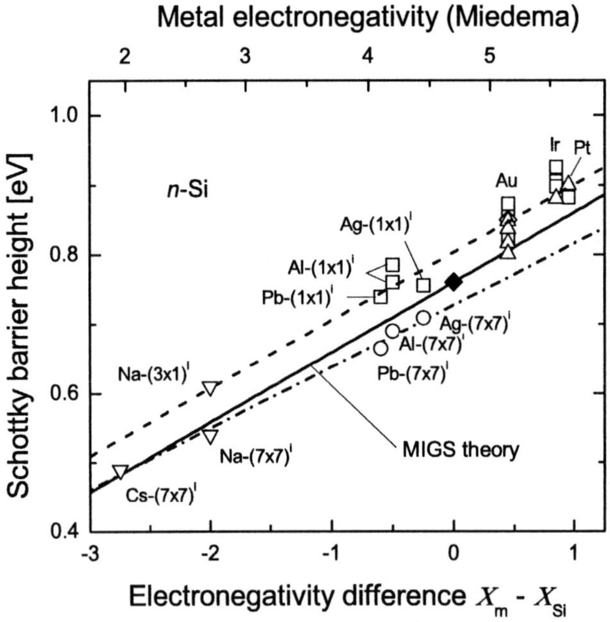

Fig. 6.1. Barrier heights of laterally homogeneous *n*-type silicon Schottky contacts versus the difference of the *Miedema* electronegativities of the metal and silicon. The ○ and □, ◇, Δ, and ∇ symbols differentiate the data from *I*/*V*, BEEM, IPEYS, and PES measurements, respectively, as compiled in Tables 3.3 to 3.6. The dashed and the dash-dotted lines are linear least-squares fits to the data of diodes with $(1\times1)^i$-unreconstructed and $(7\times7)^i$-reconstructed interfaces, respectively. The solid MIGS line is drawn with $S_X = 0.101$ eV/Miedema-unit and $\Phi_{bp}^{p} = 0.36$ eV [*Tersoff* 1986c]

Silicide–silicon contacts also show specific peculiarities. First, the epitaxial $NiSi_2$/Si(111) and $CoSi_2$/Si(001) contacts may be prepared such that they posses either of two different interface structures, which are labeled type-A and type-B. The differences of the respective barrier heights are again caused by extrinsic interfacial electric-dipole layers. This behavior is explained in detail in Sect.10.2.2. Like the barrier heights of type-A $NiSi_2$/Si and the $CoSi_2$/Si contacts, the barrier heights of the $MoSi_2$-silicon and the $CrSi_2$–silicon diodes deviate towards lower values. Presently, it is not clear whether this behavior is also caused by extrinsic electric-dipole layers, which are induced by specific interface-structures, or by interface defects. Epitaxial $GdSi_2$ films on silicon, on the other hand, will show either an hexagonal or an orthorhombic lattice structure. The difference of the respective barrier heights is again attributed to an extrinsic electric-dipole layer at one the interfaces. If the type-A $NiSi_2$- and $CoSi_2$- and the hex-$GdSi_2$/Si contacts are not considered, then the linear least-squares fit

$$\Phi_{Bn}^{sili} = 0.785 + 0.115 \cdot (X_m - X_{Si}) \text{ [eV]} \tag{6.3}$$

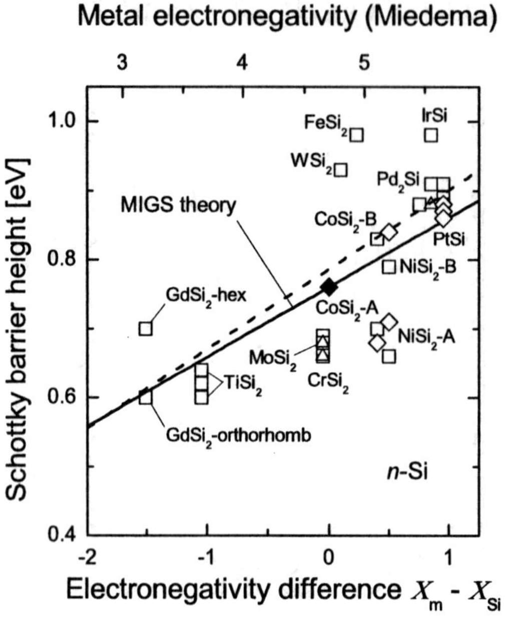

Fig. 6.2. Barrier heights of laterally homogeneous silicide–silicon contacts versus the difference of the *Miedema* electronegativities of the metal and silicon. The □, ◇, and Δ symbols differentiate the data from *I/V*, BEEM, and IPEYS measurements, respectively, as compiled in Tables 3.3 to 3.5. The dashed line is a linear least-squares fit to part of the data; for details see text. The solid MIGS line is drawn with the same parameters as in Fig. 6.1

is obtained from the remaining data points of silicide–silicon contacts plotted in Fig. 6.2.

The large scatter of the silicide–silicon barrier heights may be caused by interface defects. Their densities might depend on the specific conditions under which the silicides were grown. *Fujitani* and *Asano* [1994] found that at such defected interfaces transition-metal atoms may show dangling bonds. The respective extrinsic interface states would exist in addition to the intrinsic MIGS and might then alter the barrier heights from the values of the defect-free interfaces.

The MIGS-and-electronegativity theory predicts the barrier heights of ideal silicon Schottky contacts to vary as

$$\Phi_{Bn}^{\text{MIGS}} = 0.76 + 0.101 \cdot (X_{\text{m}} - X_{\text{Si}}) \text{ [eV]} \quad . \tag{6.4}$$

The solid lines shown in Figs. 6.1 and 6.2 represent this theoretical relationship. Within the margins of experimental error the MIGS-and-electronegativity theory obviously explains the experimental barrier heights of silicon Schottky contacts irrespective of whether pure metals or metal silicides are used.

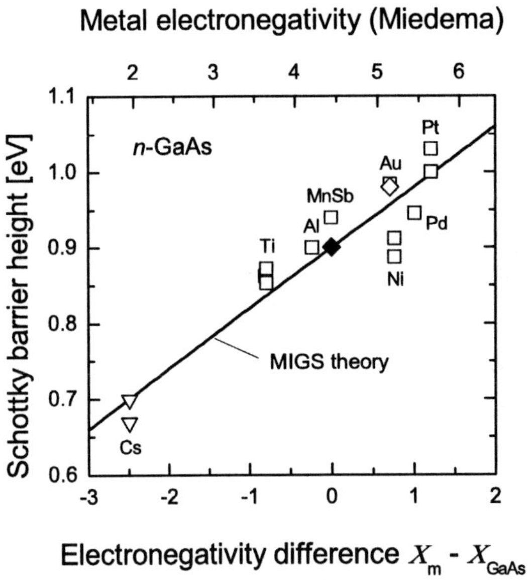

Fig. 6.3. Barrier heights of laterally homogeneous *n*-GaAs Schottky contacts versus the difference of the *Miedema* electronegativities of the metal and gallium arsenide. The □, ◇, Δ, and ∇ symbols differentiate the data from *I/V*, BEEM, IPEYS, and PES measurements, respectively, as compiled in Tables 3.3 to 3.6. The solid MIGS line is drawn with $\Phi_{bp}^{p} = 0.5$ eV [*Mönch* 1996b] and $S_X = 0.08$ eV/Miedema-unit

6.1.3 GaAs Schottky Contacts

The barrier heights of intimate, clean, and laterally homogeneous *n*-GaAs Schottky contacts which are shown in Tables 3.3, 3.4, and 3.6 are plotted versus the electronegativity difference of the metal and GaAs in Fig. 6.3. The solid line

$$\Phi_{Bn}^{MIGS} = 0.92 + 0.08 \cdot (X_{\mathrm{m}} - X_{\mathrm{GaAs}})\ [\mathrm{eV}] \tag{6.5}$$

of the MIGS-and-electronegativity theory excellently reproduces the experimental results.

The transition metals are known to react with semiconductors. While Pt shows no evidence for reaction with GaAs below approximately 300°C the other two transition metals, Ni and Pd, for which data of laterally homogeneous Schottky contacts are available, decompose GaAs even at room temperature. The slight deviations of the Ni and Pd data points from the MIGS line towards lower values may thus be attributed to defected and non-abrupt interfaces.

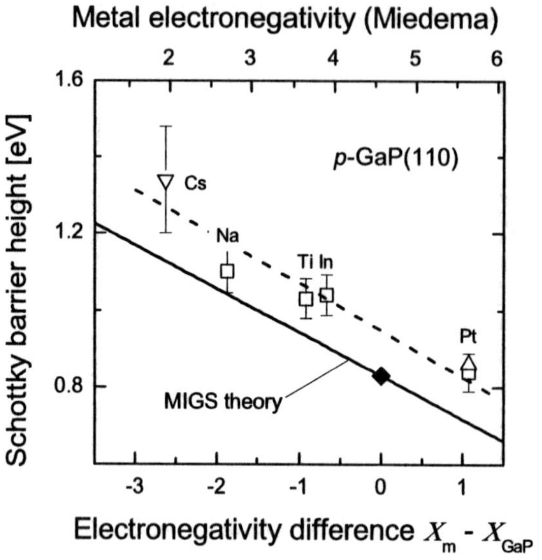

Fig. 6.4. Barrier heights of laterally homogeneous *p*-GaP(110) Schottky contacts versus the difference of the *Miedema* electronegativities of the metal and gallium phosphide. The □, ∇ and Δ symbols differentiate the data from *I*/*V*, IPEYS, and PES measurements, respectively, as compiled in Tables 3.3, 3.5, and 3.6. The dashed line is a linear least-squares fit to the experimental data. The solid MIGS line is drawn with $S_X = 0.113$ eV/Miedema-unit and $\Phi^p_{bp} = 0.83$ eV [*Mönch* 1996b]

6.1.4 GaP Schottky Contacts

The experimental barrier heights of intimate, abrupt, clean, and laterally homogeneous Schottky contacts prepared on cleaved (110) surfaces of *p*-GaP are displayed in Fig. 6.4. The experimental data are taken from Tables 3.3 and 3.6. The *Miedema* electronegativity of GaP, $X_{GaP} = 4.57$ Miedema-units, is estimated by inserting the *Miedema* electronegativity of phosphorous, $X_P = 5.1$ Miedema-units, which was obtained from its *Pauling* value by applying the conversion (5.16), into relation (5.3). The dashed line is the linear least-squares fit

$$\Phi_{Bp} = 0.95 - 0.122 \cdot (X_m - X_{GaP}) \text{ [eV]} \tag{6.6}$$

to the experimental data. The slope parameters of this fit and of the MIGS line

$$\Phi_{Bp} = 0.83 - 0.113 \cdot (X_m - X_{GaP}) \text{ [eV]} \tag{6.7}$$

agree within the margins of experimental error. However, the experimental fit is set off from the theoretical MIGS line towards larger barrier heights by approximately 120 meV. This value slightly exceeds the margins of error. Although Fig. 6.4 displays barrier heights of laterally homogeneous GaP Schottky contacts this shift obviously indicates the existence of an extrinsic mechanism in addition to the intrinsic MIGS. A most reasonable explanation is the presence of interfacial phosphorous vacancies, i.e., interface defects which are created during the fabrication of the metal contacts on the cleaved GaP surfaces as presented in Sect. 9.2.

6.1.5 GaN Schottky Contacts

The barrier heights of intimate, abrupt, clean, and laterally homogeneous *n*-GaN Schottky contacts which are compiled in Tables 3.3 to 3.6 are plotted versus the electronegativity difference of the metal and gallium nitride in Fig. 6.5. Some of the materials and sample parameters required to calculate the image-force lowering are not well established. Therefore, neither the barrier heights obtained by extrapolation of Φ_{Bn}^{eff} versus n plots to $n = 1$ nor the center value of the Gaussian distribution of the Au/*n*-GaN BEEM barrier heights nor the IPEYS threshold of the Ni/*n*-GaN diode were corrected for the image-force effect. However, the error introduced by neglecting the image-force lowering amounts to 20 meV at maxi-

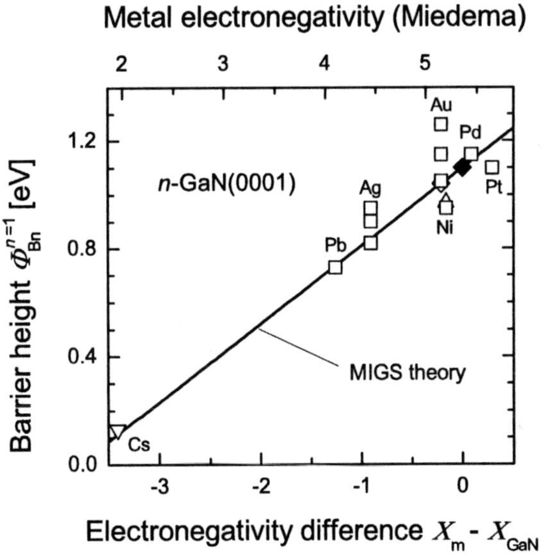

Fig. 6.5. Barrier heights of laterally homogeneous *n*-GaN Schottky contacts versus the difference of the *Miedema* electronegativities of the metal and gallium nitride. The □, ◇, Δ, and ∇ symbols differentiate the data from *I*/*V*, BEEM, IPEYS, and PES measurements, respectively, as compiled in Tables 3.3 to 3.6. The solid MIGS line is drawn with $\Phi_{bp}^{p} = 2.37$ eV [*Mönch* 1996b] and $S_X = 0.29$ eV/Miedema-unit

mum and is less than the margins of error of the individual experimental values. The MIGS line

$$\Phi_{Bn}^{MIGS} = 1.1 + 0.29 \cdot (X_{\mathrm{m}} - X_{\mathrm{GaN}}) \text{ [eV]} \tag{6.8}$$

excellently reproduces the experimental results.

Mönch [1996b] calculated the branch-point energies $W_{bp} - W_v(\Gamma)$, which are displayed in Table 5.1, and from this the zero-charge-transfer barrier heights Φ_{bp}^{p} and $\Phi_{bp}^{n} = W_g - \Phi_{bp}^{p}$ for zincblende- rather than wurtzite-structure compounds. However, the MIGS line in Fig. 6.5 was drawn by assuming that the energy position $\Phi_{bp}^{p} = W_{bp} - W_v(\Gamma)$ of the MIGS branch point above the valence-band maximum is the same irrespective of whether the lattice structure of gallium nitride is zincblende or wurtzite. The experimental barrier heights apparently corroborate this assumption. Further support of this concept comes from DFT-LDA calculations performed by *Stampfl* and *Van der Walle* [1998]. They studied stacking faults in wurtzite GaN which can be considered as thin layers of zincblende material embedded in the wurtzite lattice. As was to be expected, they found no defect-induced localized states in the band gap. Furthermore, they estimated a valence-band offset of only ≈ 70 meV at the zincblende–wurtzite or, to use *Ramsdell*'s notation, the 3*C*–2*H* interfaces of gallium nitride. Considering the margins of computational error, the branch-point energies of cubic 3*C*-GaN and hexagonal 2*H*-GaN may thus be taken as equal. Further support for this assumption comes from Schottky contacts on different SiC polytypes.

6.1.6 SiC Schottky Contacts

The experimental barrier heights of intimate, abrupt, clean, and laterally homogeneous Schottky contacts on the three SiC polytypes 3*C*, 6*H*, and 4*H* are plotted versus the difference $X_{\mathrm{m}} - X_{\mathrm{SiC}}$ of the *Miedema* electronegativities of the metals and SiC in Fig. 6.6. The experimental data are taken from Tables 3.3, 3.4, and 3.6. The MIGS lines of the three SiC polytypes were again drawn with the assumption that the energy position of the branch point relative to the valence-band maximum is the same irrespective of the polytype considered. Consequently, the *n*-type zero-charge-transfer barrier heights Φ_{bp}^{n} are the differences between the band gap of the specific polytype considered and the *p*-type zero-charge-transfer barrier height Φ_{bp}^{p}(3C) calculated for the zincblende-structure or 3*C* polytype. The MIGS lines excellently reproduce the experimental barrier heights of metal contacts on *n*-type 3*C*-, 6*H*-, and 4*H*-SiC. Most importantly, they quantitatively explain the barrier-height differences between the three polytypes, although the barrier heights of the

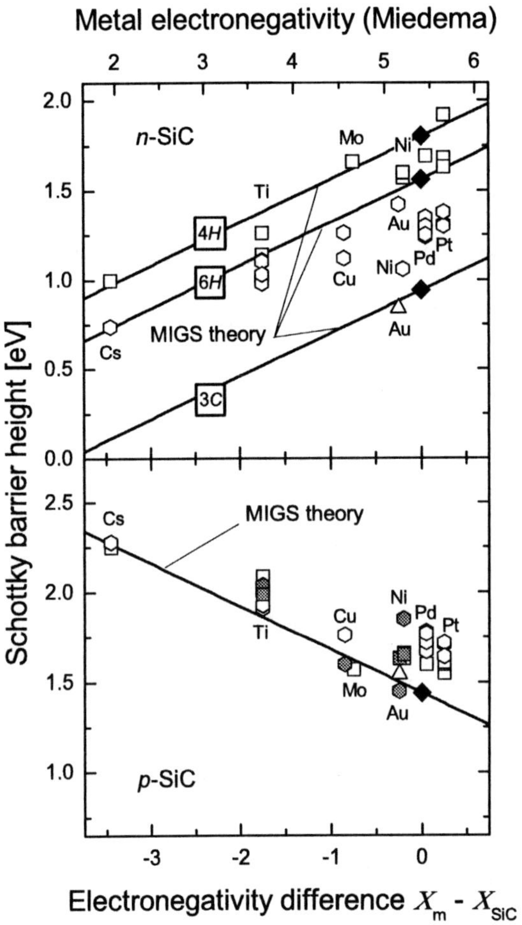

Fig. 6.6. Barrier heights of laterally homogeneous 3*C*-, 6*H*-, and 4*H*-SiC *Schottky* contacts versus the difference of the *Miedema* electronegativities of the metal and silicon carbide. The □, ○, and ∇ symbols differentiate data of 4*H*-, 6*H*- and 3*C*-SiS Schottky contacts, respectively. The upper diagram displays experimental data of n-type contacts as compiled in Tables 3.3, 3.4, and 3.6. The gray and the open data points of the lower diagram represent true experimental and hypothetical (see text) barrier heights of laterally homogeneous *p*-type contacts, respectively. The solid MIGS lines are drawn with Φ^{p}_{bp} = 1.44 eV of cubic 3*C*-SiC [*Mönch* 1996b], the band gaps of the polytypes, and S_X = 0.24 eV/Miedema-unit (see text)

Pd and Pt contacts on 6*H*-and 4*H*-SiC are smaller than theoretically predicted by approximately 100 meV.

Table 6.1. Theoretical valence-band offsets, in meV, at interfaces between different modifications of one and the same semiconductor

Semiconductor	Interface	ΔW_v	Reference
SiC	3*C*/2*H*	130	*Qteish* et al. [1992]
	3*C*/2*H*	130	*Käckell* et al. [1994]
	3*C*/6*H*	20	*Käckell* et al. [1994]
	3*C*/4*H*	50	*Käckell* et al. [1994]
	4*H*/6*H*	30	*Käckell* et al. [1994]
	3*C*/2*H*	140	*Ke* et al. [1996a, b]
	3*C*/2*H*	40	*Binggeli* et al. [2001]
BN	3*C*/2*H*	160	*Ke* et al. [1996b]
AlN	3*C*/2*H*	100	*Ke* et al. [1996a]
	3*C*/2*H*	20	*Majewski & Vogl* [1998]
	3*C*/2*H*	40	*Binggeli* et al. [2001]
GaN	3*C*/2*H*	10	*Ke* et al. [1996b]
	3*C*/2*H*	70	*Stampfl & Van de Walle* [1998]
	3*C*/2*H*	20	*Majewski & Vogl* [1998]
CdS	3*C*/2*H*	46	*Wei & Zhang* [2000]
CdSe	3*C*/2*H*	35	*Wei & Zhang* [2000]
CdTe	3*C*/2*H*	18	*Wei & Zhang* [2000]

The assumption of identical energy positions $\Phi_{bp}^{p} = W_{bp} - W_v(\Gamma)$ of the branch points above the valence-band maximum for all SiC polytypes is further illustrated by the lower diagram of Fig. 6.6. It displays the hypothetical *p*-type barrier heights $\Phi_{Bp}^{hom} = W_g - \Phi_{Bn}^{hom}$ which are obtained by subtracting the experimental *n*-type barrier heights shown in the diagram above from the band gaps of the corresponding SiC polytypes and IPEYS barrier heights of Au-, Ti-, Ni-, and Cu/*p*-6*H*-SiC contacts [*Aboelfotoh* et al. 2003] as well as barrier heights of laterally homogeneous Au-, Ti-, and Ni/*p*-4*H*-SiC contacts obtained from Φ_{Bp}^{eff} versus *n* diagrams [*Lee* et al. 2001]. First of all, the hypothetical and the true experimental data excellently agree. The deviations of the Pd and Pt data points towards larger values indicate extra dipole contributions which might be caused by interfacial reactions. On the whole,. the data points excellently agree with the straight line

$$\Phi_{Bp}^{MIGS} = 1.44 - 0.24 \cdot (X_{\mathrm{m}} - X_{\mathrm{SiC}}) \text{ [eV]} \tag{6.9}$$

of the MIGS-and-electronegativity concept drawn with the branch-point energy Φ_{bp}^{p} = 1.44 eV calculated for zincblende-structure or 3*C* silicon carbide. Together with the GaN data these experimental results strongly support the assumption that the energy position of the branch point relative to the valence-band maximum is the same irrespective of the specific lattice structure of the semiconductor, i.e., the branch-point energies Φ^p_{bp} are ground-state properties. [1] The band-edge discontinuities calculated for heterostructures between different SiC polytypes as well as between wurtzite- and zincblende-structure GaN and Cd chalcogenides confirm these conclusions.

The valence-band offsets at interfaces between the cubic zincblende or $3C$ and the hexagonal wurtzite or $2H$ modifications of SiC, GaN, and Cd chalcogenides as well as the $4H$ and $6H$ polytypes of SiC were calculated by applying LDA-DF theory and using *ab-initio* pseudopotentials. Table 6.1 displays the results obtained. For the SiC polytypes of interest here, *Käckell* et al. [1994] yielded valence-band offsets of up to 50 meV for $3C/6H$-, $3C/4H$-, and $4H/6H$-SiC. *Stampfl* and *Van de Walle* [1998] and *Majewski* and *Vogl* [1998] reported valence-band offsets between 20 to 70 meV for $3C/2H$-GaN and $3C/2H$-AlN interfaces while *Wei* and *Zhang* [2000] calculated ΔW_v values varying between 18 and 46 meV for the $3C/2H$ interfaces of the Cd chalcogenides. At interfaces between semiconductors with the same electronegativities and, therefore, also between chemically identical semiconductors the branch points of the two semiconductors in contact will align; see Sect. 5.3.3. The theoretical results just mentioned thus imply that

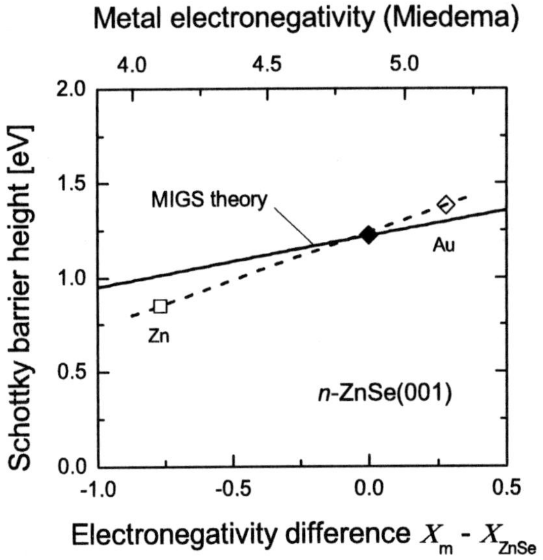

Fig. 6.7. Barrier heights of laterally homogeneous *n*-ZnSe(001) *Schottky* contacts versus the difference of the *Miedema* electronegativities of the metal and zinc selenide. The □ and ◇ symbols differentiate the data from *I/V* and BEEM measurements, respectively, as compiled in Tables 3.3 and 3.4. The solid MIGS line is drawn with $S_X = 0.232$ eV/Miedema-unit and $\Phi^p_{bp} = 1.46$ eV [*Mönch* 1996b]

[1] *Mönch* [1994a] already reached this conclusion from barrier heights of single but laterally inhomogeneous *Schottky* contacts on $3C$- and $6H$-SiC.

the branch-point energies of the SiC polytypes 3*C*, 6*H*, and 4*H*, of hexagonal wurtzite-structure and cubic zincblende-structure GaN and Cd chalcogenides, respectively, are almost identical. Within the margins of error of both theory and experiment, these findings corroborate the conclusions drawn above from the experimental barrier heights of laterally homogeneous SiC and GaN *Schottky* contacts.

6.1.7 ZnSe Schottky Contacts

Experimental barrier heights of intimate, abrupt, clean, and laterally homogeneous *n*-ZnSe *Schottky* contacts are shown as a function of the difference $X_m - X_{ZnSe}$ of the metal and ZnSe electronegativities in Fig. 6.7. The experimental data are taken from Tables 3.3 and 3.4. The *Miedema* electronegativity of ZnSe, $X_{ZnSe} = 4.87$ Miedema -units, is estimated by inserting the *Miedema* electronegativity of selenium, $X_{Se} = 5.79$ Miedema-units, which was obtained from its *Pauling* value by applying the conversion (5.16), into relation (5.3). The MIGS line

$$\Phi_{Bn}^{MIGS} = 1.23 + 0.27 \cdot (X_m - X_{ZnSe}) \text{ [eV]} \tag{6.10}$$

describes the chemical trend of the experimental data, even if there are only two points, quite well.

6.1.8 Schottky Contacts on Layered Ga-Chalcogenides

Many of the chalcogenides are layered compounds. Typical examples are the III–VI compounds and the transition-metal dichalcogenides TM–VI_2. Within the layers, the metal atoms are sandwiched between two planes of chalcogen atoms and the bonds are covalent. Since the layers possess no dangling bonds they are themselves bound by van der Waals forces. However, at Schottky contacts and at heterostructures of such materials the wave functions of electrons will also tunnel into layered semiconductors in energy regions where their band gaps overlap the conduction band of a metal or the valence band of another semiconductor. Hence, the IFIGS theory should also apply to interfaces of layered compounds [*Mönch* 1998b].

Figure 6.8 displays barrier heights of Schottky contacts on *p*-type gallium chalcogenides as a function of the metal electronegativity [*Kurtin* and *Mead* 1969]. The metal electronegativities rather than the electronegativity differences $X_m - X_{Ga-VI}$ are taken since no *Miedema* values are available for the group-VI atoms. The barrier heights were determined using internal photoemission yield spectroscopy. The full data points for Al- [*Bose* and *Pal* 1997] and Cd/*p*-GaTe [*Abay* et al. 2003], on the other hand, were obtained by linear extrapolation of $\Phi_{Bp}^{eff}(T)$ versus $n(T)$ data to n_{if}, the ideality factor determined by the image-force effect only. The *I*/*V* value of the Cd contact exactly agrees with the corresponding

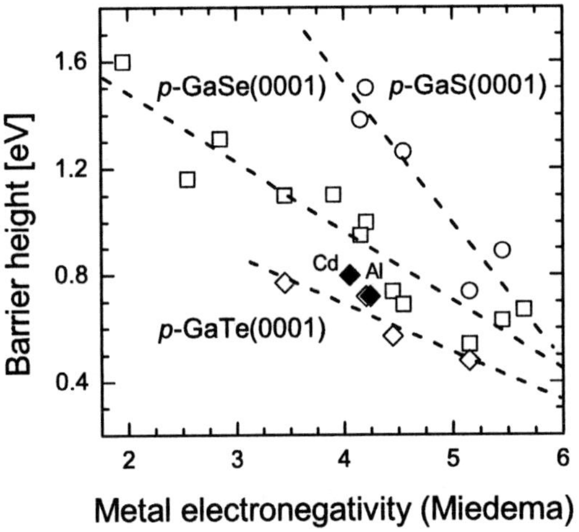

Fig. 6.8. IPEYS barrier heights of metal contacts on *p*-type GaS, GaSe, and GaTe layered compounds versus the *Miedema* electronegativities of the metals; data from *Kurtin* and *Mead* [1969]. The dashed lines are linear least-squares fits to the experimental data. The filled data points were obtained from plots of $\Phi_{Bp}^{eff}(T)$ versus $n(T)$ determined from I/V characteristics of one Al- and Cd/*p*-GaTe diode each; data from *Bose* and *Pal* [1997] and *Abay* et al [2003], respectively

IPEYS data point and the Cd data point fits in the general trend. These findings again support the conclusion that internal photoemission yield spectroscopy provides the barrier heights of the laterally homogeneous regions of real and patchy Schottky contacts. The data shown in Fig. 6.8 are thus for laterally homogeneous GaS, GaSe, and GaTe Schottky contacts.

The experimental GaSe data displayed in Fig. 6.8 give a slope parameter $S_X^{\exp} = \partial \Phi_{Bp}^{\text{hom}} / \partial X_{\text{m}}$ of 0.26 ± 0.03 eV/Miedema-unit. This value agrees with the prediction of the MIGS model. Relation (5.65) relates the slope parameter S_X and the optical dielectric constant ε_∞ of the respective semiconductor. With the optical dielectric constant $\varepsilon_{\infty\|} = 5.76$ of GaSe [*Bucher* 1992] one obtains $S_X^{\text{theo}}(\text{GaSe}) = 0.26$ eV/Miedema-unit, in excellent agreement with the experimental value. The same conclusion holds for the slope parameter of the GaTe Schottky contacts while there is a somewhat larger deviation for the GaS contacts.

For group-III chalcogenides, no branch-point energies have been calculated. The reasons are twofold. Relation (5.59) can not be applied since the mean-value $\underline{k}$-point has not been evaluated for the layered lattice-structure and the universal coefficients (5.58) of the inter-orbital matrix elements $V_{ll'\lambda}$, which are required for the ETB calculations of the dispersion $W_v(\Gamma) - W_v(\underline{k}_{mv})$ of the valence band from

Table 6.2. Effective barrier heights of Schottky contacts, in eV, on I–III–VI_2 chalcopyrites

Semiconductor	Metal	Φ_{Bn}	Φ_{Bp}	Method	Reference
$CuInS_2$	Au	0.83		*C*/*V*	*Kościelniak-Mucha & Opanowicz* [1992]
$CuInSe_2$	Al		0.48	*C*/*V*	*Rao* et al. [1986]
	Al		0.80	*C*/*V*	*Chan & Shih* [1990]
	Au	0.36		*C*/*V*	*Parkes* et al. [1973]
	Au	0.47		*C*/*V*	*Prasad* et al. [1985]
	Au	0.50	0.60	SXPS	*Nelson* et al. [1991a]
	Au	0.35		*C*/*V*	*Opanowicz* et al. [1988]
		0.83			*Opanowicz* et al. [1988]
		0.60		IPEYS	*Opanowicz* et al. [1988]
	In		0.62	*I*/*V*	*Kościelniak-Mucha & Opanowicz* [1994]
	Mo		0.80		*Löher* et al. [1994]
$AgInSe_2$	Ni	0.60		IPEYS	*Ramesh* et al. [1997]

its maximum in the middle of the Brillouin zone towards the mean-value $\underline{k}$-point, were derived for zincblende-structure rather than layered compounds. However, the branch-point energies may be estimated. For this purpose, one has to switch from *Miedema*'s to *Pauling*'s electronegativities. The experimental slope parameter, $S_X^{\text{theo}} = 0.26 \pm 0.03$ eV/Miedema-unit, of the GaSe Schottky contacts in Fig. 6.7 converts into 0.52 ± 0.07 eV/Pauling-unit if in relation (5.23) the A_X value of 1.79 eV/Pauling-unit is exchanged for 0.86 eV/Miedema-unit. Gold has an electronegativity of 2.54 Pauling-units and the average electronegativity of the group-III chalcogenides amounts to 2.20 ± 0.18 Pauling-units. Therefore, the dipole term $S_X = X_{\text{Au}} - X_{\text{III-VI}}$ in the barrier-height relation (5.22) is small for these compounds and may be neglected. The (interpolated) barrier heights of Au contacts are thus close to the branch-point energies Φ_{bp}^{p} of the layered group-III chalcogenides.

6.1.9 $CuInSe_2$ Schottky Contacts

The I–III–VI_2 chalcopyrites play an important role in thin-film solar cells. However, there exist only very few studies on Schottky contacts with these technologically interesting compounds. They are compiled in Table 6.2. Unfortunately, no sets of $\Phi_{Bn}(n)$ or $\Phi_{Bp}^{eff}(n)$ data have been published. The experimental $CuInSe_2$ data available and the predictions of the MIGS theory for ideal $CuInSe_2$ Schottky

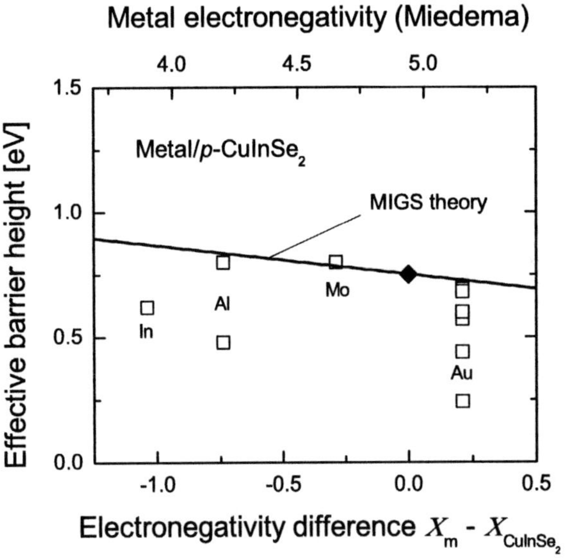

Fig. 6.9. Barrier heights of p-$CuInSe_2$ Schottky contacts versus the difference of the *Miedema* electronegativities of the metal and $CuInSe_2$. The data are from Table 6.1. The solid MIGS line is drawn with $S_X = 0.37$ eV/Pauling-unit and $\Phi^p_{bp} = 0.75$ eV [*Mönch*, unpublished]

contacts are plotted in Fig. 6.9. The experimental data points are on or below the MIGS line. The latter observation indicates non-ideal behavior, i.e., these contacts contain nm-size patches of lowered barrier height.

6.1.10 Schottky Contacts on Ternary III–V Alloys

The branch-point energies of binary compound semiconductors are slightly below the middle of the dielectric band gap at the mean-value $\underline{k}$-point of the Brillouin zone so that the energy position of the branch point relative to the valence-band maximum results as

$$W_{bp} - W_v(\Gamma) = 0.449 \cdot W_{dg} - [W_v(\Gamma) - W(\underline{k}_{mv})] \tag{6.11}$$

(see Sect. 5.5.2). The dispersion $W_v(\Gamma) - W_v(\underline{k}_{mv})$ of the valence band from its maximum in the middle of the Brillouin zone towards the mean-value $\underline{k}$-point may be calculated in the empirical tight-binding approximation (ETB). *Mönch* [1996b] applied this concept also to ternary semiconductor alloys. Both the width W_{dg} of the dielectric band gap and the energy dispersion $W_v(\Gamma) - W_v(\underline{k}_{mv})$ vary with alloy composition. If the alloys are assumed to be randomly disordered, then their band structures may be described in the virtual-crystal approximation. Hence, the atomic term values of *virtual* cations (c) or anions (a) in, for example, ternary $A_{1-x}B_xC$ alloys are then given by

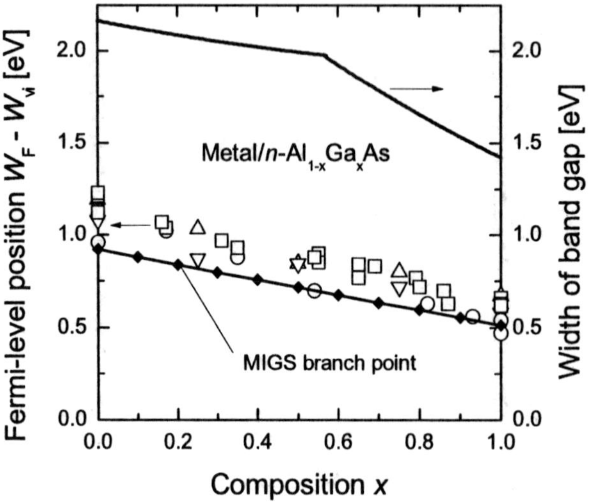

Fig. 6.10. Branch-point energies and energy positions of the Fermi level relative to the valence-band maxima at n-$Al_{1-x}Ga_xAs$ Schottky contacts and width of band gap. O data for Au from *Best* [1979], □ data for Al from *Okamoto* et al. [1981], *Missous* et al. [1990] and *Revva* et al. [1993], Δ and ∇ data for CoGa and CoAs, respectively, from *Palmstrøm* et al. [1989]. The –◆– data are ETB branch-point energies calculated in the virtual-crystal approximation [*Mönch* 1996b]

$$W_l^{c,a}(x) = (1-x) \cdot W_l^{\mathrm{A}} + x \cdot W_l^{\mathrm{B}} \quad . \tag{6.12}$$

The subscript l distinguishes between atomic s and p levels. The letters A and B stand for either cations or anions. The bond lengths $d_{nn}(x)$ and, by this, the interatomic matrix elements $V_{ll'\lambda}$ and the energies $\hbar\omega_p(x)$ of the valence-electron plasmons as well as the optical dielectric constants $\varepsilon_\infty(x)$ vary as a function of composition x. The nearest-neighbor distances are calculated from the experimental lattice parameters, while the optical dielectric constants are linearly interpolated between the values of the corresponding binary compounds. Figures 6.10 to 6.13 display the calculated branch-point energies $[W_{bp} - W_v(\Gamma)]_{\mathrm{ETB}}$ of the ternary systems $Al_{1-x}Ga_xAs$, $GaAs_{1-x}P_x$, $Ga_{1-x}In_xP$, $Al_{1-x}In_xAs$, and $In_{1-x}Ga_xAs$. The bowing parameters of the branch-point energies are virtually zero. The branch-point energies of ternary $A_{1-x}B_xC$ alloys thus vary linearly between the respective values of the corresponding two binary compound semiconductors, AC and BC, i.e.,

$$\Phi_{bp}^{p}(x) = (1-x) \cdot \Phi_{bp}^{p}(\mathrm{AC}) + x \cdot \Phi_{bp}^{p}(\mathrm{BC}) \quad . \tag{6.13}$$

In Fig. 6.13 this relation is assumed to be valid also for $Al_{1-x}Ga_xN$, $Ga_{1-x}In_xN$, and $Al_{1-x}In_xN$, the alloys of the group-III nitrides. The same should be also true for II–VI alloys.

Within a specific alloy system the branch point shifts linearly with composition, but the width of the band gap shows either bowing when it is direct in the en-

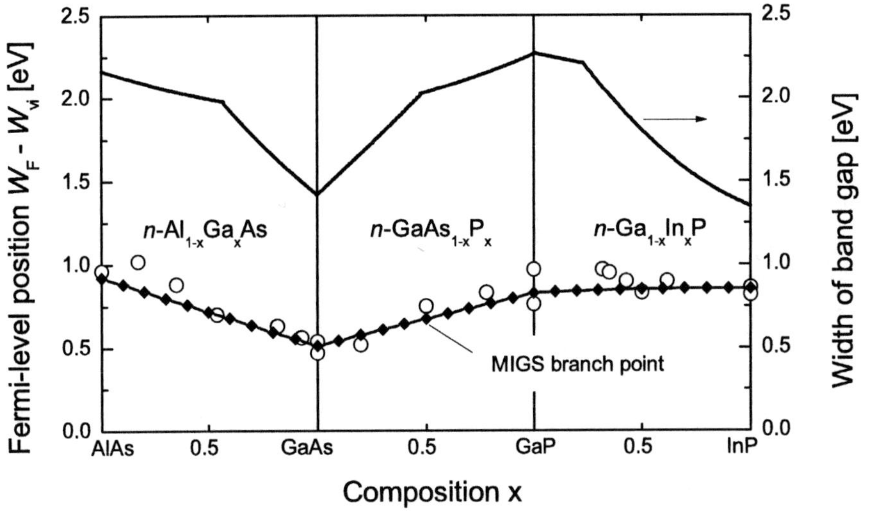

Fig. 6.11. Branch-point energies and energy positions of the Fermi level relative to the valence-band maxima at n-$Al_{1-x}Ga_xAs$, n-$GaAs_{1-x}P_x$, and n-$Ga_{1-x}In_xP$ Schottky contacts and band-gap widths of the alloys. Au/n-$Al_{1-x}Ga_xAs$ data from *Best* [1979], Au/n-$GaAs_{1-x}P_x$ data from *Spitzer* and *Mead* [1964] and *Neamen* and *Grannemann* [1971], Au/n-$Ga_{1-x}In_xP$ data from *Kuech* and *McCaldin* [1980]. The –◆– data are ETB branch-point energies calculated in the virtual-crystal approximation [*Mönch* 1996b]

tire composition range or exhibits a kink at that composition where its character changes from direct to indirect. These observations suggest to consider the ener-position $W_F(x) - W_{vi}$ of the Fermi level relative to the valence-band maximum at the interface or, in other words, the barrier height $\Phi_{Bp}(x)$ not only for p-type but also for n-type semiconductor-alloy Schottky contacts. For ternary $A_{1-x}B_xC$ alloys one obtains

$$\begin{aligned} \Phi_{Bp}(x) &= \Phi_{bp}^{p}(x) - S_X(x)[X_m - X_s(x)] \\ &\approx \Phi_{bp}^{p}(\mathrm{AC}) - S_X(\mathrm{AC})(X_m - \langle X_s \rangle) \\ &\quad - x\{\Phi_{bp}^{p}(\mathrm{AC}) - \Phi_{bp}^{p}(\mathrm{BC}) - [S_X(\mathrm{AC}) - S_X(\mathrm{BC})](X_m - \langle X_s \rangle)\}. \end{aligned} \tag{6.14}$$

The electronegativities of the binary III–V and II–VI compound semiconductors differ by only ± 4%. Therefore, the average electronegativity $\langle X_s \rangle$ is substituted for the electronegativities X_{AC} and X_{BC} of the corresponding binary compounds AC and BC, respectively, in relation (6.14). For most ternary alloy systems one finds

$$\Phi_{bp}^{p}(AC) - \Phi_{bp}^{p}(BC) > [S_X(AC) - S_X(BC)](X_m - \langle X_s \rangle) \ ,$$

so that the second term in the braces on the right-hand side of (6.11) may be omitted in most cases.

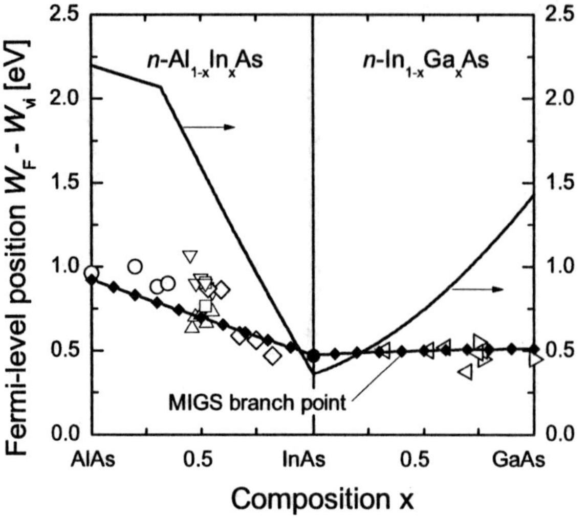

Fig. 6.12. Branch-point energies and energy positions of the Fermi level relative to the valence-band maxima at n-$Al_{1-x}In_xAs$ and n-$In_{1-x}Ga_xAs$ Schottky contacts and band-gap widths of the alloys. n-$Al_{1-x}In_xAs$: O data from *Chyi* et al. [1995], Δ data from *Lin* et al. [1986], ∇ data from *Chu* et al. [1988], □ data from *Sadwick* et al. [1991], ◇ data from *Schramm* et al. [1991]; n-$In_{1-x}Ga_xAs$: ◁ and ▷ data from *Kajiyama* et al. [1973]. The –◆– data are ETB branch-point energies calculated in the virtual-crystal approximation [*Mönch* 1996b]

Figures 6.10 to 6.13 show the energy positions $W_F(x) - W_{vi} = W_g(x) - \Phi_{Bn}^{eff}(x)$ of n-type Schottky contacts as a function of alloy compositions rather than the experimental barrier heights $\Phi_{Bn}^{eff}(x)$ themselves. Unfortunately, no data sets of effective barrier heights and ideality factors n are available for such Schottky contacts. Therefore, the effective barrier heights rather than the barrier heights of laterally homogeneous contacts are plotted. This limited data base is responsible for the scatter of the data points in the figures. Nevertheless, all the experimental results not only confirm the theoretically predicted linear variation of the p-type zero-charge-transfer barrier heights $\Phi_{bp}^{p}(x)$ as a function of alloy composition but also follow closely the calculated values.

The slopes $\partial\Phi_{Bp}/\partial x$ of Al, Au, CoGa, and CoAs Schottky contacts on n-type $Al_{1-x}Ga_xAs$ are slightly larger than predicted by relation (6.11). This means that the zero-charge-transfer barrier height Φ_{bp}^{p} of AlAs might be larger by approximately 0.1 eV. The branch-point energy calculated by *Tersoff* [1986c] for AlAs supports this conclusion, see Table 5.1.

Experimental investigations of Schottky contacts on alloys of the group-III nitrides are very rare. The data of *Khan* et al. [1997] for Au/n-$Al_{1-x}Ga_xN$ contacts and of *Yu* et al. [1998] for Ni/n-$Al_{1-x}Ga_xN$ contacts are the only ones available. Figure 6.13 displays the homogeneous barrier heights of the Ni/n-$Al_{1-x}Ga_xN$ con-

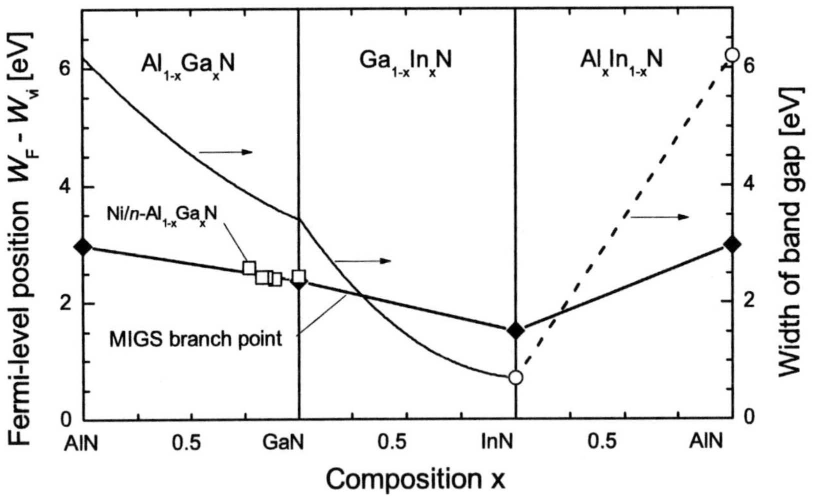

Fig. 6.13. Branch-point energies relative to the valence-band maxima and band-gap widths of $Al_{1-x}Ga_xN$, $Ga_{1-x}In_xN$, and $In_{1-x}Al_xN$ alloys and energy positions of the Fermi level relative to the valence-band maximum at Ni/*n*-$Al_{1-x}Ga_xN$ *Schottky* contacts. IPEYS barrier heights from *Qiao* et al. [2000]. The straight MIGS lines are linear interpolations between the ETB branch-point energies of the binary nitrides [*Mönch* 1996b]; for details see text

tacts [*Lau* and *Qiao* 2003]. They are the only ones available. As Fig. 6.13 demonstrates, they definitely agree with the branch-point energies of the MIGS which were theoretically predicted by applying relation (6.10) to $Al_{1-x}Ga_xN$ alloys [*Mönch* 1998a].

The theoretical and the experimental data displayed in Figs. 6.10 to 6.13 clearly demonstrate that ohmic metal contacts on moderately doped III–V compounds and alloys are *naturally* obtained only on *n*-InAs and *n*-InN. The branch-point energies given in Table 5.1 further indicate that the same also holds for *p*-GaSb and *p*-InSb. Contacts of the common metals on all other semiconductors are rectifying so that ohmic behavior requires some *manipulations* on the semiconductor side of the contacts. The most well-known approach are contacts where the current transport occurs via tunneling through rather than by thermionic emission over the potential barrier. The interface regions of such tunnel-contacts have to be highly doped n^+- or p^+-type in order to achieve thin depletion layers. Ohmic contacts are briefly discussed in Chap. 11.[2]

There is great interest in ternary III–V alloys such as, for example, $Ga_{0.51}In_{0.49}P$ that are lattice-matched to GaAs. Schottky contacts on such layers have been widely investigated. However, no sets of $\Phi_{Bn}^{eff}(n)$ data are available. Therefore, effective barrier heights rather than barrier heights of laterally homogeneous

[2] The formation of ohmic contacts to wide-band gap semiconductors has been reviewed by *Holloway* et al. [1997].

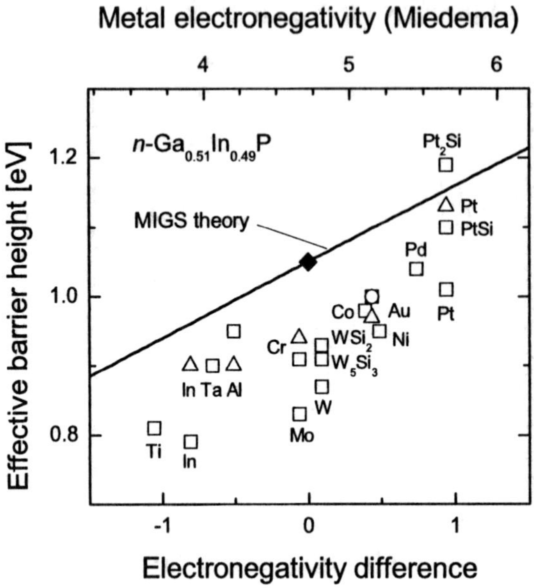

Fig. 6.14. Effective barrier heights of n-$Ga_{0.51}In_{0.49}P$ *Schottky* contacts versus the differences of the *Miedema* electronegativities of the metals and the alloy. O data from *Nanda* et al. [1992], □ data from *Chang* et al. [1993], Δ data from *Kwon* et al. [1995]. The solid MIGS line is drawn with $\Phi^p_{bp} = 0.845$ eV and $S_X = 0.11$ eV/Miedema-unit (see text)

metal/n-$Ga_{0.51}In_{0.49}P(001)$ contacts are plotted in Fig. 6.14 as a function of the electronegativity difference. The experimental data are bordered by the MIGS line. The zero-charge-transfer barrier height Φ^n_{bp} equals the difference between the width of the band gap, which amounts to 1.9 eV for $Ga_{0.51}In_{0.49}P$, and the energy position of the branch point relative to the valence band. Relation (6.12) gives $\Phi^p_{bp}(Ga_{0.51}In_{0.49}P) = 0.845$ eV, which is the mean value of the branch-point energies of GaP and InP; see Table 5.1. The slope parameter is estimated by inserting the average of the optical dielectric constants ε_∞ of GaP and InP, which differ by only 5%, into relation (5.65). Finally, the MIGS theory predicts the barrier heights of laterally homogeneous *Schottky* contacts on n-$Ga_{0.51}In_{0.49}P$ to vary as

$$\Phi_{Bn}(Ga_{0.51}In_{0.49}P) = 1.05 + 0.11(X_m - X_{GaInP})[eV] \quad . \tag{6.15}$$

The experimental data plotted in Fig. 6.14 are all below the MIGS line, as is to be expected for non-ideal and, therefore, patchy Schottky contacts. However, the experimental results clearly follow the chemical trend predicted by the MIGS theory.

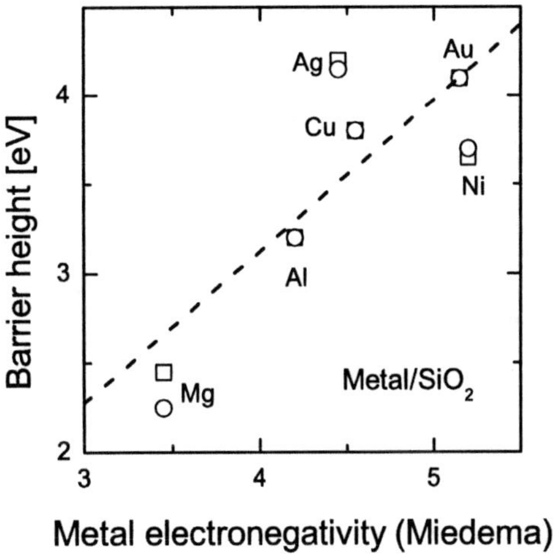

Fig. 6.15. Metal–SiO_2 barrier heights as a function of the metal electronegativity. The dashed line is a linear least-squares fit to the data and has a slope parameter of 0.85 eV/Miedema-unit. Data from *Deal* et al. [1966]

6.1.11 Metal–Insulator Contacts

Metal–insulator–semiconductor structures are of essential importance in semiconductor devices. Such MIS structures with very thin insulating interlayers behave electrically like Schottky contacts. The technologically most important examples are metal–oxide–silicon (MOS) and metal–nitride–silicon (MIS) structures. The barrier heights of such devices depend on the nature of the insulator, its thickness, and the specific metal used. Experimentally, metal–SiO_2 barrier heights in MOS structures were determined by using internal photoemission yield spectroscopy and capacitance–voltage (C/V)) measurements [*Deal* et al. 1966]. Si_3N_4–Si barrier heights were evaluated from C/V characteristics measured with MNS contacts [*Sobolewski* and *Helms* 1989]. Figures 6.15 and 6.16 display the MOS and MNS data, respectively. They will be analyzed in the following [*Mönch* 1996c].

Insulators may be considered as semiconductors with large band gaps. Therefore, the concept of metal-induced gap states and, consequently, relation (5.22) should also apply to metal–insulator contacts. Figure 6.15 displays the barrier heights of metal–SiO_2 contacts as a function of the metal electronegativity X_m. The dashed line is a linear least-squares fit to the data. Its slope parameter S_X amounts to 0.85 ± 0.2 eV per Miedema-unit. In the MIGS-and-electronegativity theory of metal–semiconductor contacts, the slope parameters $S_X = \partial \Phi_B / \partial X_m$ are determined by the optical dielectric constant ε_∞ of the corresponding semiconductor; see relation (5.65). In the infrared spectral region, amorphous SiO_2 has an index of refraction $n = 1.45$, so that one obtains $\varepsilon_\infty = n^2 = 2.1$. Through relation (5.65) the MIGS concept then predicts a slope parameter $S_X(SiO_2) = 0.77$ eV per

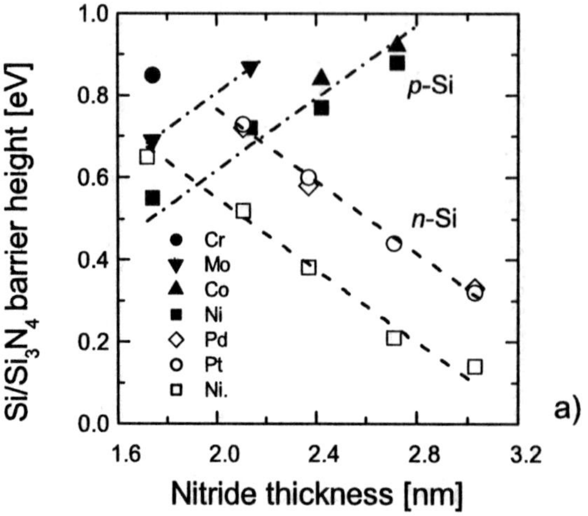

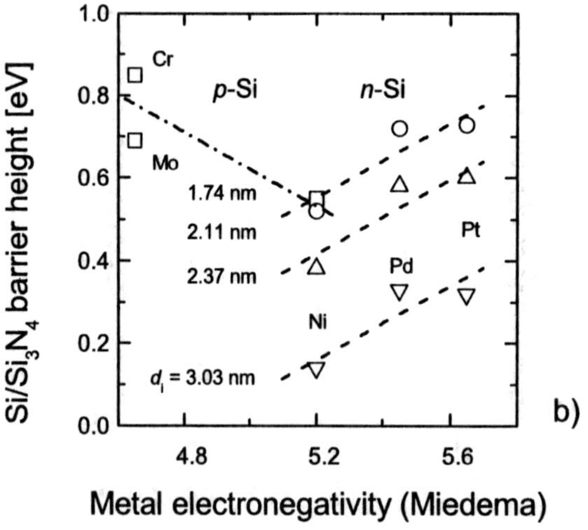

Fig. 6.16. Barrier heights of metal/Si_3N_4/Si contacts as a function of the nitride thickness for a variety of metals (a) and as a function of the metal electronegativity for the same nitride films (b). Data from *Sobolowski* and *Helms* [1989]

Miedema-unit for SiO_2 Schottky contacts. This theoretical value agrees remarkably well with what resulted from the above fit to the experimental results plotted in Fig. 6.15, although these data points exhibit a large scatter.

Figure 6.16a displays the barrier heights at Si_3N_4–Si interfaces of MNS structures. Irrespective of the contact metal used, the barrier heights vary linearly as a function of the insulator thickness. The sign of the corresponding slope parameters is negative for *n*-type and positive for *p*-type silicon. However, the linear least-squares fits to the data give identical absolute values of the slope parameters irrespective of the type of doping and of the specific metal considered.

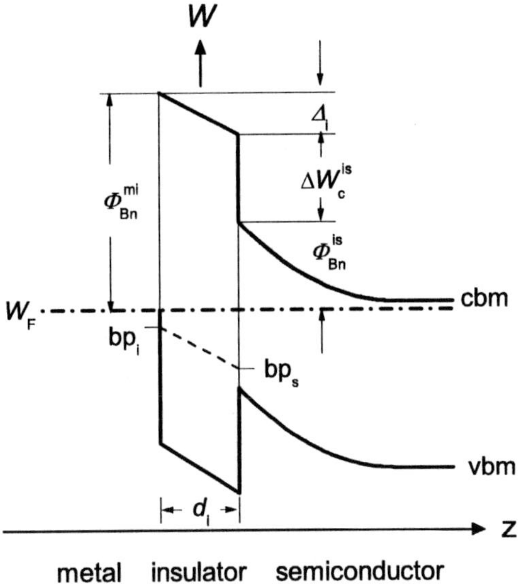

Fig. 6.17. Band diagram of metal–insulator–semiconductor structures (schematically)

MIS structures may be described as a metal–insulator and an insulator–semiconductor contact in series. Thus, the concept of interface-induced gap states outlined in Chap. 5 may be applied to MIS structures as well. Figure 6.17 shows schematically the band diagram of MIS structures in thermal equilibrium. At the two insulator interfaces the different energy contributions add up as

$$\Phi_{Bn}^{mi} = \Delta_i + \Delta W_c^{is} + \Phi_{Bn}^{is} \quad . \tag{6.16}$$

As in semiconductor heterostructures, the branch points of an insulator and a semiconductor in contact line up. The conduction-band offsets at such interfaces are then given by

$$\Delta W_c^{is} = \Phi_{bp}^{ni} - \Phi_{bp}^{ns} - D_X \cdot (X_i - X_s) \quad , \tag{6.17}$$

where the last term on the right-hand side accounts for the dipole contribution due to a nonvanishing difference $X_i - X_s$ of the insulator and the semiconductor electronegativities; see relations (5.6) and (5.25). The potential drop Δ_i across the insulator accounts for a fixed charge

$$Q_i^f = \frac{\varepsilon_{bi}\varepsilon_0}{e_0}\frac{\Delta_i}{d_i} \quad , \tag{6.18}$$

where ε_{bi} and d_i are the static dielectric constant and the thickness of the insulator, respectively. By combining (6.16) to (6.18) and considering relation (5.22), one obtains

$$\Phi_{Bn}^{is} = (\Phi_{Bn}^{mi} - \Phi_{bp}^{ni}) + \Phi_{bp}^{ns} - \frac{e_0}{\varepsilon_{bi}\varepsilon_0} Q_i^f d_i + D_X \cdot (X_i - X_s)$$
$$= S_{Xi} \cdot (X_i - X_s) - \frac{e_0}{\varepsilon_{bi}\varepsilon_0} Q_i^f d_i + \Phi_{bp}^{ns} + D_X \cdot (X_i - X_s) \ . \quad (6.19)$$

For p-type semiconductors the first two terms of the preceding relation have opposite sign.

The barrier height Φ_{Bn}^{is} at the insulator–semiconductor interface will thus decrease proportionally to the thickness d_i of the insulator, provided that the charge density Q_i^f is fixed. Relation (6.19) thus explains the experimentally observed linear variation of the barrier heights Φ_{Bn}^{is} as a function of the nitride thickness; see Fig. 6.16a. This result is well-known and by no means surprising.

For constant thickness of the insulator, relation (6.19) predicts the barrier height Φ_{Bn}^{is} at the Si_3N_4–Si interface of MIS structures to increase proportionally to the electronegativity of the metal. Therefore, the barrier heights displayed in Fig. 6.16a are replotted in Fig. 6.16b, but now versus the electronegativities of the metals with the nitride thickness as a parameter. As is to be expected, the barrier heights increase for n-type but decrease for p-type doping of the silicon substrate. The straight lines show linear least-squares fits to the data. The average of the slope parameters amounts to $\langle S_{Xi} \rangle = 0.45$ eV/Miedema-unit. In the MIGS model, the slope parameters $S_X = \partial\Phi_B/\partial X_m$ are determined by the optical dielectric con-

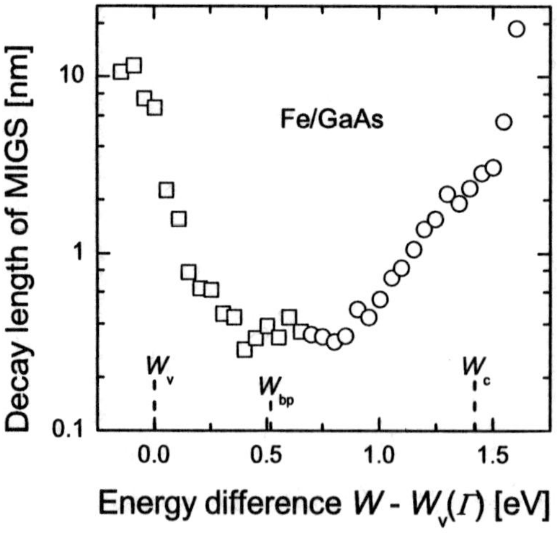

Fig. 6.18. Energy dependence of the decay length of gap states tailing away from metallic Fe islands on GaAs(110) surfaces as detected by using a scanning tunneling microscope: O and □ data recorded with samples doped p- and n-type, respectively. After *First* et al. [1989]

stants ε_∞ of the corresponding insulators; see relation (5.65). In the infrared spectral region, Si_3N_4 has an index of refraction $n = 2.05$, so that the optical dielectric constant results as $\varepsilon_\infty = n^2 = 4.2$. Through relation (5.65) the MIGS concept then predicts the slope parameter $S_X(Si_3N_4) = 0.42$ eV/Miedema-unit. Again, the MIGS concept finally confirms and explains the experimental data.

6.1.12 Direct Observations of MIGS

The direct experimental proof of the existence of the continuum of metal-induced gap states requires probes with sub-nm resolution. *First* et al. [1989] approached this problem by using a scanning tunneling microscope and *Muller* et al. [1998] applied a 100-kV scanning transmission electron microscope.

In their STM study, *First* et al. [1989] evaporated submonolayer quantities of iron onto cleaved GaAs(110) surfaces at room temperature. They found the deposit to coagulate in small epitaxial clusters. Islands with volumes larger than 1 nm^3 were definitely observed to have a metallic character. Around such particles, *First* et al. detected metallic states on such particles, which have energies within the semiconductor band gap, to propagate along the semiconductor surface. These gap states decay exponentially as a function of distance away from the metallic Fe islands. Figure 6.18 shows that the decay lengths of these gap states varies as a U-shape across the energy gap with a minimum value of 0.34 nm in the lower half of the gap. This experimental value excellently verifies the results of

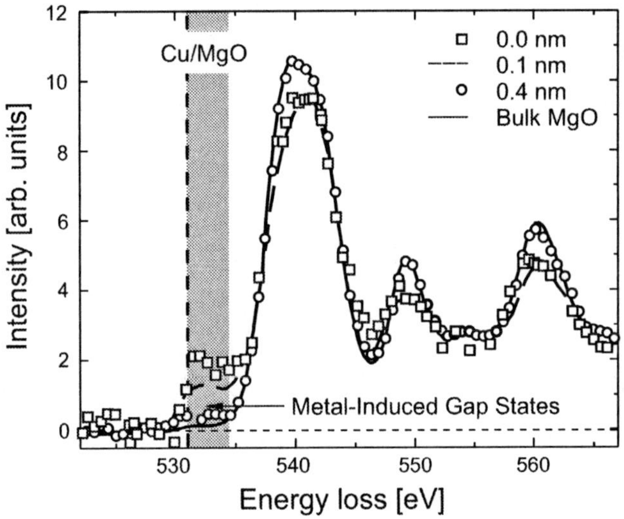

Fig. 6.19. Oxygen *K*-edge electron energy-loss spectra recorded at different distances from a Cu/MgO interface. The hatching marks the approximate position of the unoccupied portion of the MgO band gap. After *Muller* et al. [1998]

the theoretical computations; see Table 5.1. The broad minimum of the tailing states is centered around the calculated energy position, $W_{bp} - W_v(\Gamma) = 0.52$ eV, of the MIGS branch-point above the valence-band maximum, see Table 5.1. These gap states observed around metallic Fe particles on GaAs(110) surfaces thus exhibit all the features characteristic of the continuum of metal-induced gap states.

In their high-resolution transmission microscopy study, *Muller* et al. [1998] investigated Cu/MgO interfaces. They produced small MgO precipitates in a Cu matrix by annealing a 2.5-at.% Cu/Mg alloy in oxygen at 1223 K. The MgO particles had diameters between 10 and 30 nm. The Cu/MgO interface was examined using electron energy-loss spectroscopy (EELS). For that purpose the interfaces were oriented parallel to the probing electron beam, which had a width of 0.22 nm.

Figure 6.19 displays four EEL spectra of the oxygen *K*-edge recorded at different locations across a Cu/MgO interface and at its center, respectively. The spectra taken directly at the interface and at 0.1 nm and 0.4 nm into the MgO grain show a well-pronounced and 6 eV wide prepeak just at the onset of the bulk edge. This signal images a continuum of unoccupied states below the conduction-band minimum, i.e., within the MgO band gap. The band gap of MgO is 7.7 eV wide so that the barrier height of the Cu/MgO contacts amounts to approximately 1.7 eV. The intensity of the prepeak decays rapidly with increasing distance from the interface. By fitting its intensity to an exponential function *Muller* et al. obtained an effective decay length of 0.27 nm. Deconvolving the inelastic point spread function gives a decay length of 0.13 ± 0.05 nm. In their DFT-LDA calculations performed for Cu/MgO contacts *Blöchl* et al. [1990] found a decay length of the MIGS of approximately 0.11 nm. Within the margins of error, the experimental result confirms this theoretical value. Again, the gap states observed at Cu/MgO interfaces exhibit all the features characteristic of the continuum of metal-induced gap states.

6.2 Heterostructures

6.2.1 Introductory Remarks

Semiconductor heterostructures generally grow in a layer-by-layer fashion, at least initially. In the bulk of the sp^3-coordinated semiconductors, the chemical bonds are covalent and, naturally, the same holds for their interfaces. The most simple case is *lattice-matched* heterostructures where the two semiconductors have identical lattice parameters. However, if the bond lengths differ then the interface will respond by tetragonal distortions. Such *pseudomorphic* interfaces are under compressive or tensile stress. If the strain energy becomes too large it is energetically more favorable to release the stress by the formation of misfit dislocations which accommodate the lattice mismatch. Such *metamorphic* interfaces are almost relaxed.

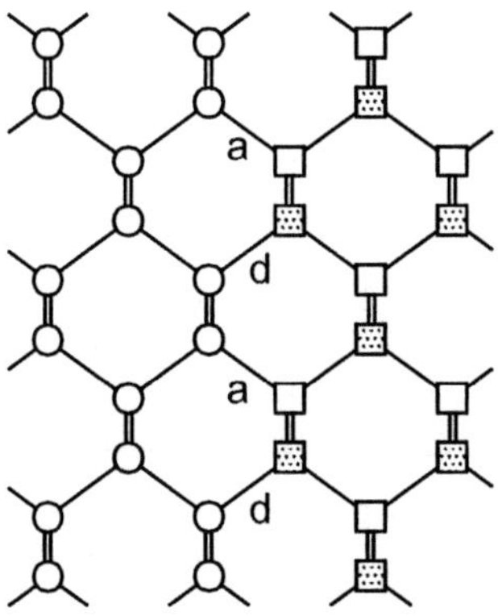

Fig. 6.20. Atomic arrangement at abrupt Ge/GaAs(110) heterostructures (schematic); O Ge atom; □ Ga atom; ■ As atom; a acceptor bond;, d donor bond

In contrast to isovalent heterostructures, the chemical bonds at heterovalent interfaces require special attention.[3] This will be explained by discussing the bond configurations at (110)- and (001)-oriented Ge/GaAs interfaces. Figure 6.20 illustrates schematically the atomic arrangement at an abrupt Ge/GaAs(110) heterostructure. Such interfaces are almost lattice-matched. Right at this non-polar interface there exist Ge–Ga and Ge–As bonds. In the most simple picture, atoms with z valence electrons contribute $v_z = z/4$ electrons per tetrahedral bond. Such electron counting gives $4/4 + 3/4 = 2 - 1/4$ electrons per Ge–Ga and $4/4 + 5/4 = 2 + 1/4$ electrons per Ge–As bond. Covalent bonds always contain two electrons

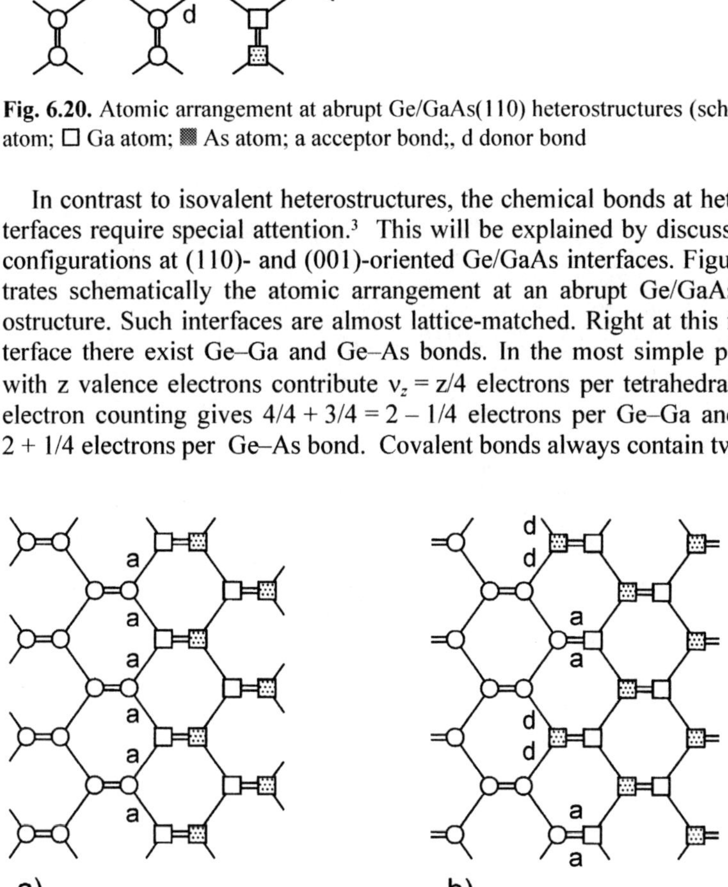

Fig. 6.21. Atomic arrangements at Ge/GaAs(001) heterostructures (schematic): a) bulklike terminated interface, b) mixed interface with one interface layer consisting of 50% Ge and As atoms each; O Ge atom; □ Ga atom; ■ As atom; a acceptor bond; d donor bond

[3] Such effects were first discussed and analyzed in a pioneering article by *Harrison* et al. [1978].

and, therefore, charge is transferred from the Ge–As donor to the Ge–Ga acceptor bonds at the interface. Thus, As^{+}–Ga^{-} interface dipoles will exist at such interfaces. They are oriented parallel to the interface and, therefore, add no *extrinsic* dipole contribution to the valence-band offset (5.25).

Polar interfaces behave quite differently. As an example, Fig. 6.21a illustrates schematically the atomic arrangement at a Ge/GaAs(001) interface. Bulk {100} lattice planes of binary zincblende-structure compounds are either occupied by anions or cations. Thus, either Ge–Ga acceptor-bonds or Ga–As donor-bonds would exist at such interfaces. Figure 6.21a assumes the GaAs to be terminated by a Ga layer so that Ge–Ga acceptor-bonds are present at the interface only. Provided that no defects form, the charge required to complete the bond charges would have to be supplied by atoms at the free GaAs surface. As a consequence, a macroscopic electric field would exist across the GaAs film as a whole. This unfavorable situation is avoided when in addition to Ge–Ga acceptor-bonds an equal number of Ge–As donor-bonds per unit area is present at the interface. This may be achieved, for example, when As atoms are substituted for half of the Ge atoms of the interfacial Ge layer. Figure 6.21b illustrates schematically the atomic arrangement at such a mixed interface. The interface as a whole is neutral and the corresponding charge transfer from Ge–As donor-bonds to Ge–Ga acceptor-bonds now occurs right at the interface. The Ga^{-} and As^{+} species are now placed in adjacent planes parallel to the interface. The corresponding electric dipoles are thus oriented normal to the interface plane and will add an *extrinsic* dipole contribution to the valence-band offset (5.25).

Layered semiconductors behave somewhat differently. Within the layers the bonds are also covalent, while the layers themselves are bound by van der Waals forces. Hence, no covalent bonds exist between two layered compounds. Because of the van der Waals bonds between layers, interfaces of layered semiconductors are unstrained irrespective of how large is the lattice mismatch.[4] The same applies to interfaces of layered compounds with sp^3-coordinated semiconductors.

6.2.2 Non-Polar Heterostructures

The valence-band offsets at non-polar (110) heterostructures should equal the difference of branch-point energies $\Phi_{bp}^{p} = W_{bp} - W_{v}(\Gamma)$ of the two semiconductors in contact if the intrinsic dipole contribution $\Delta_{ss}^{\text{int}} = D_X(X_{\text{sr}} - X_{\text{sl}})$ is neglected; see (5.25). Figure 6.22 displays experimental results as a function of the differences of the branch-point energies $\Phi_{bp}^{p} = [W_{bp} - W_{v}(\Gamma)]_{\text{ETB}}$ given in Table 5.1. The experimental values were all obtained by applying the XPS method outlined in Sect. 3.12.2. The dashed line clearly demonstrates that the experimental data are excellently explained by the theoretical branch-point energies or, in other words, by the IFIGS theory.

---

[4] This was first pointed out by *Koma* et al. [1985, 1991] and *Ueno* et al. [1991].

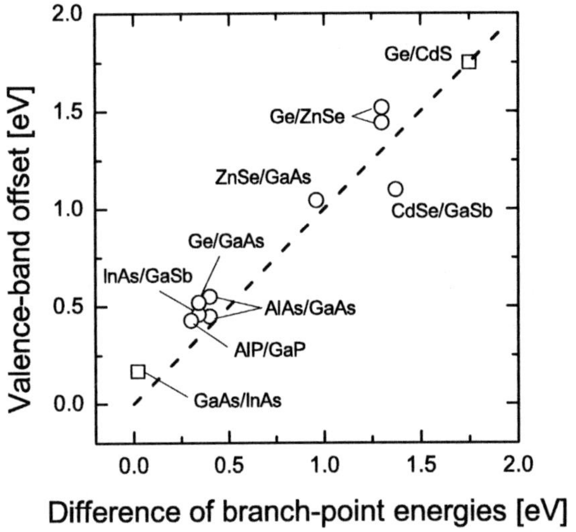

Fig. 6.22. Valence-band offsets at non-polar (110) semiconductor heterostructures against the differences of the branch-point energies relative to the corresponding valence-band maxima of the semiconductors in contact. Data from *Sakaki* et al. [1977] (InAs/GaSb), *Kraut* et al. [1980] (Ge/GaAs), *Kowalczyk* et al [1982a] (ZnSe/Ge, ZnSe/GaAs), *Kowalczyk* et al [1982b] (GaAs/InAs), *Wang* and *Stern* [1985] and *Duggan* [1985] (AlAs/GaAs), *Duc* et al. [1987] (ZnTe/CdTe), *Niles* et al. [1988] (Ge/ZnSe, Ge/CdS), *Neuhold* et al. [1995] (CdSe/GaSb), *Ding* et al. [1997] (GaN/GaAs), *Ban* et al. [1998] (Ge/ZnS), *Rizzi* et al. [1999] (AlN/SiC, GaN/AlN, GaN/SiC/). Branch-point energies from Table 5.1

6.2.3 Lattice-Matched Ternary and Quaternary III–V Alloys

Semiconductor heterostructures which contain ternary and quaternary III–V alloys are an integral part of the design of semiconductor devices. By varying the alloy compositions not only may the width of the band gap be adjusted to the device requirements but also the lineup of the bands at the interface may be chosen over a wide range. Straddling or type-I, staggered or type-II, and broken-gap or type-III lineups may be distinguished. They are explained in Fig. 6.23.

Another design parameter in epitaxial heterostructures is their growth mode. It depends on the lattice parameters of the substrate and the overlayer. The most simple case is lattice-matched (LM) systems such as, for example, $Al_{0.2}Ga_{0.8}As/Al_{0.5}Ga_{0.5}As$, $Al_{1-x}Ga_xAs$- and $Ga_xIn_{1-x}As_{1-y}P_y/GaAs$ and $Ga_xIn_{1-x}As_{1-y}P_y$- and $Al_yGa_{0.47-y}In_{0.53}As/InP$ heterostructures. The compositional parameters x and y of the quaternary alloys $Ga_xIn_{1-x}As_{1-y}P_y$, for example, are related to the lattice parameters a_0(III–V) of the corresponding four binary compounds as

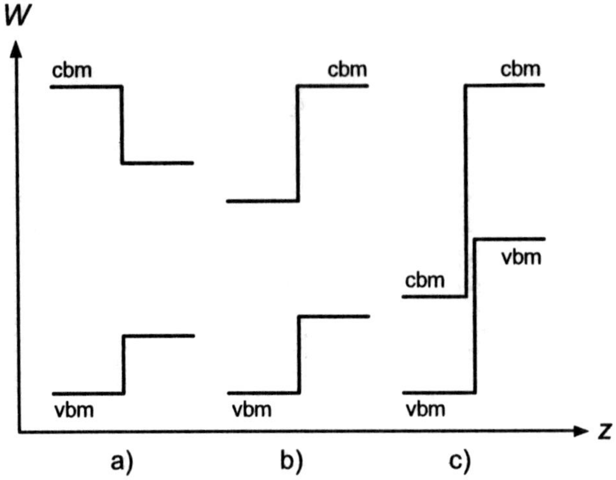

Fig. 6.23. Types of energy-band lineups in semiconductor heterostructures: (a) straddling or type-I,(b) staggered or type-II, and (c) broken-gap or type-III lineup

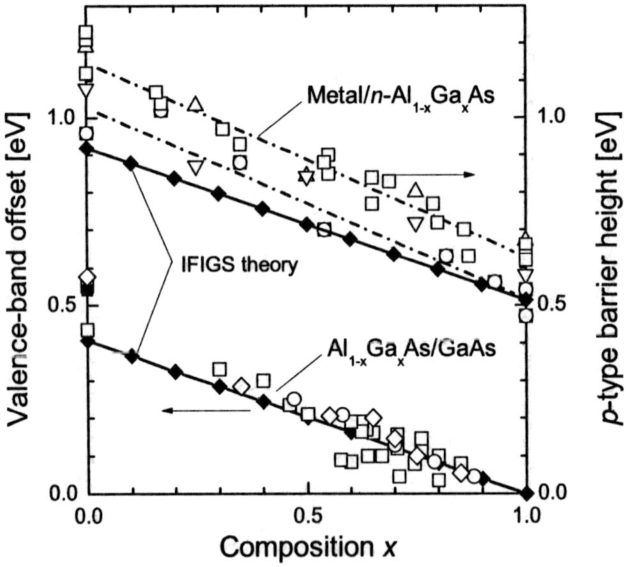

Fig. 6.24. Valence-band offsets at $Al_{1-x}Ga_xAs/GaAs$ heterostructures and barrier heights of metal/*n*-$Al_{1-x}Ga_xAs$ contacts. Heterostructures: experimental data from *Duggan* [1985] and *Wang* and *Stern* [1985] (□), *Forchhammer* et al. [1995] (○), and *Leuther* et al. [1996] (◇); theoretical IFIGS data from *Mönch* [1996b] (–◆–). Schottky contacts: experimental and theoretical data are the same as in Fig. 6.10

$$x = \frac{[a_0(\text{InP}) - a_0(\text{sub})] + [a_0(\text{InP}) - a_0(\text{InAs})](1-y)}{[a_0(\text{InAs}) - a_0(\text{GaAs})] - [a_0(\text{GaAs}) + a_0(\text{InP}) - a_0(\text{GaP}) - a_0(\text{InAs})](1-y)} = \frac{[0.58687 - a_0(\text{sub})] + 0.01896(1-\text{y})}{0.0451 - 0.00125(1-\text{y})} , \tag{6.20}$$

for lattice matching on GaAs or InP substrates with the corresponding lattice parameter a_0(sub) in nm.

The IFIGS theory predicts the branch-point energies $\Phi_{bp}^{p}(x)$ of ternary alloys $A_{1-x}B_xC$ to vary linearly between the corresponding values of the two binary compounds AC and BC, i.e.,

$$\Phi_{bp}^{p}(x) = (1-x) \cdot \Phi_{bp}^{p}(\text{AC}) + x \cdot \Phi_{bp}^{p}(\text{BC}) \quad ; \tag{6.21}$$

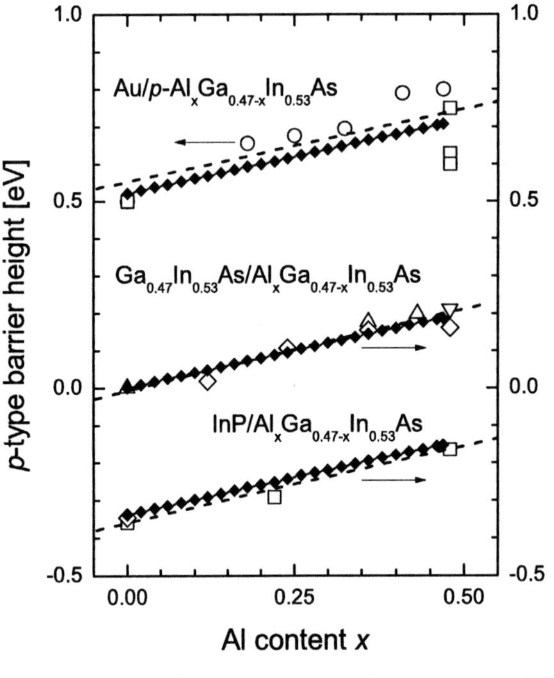

Fig. 6.25. Barrier heights of Au/p-$Al_xGa_{0.47-x}In_{0.53}As$ *Schottky* contacts and valence-band offsets of lattice-matched $Ga_{0.47}In_{0.53}As/Al_xGa_{0.47-x}In_{0.53}As$ and $InP/Al_xGa_{0.47-y}In_{0.53}As$ heterostructures. Data for *Schottky* contacts from *Kajiyama* et al. [1973], *Lin* et al. [1986], *Schramm* et al. [1991], *Sadwick* et al. [1991], and *Clark* et al. [1994]. Data for heterostructures from *Chan* et al. [1997], *People* et al. [1983], *Hiyamizu* et al. [1987], *Lang* et al. [1987] and *Böhrer* et al. [1993a]. The dashed lines are linear least-squares fits to the experimental data. The –◆– data are ETB branch-point energies and valence-band offsets calculated in the virtual-crystal approximation [*Mönch* 1996b]

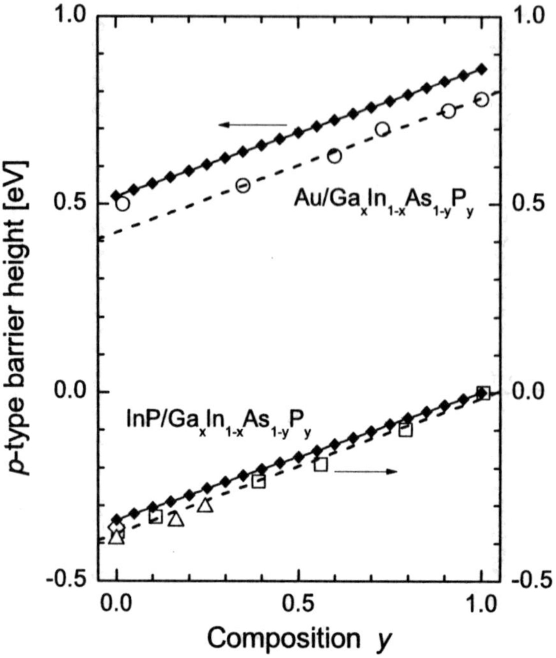

Fig. 6.26. Barrier heights of Au/p-$Ga_xIn_{1-x}As_{1-y}P_y$ contacts and valence-band offsets of lattice-matched InP/$Ga_xIn_{1-x}As_{1-y}P_y$ heterostructures. Data for *Schottky* contacts from *Escher* et al. [1976] and of heterostructures from *Forrest* et al. [1984], *Zachau* et al. [1989], and *Böhrer* et al. [1993b]. The dashed lines are linear least-squares fits to the experimental data. The –◆– data are ETB branch-point energies and valence-band offsets calculated in the virtual-crystal approximation [*Mönch* 1996b]

see (6.13) in Sect. 6.1.10. This relationship is verified by the variation of the barrier heights of Schottky contacts on ternary and also on quaternary III–V alloys as a function of composition; see Figs. 6.10 to 6.13 and 6.24 to 6.27, respectively. Relation (6.21) immediately predicts the valence-band offsets of, for example, $A_{1-x}B_xC$/AC heterostructures also to vary linearly as a function of composition x, i.e., one obtains [*Mönch* 1996b]

$$\Delta W_v(x) = (1-x)\cdot[\Phi^p_{bp}(\mathrm{BC}) - \Phi^p_{bp}(\mathrm{AC})] \quad , \tag{6.22}$$

provided that the intrinsic dipole term is again neglected. Linear relationships of the valence-band offsets as a function of composition are also found for heterostructures with quaternary alloys as verified by the calculated data shown in Figs. 6.24 to 6.27.

The most simple example of a lattice-matched heterostructure is the isotypic p-$Al_{0.2}Ga_{0.8}As$/p-$Al_{0.5}Ga_{0.5}As$ heterojunction. *Zubkov* et al. [1999] measured and simulated the *C/V* characteristics of such heterostructures. They obtained a valence-band offset of 0.113 eV. Using the ETB branch-point energies of GaP and GaAs, 0.92 and 0.52 eV, respectively, as shown in Table 5.1 relation (6.22) then

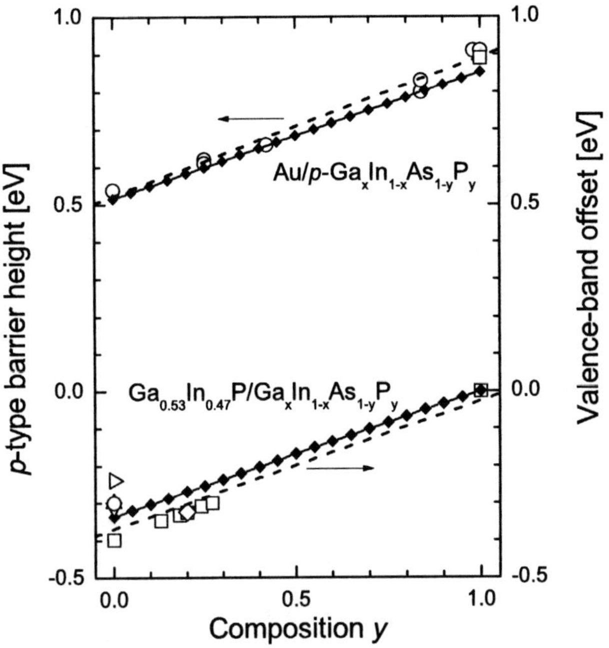

Fig. 6.27. Barrier heights of Au/p-Ga$_x$In$_{1-x}$As$_{1-y}$P$_y$ contacts and valence-band offsets of Ga$_{0.5}$In$_{0.5}$P/Ga$_x$In$_{1-x}$As$_{1-y}$P$_y$ heterostructures lattice-matched to GaAs. Data of Schottky contacts from *Kwon* et al. [1995] and *Lee* et al. [1997], and of heterostructures from *Rao* et al. [1987], *Watanabe* and *Ohba* [1987], *Cho* et al. [1995, 1996], and *Kuo* et al. [1997]. The dashed lines are linear least-squares fits to the experimental data. The –◆– data are ETB branch-point energies and valence-band offsets calculated in the virtual-crystal approximation [*Mönch* 1996b]

gives a valence-band offset of 0.12 eV. The IFIGS theory thus excellently explains the experimentally observed valence-band offset result of this most simple example of lattice-matched and isovalent heterostructures.

Experimental valence-band offsets of lattice-matched Al$_{1-x}$Ga$_x$As/GaAs heterostructures are displayed in Fig. 6.24. The experimental data [*Duggan* 1985, *Wang* and *Stern* 1985, *Forchhammer* et al. 1995, *Leuther* et al. 1996] are excellently reproduced by the results of IFIGS theory [*Mönch* 1996b]. The compositional variation of the branch-point energies $\Phi^p_{bp}(x)$ was calculated using the virtual-crystal approximation (6.12). The results of *ab-initio* theories are discussed in Sect. 9.2.1.

The experimental barrier heights Φ_{Bp} of n-Al$_{1-x}$Ga$_x$As Schottky contacts and the corresponding ETB branch-point energies $\Phi^p_{bp}(x)$ of Fig. 6.10 are replotted in Fig. 6.24. The compositional variations of valence-band offsets and of the Schottky barrier heights are well described by the difference $\Phi^p_{bp}(x) - \Phi^p_{bp}(\text{GaAs})$ of the branch points and by the branch points $\Phi^{p}_{bp}(x)$ themselves, respectively. The same observation applies to with quaternary III–V alloys.

Figures 6.25 to 6.27 display experimental barrier heights of Au *Schottky* contacts and valence-band offsets for some quaternary III–V alloys lattice-matched to the compounds GaAs or InP. These substrates are commonly used for epitaxial growth in III–V semiconductor technology. All the data for $In_{1-x}Ga_xAs_{1-y}P_y$ lattice-matched to GaAs or to InP and $Al_xGa_{0.47-x}In_{0.53}As$ lattice-matched to InP are again excellently explained by the IFIGS branch-point energies calculated in the virtual-crystal approximation and the corresponding valence-band offsets. These results again and most directly indicate that the continua of the intrinsic interface-induced gap states or, in other words, the tunneling of valence electrons at semiconductor interfaces determines the band lineup of the solids in contact irrespective of whether another elemental, compound or alloy semiconductor or a metal is involved.

6.2.4 I–III–VI$_2$ Chalcopyrites

The $Cu(Ga,In)(S,Se)_2$ chalcopyrites are important absorber materials in the research and development of highly efficient thin-layer solar cells. One of the crucial parameters of such devices is the band-structure lineup at the interface between the absorber and the window layer. Table 6.3 summarizes experimental valence-band offsets at various $Cu(GaIn)(S,Se)_2$ heterostructures. The data were

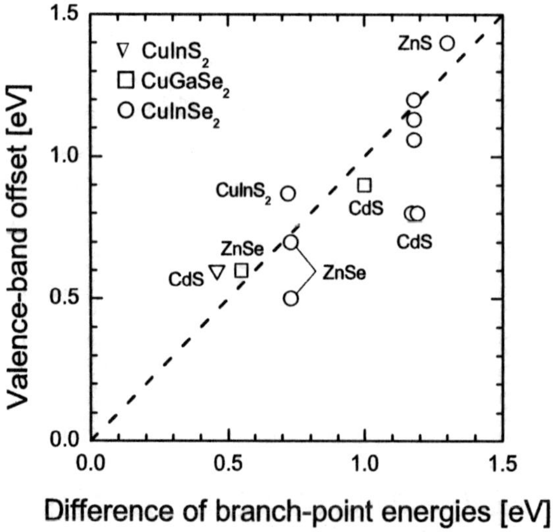

Fig. 6.28. Experimental valence-band offsets of I–III–VI$_2$ heterostructures versus the difference of the IFIGS branch-point energy. Data from Table 6.3. The dashed line indicates complete agreement between experimental and theoretical data

Table 6.3. Valence-band offsets of I–III–VI_2 chalcopyrites, in eV. The experimental values were determined by using XPS. The ETB values are from Table 5.1. The LAPW values in brackets assumed the transitivity rule to hold.

Heterostructure Layer 1	Layer 2	ETB	LAPW	Experiment	Reference
$CuInS_2$	ZnS	0.58	(095)[b]	2.3	*Wörz* et al. [1998]
	CdS	0.46	(0.79)[b]	0.6	*Klein* et al. [1997]
$CuGaSe_2$	ZnSe	0.55	(1.19)[b]	0.6	*Bauknecht* et al. [1999]
	CdS	1.00	(1.03)[b]	0.9	*Nadenau* et al. [1998]
$CuInSe_2$	ZnS	1.30	(1.23)[b]	1.4	*Wörz* et al. [1998]
	ZnSe	0.73	0.70[a]	0.7	*Nelson* et al. [1993]
				0.5	*Wörz* et al. [1998]
	CdS	1.18	1.07[a]	1.13	*Robinson* et al. [1977]
				0.8	*Löher* et al. [1995]
				1.06	*Niles* et al. [1996]
				1.2	*Kuhaimi* [1998]
				0.8	*Morkel* et al. [2001]
	$CuInS_2$	0.72	0.28[b]	0.87	*Wörz* et al. [1998]
	$CuInTe_2$	0.2	0.50[b]	0.85	*Wörz* et al. [1998]

[a] *Wei* and *Zunger* [1993].
[b] *Wei* and *Zunger* [1995].

obtained by applying core-level XPS. The table also gives the corresponding results of two different calculations. In their computations, *Wei* and *Zunger* [1993, 1995] used the density-functional formalism as implemented by the general potential, relativistic, all electron, linearized augmented plane wave (LAPW) method. *Mönch* [2002] obtained the branch-point energies of the I–III–VI_2 chalcopyrites using relation (5.59). He again calculated the valence-band dispersion $[W_v(\Gamma) - W_v(\underline{k}_{mv})]_{ETB}$ using the empirical tight-binding approach and the virtual crystal approximation, i.e., the atomic term values of the cations were taken as $W_l^c = (W_l^{Cu} + W_l^{III})/2$. The subscript l distinguishes between atomic *s*- and *p*-levels. Table 5.1 displays these branch-point energies of the I–III–VI_2 chalcopyrites.

The experimental valence-band offsets of I–III–VI_2 heterostructures are plotted versus the corresponding differences of the ETB branch-point energies in Fig. 6.28. The dashed line indicates that the IFIGS theory also explains the experimental data of the I–III–VI_2 heterostructures. The ZnS/$CuInS_2$ and $CuInTe_2$/$CuInSe_2$ results are not considered since they strongly deviate from the general trend. One of the two data points for CdS/$CuInSe_2$ which deviate by approximately 0.3 eV to lower values was obtained with technical devices [*Morkel* et al. 2001]. Additional investigations [*Heske* et al. 1999] using a combination of X-ray emission and X-ray photoemission spectroscopy revealed an anion intermixing at the interface of these cells. This effect reduces the difference of the corresponding branch-point energies and brings the experimental valence-band offset in closer agreement with the prediction of the IFIGS theory.

For some of the I–III–VI_2 heterostructures, the ETB and LAPW valence-band offsets differ. However, and most remarkably, when both theoretical approaches yielded almost identical results they were also close to the experimental values. In the remaining examples, the agreement is better between the experimental data and the ETB results.

6.2.5 Layered Semiconductors

The experimental valence-band offsets of GaSe/InSe and InSe/GaSe heterostructures are displayed in Fig. 6.29 [*Lang* et al. 1996]. The underlining indicates the respective substrate. The data clearly confirm the commutativity and transitivity of the valence-band offsets. The figure also shows the barrier heights $\Phi_{Bp} = W_F - W_{vi}$ of Au contacts on GaSe and InSe [*Kurtin* and *Mead* 1969, *Mamy* et al. 1990], see Fig. 6.8. Obviously, the energy positions of the Fermi level line up. This behavior is easily explained since, as concluded in Sect. 6.1.8, the barrier heights of Au contacts on *p*-type group-III chalcogenides are close to the branch-point energies Φ^{p}_{bp} of these compounds.

While Fig. 6.29 considers two layered compounds only, Fig. 6.30 displays valence-band offsets at interfaces between two II–VI compounds and $CuInSe_2$ [*Nelson* et al. 1994], between CdS and layered chalcogenides [*Löher* 1995], and between $SnSe_2$ and other layered chalcogenides [*Schlaf* et al. 1997]. The heterostructures were grown on the underlined substrates. As in Fig. 6.29, the valence-band maximum of GaSe is arbitrarily chosen as the reference. The dashed

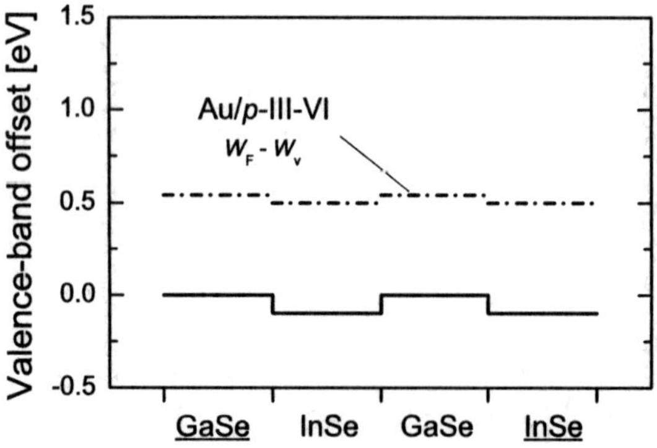

Fig. 6.29. Experimental valence-band offsets of GaSe/InSe and InSe/GaSe heterostructures (solid line) and experimental barrier heights of Au contacts on InSe and GaSe (dash-dotted lines). The substrates are underlined. Data from *Lang* et al. [1996], and *Kurtin* and *Mead* [1969] and *Mamy* et al. [1990], respectively

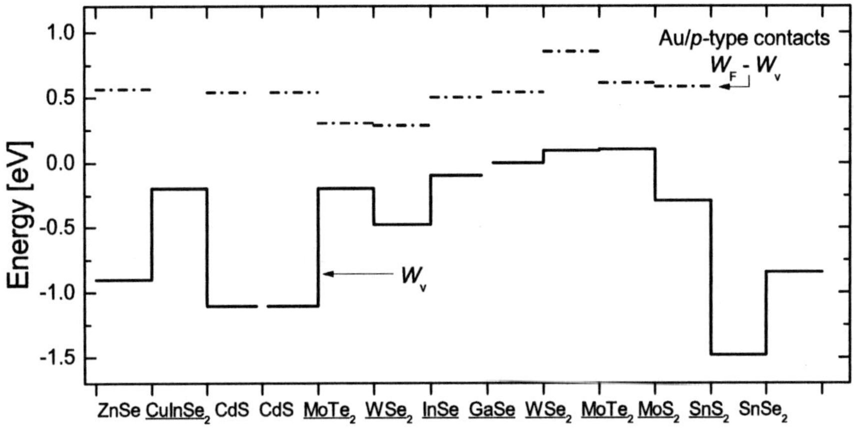

Fig. 6.30. Valence-band offsets for ZnSe and CdS on $CuInSe_2$, CdS on $MoTe_2$, WSe_2, and InSe, and $SnSe_2$ on GaSe, WSe_2, $MoTe_2$, and MoS_2 (solid lines) and barrier heights of Au contacts on the semiconductors considered (dashed lines). Data from *Nelson* et al. [1994], *Löher* [1995], *Schlaf* et al. [1997], *Mead* and *Spitzer* [1964], *Mead* [1965], *Kurtin* and *Mead* [1969], *Lince* et al. [1987], *Mamy* et al. [1990], *Bortz* et al. [1989], *Morgan* et al. [1996b], and Klein et al. [1994]

lines in Fig. 6.30 indicate the Fermi-level positions of Au contacts on the respective chalcogenides relative to their valence-band maxima [*Mead* and *Spitzer* 1964, *Mead* 1965, *Kurtin* and *Mead* 1969, *Lince* et al. 1987, *Mamy* et al. 1990, *Bortz* et al. 1989, *Morgan* et al. 1996b, *Klein* et al. 1994]. The Fermi-level positions of the Au contacts on ZnSe, CdS, InSe, GaSe, $MoTe_2$, and MoS_2 line up, as is to be ex-

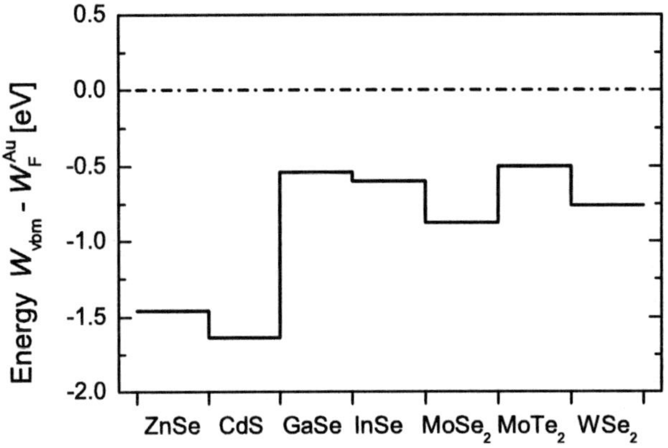

Fig. 6.31. Consistent valence-band positions of the sp^3-coordinated II–VI compounds ZnSe and CdS and of the layered metal chalcogenides GaSe, InSe, $MoSe_2$, $MoTe_2$, and WSe_2 using *Schottky* barrier heights and valence-band offsets of these compounds. From *Mönch* [1998b]

pected for the reasons outlined above. Only the data for the Au/WSe_2 and Au/$MoTe_2$ contacts deviate. This finding correlates with $MoTe_2$/WSe_2 valence-band offsets of different sign and size that result from the two sets of data obtained with CdS and $SnSe_2$ overlayers, respectively. The data are thus not transitive. This behavior is most probably caused by interfacial reactions and the formation of imperfect interfaces. The IFIGS concept, on the other hand, assumes ideal interfaces. The seeming discrepancies are easily resolved if the valence-band maxima of WSe_2 in the right-hand diagram and of both WSe_2 and $MoTe_2$ in the middle diagram are moved by approximately 0.25 eV towards lower and higher energies, respectively. These shifts adjust the valence-band offset of $MoTe_2$/WSe_2 heterostructures to the same value in both diagrams. The consistent energy positions of the valence-band maxima and the estimated branch-point energies are summarized in Fig. 6.31 for the layered chalcogenides considered and some of the sp^3-coordinated II–VI compounds.

6.2.6 Metamorphic Heterostructures

Slightly mismatched epitaxial structures are strained. However, if the difference of the lattice parameters of the epitaxial film and the substrate becomes larger the lattice mismatch is accommodated by the formation of misfit dislocations and the heterostructure is almost fully relaxed. Valence-band offsets of such so-called me-

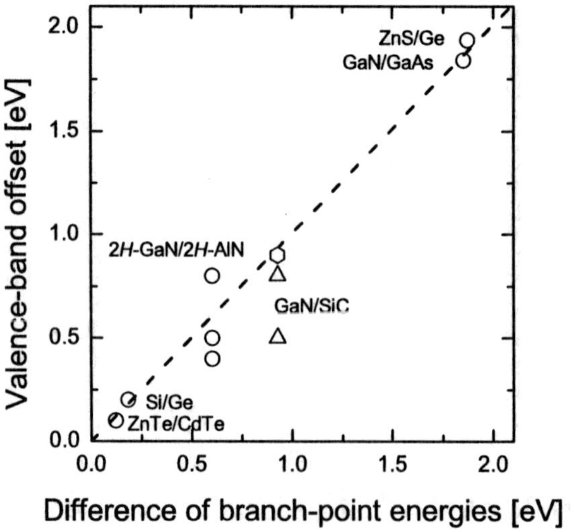

Fig. 6.32. Valence-band offsets of metamorphic heterostructures versus difference of IFIGS branch-point energies. Data from *Margaritondo* et al. [1982] (Si/Ge), *Duc* et al. [1987] (ZnTe/CdTe), *Martin* et al. [1994] (2*H*-GaN/2*H*-AlN), *King* et al. [1998] (2*H*-GaN/2*H*-AlN), *Ban* et al. [1996] (Ge/ZnS), *Ding* et al. [1997] (GaN/GaAs), *King* et al. [1999] (△ 2*H*-GaN/3*C*-SiC), *Rizzi* et al. [1999] (○ 2*H*-GaN/6*H*-SiC; 2*H*-GaN/2*H*-AlN)

tamorphic heterostructures are displayed in Fig. 6.32 versus the difference of the corresponding IFIGS branch-point energies. Only one of the two 2*H*-GaN/3*C*-SiC data points deviates from the general trend by approximately 0.3 eV. The dashed line indicates that the experimental results are excellently described by the theoretical IFIGS data. This is true not only for heterostructures of zincblende-structure or of wurtzite-structure compounds but even for films of wurtzite-structure group-III nitrides grown on both cubic 3*C*- and hexagonal 6*H*-SiC substrates. These observations suggest the following conclusions. First, all the heterostructures considered in Fig. 6.32 are only slightly, if at all, strained, although their lattice parameters differ by up to 19.8% as, for example, in the case of the GaN/GaAs heterostructures. Second, the calculations of the IFIGS branch-point energies Φ^p_{bp} assumed the zincblende-structure for the semiconductors. These values, on the other hand, reproduce the experimental valence-band offsets irrespective of whether the semiconductors have zincblende, wurtzite or, as in the case of 6*H*-SiC, another hexagonal-polytype structure. These findings again support the conclusion drawn already from the GaN and SiC Schottky barrier heights in Sects. 6.1.5 and 6.1.6 that all modifications or polytypes of any given semiconductor have identical IFIGS branch-point energies $\Phi^p_{bp} = W_{bp} - W_v(\Gamma)$. The results of *ab initio* LDA-DFT calculations which are summarized in Table 6.1 corroborate this conclusion drawn from the experimental data. Possibly, 2*H*-SiC will be an exception to the rule. Third, the agreement between the experimental valence-band offsets and the theoretical IFIGS predictions again confirms that the IFIGS branch-point energies (5.59) represent a reliable data basis for compound semiconductors.

6.2.7 Semiconductor–Insulator Interfaces

The interfaces between semiconductors and insulators are of eminent importance in semiconductor devices. The most prominent examples are SiO_2/Si and Si_3N_4/Si interfaces. However, the continuous shrinkage of the insulator thickness to below 100 nm stimulated the interest in materials with high dielectric constants such as, for example, Al_2O_3, ZrO_2, and $SrTiO_3$. The band lineup at insulator–semiconductor interfaces is one of the important design parameters in, for example, MOS field-effect transistors and dynamic random access memories (DRAMs). Valence- and conduction-band offsets have been determined using X-ray photoemission spectroscopy and internal photoemission, respectively. Tables 6.4 and 6.5 display the experimental data.

Insulators may be considered as semiconductors with very large band gaps. Consequently, the interface-induced gap states determine the band alignment at these interfaces. This means that relation (5.25) also describes the valence-band

Table 6.4. Experimental valence- and conduction-band offsets at silicon–insulator interfaces, in eV

Insulator	Orientation	ΔW_v	ΔW_c	Reference
SiO_2	–	4.45		*Williams* [1965]
	(111)	4.5		*Grunthaner & Grunthaner* [1986]
	(001)	4.3		*Himpsel* et al. [1988]
	(001)	4.53		*Afanas'ev* et al. [1996]
	(001)	4.49		*Alay & Hirose* [1997]
	(111)	4.36		*Alay & Hirose* [1997]
	(001)	4.54		*Keister* et al. [1999]
	(111)	4.35		*Keister* et al. [1999]
	(111)	4.42		*Hirose* et al. [2001]
	(111)	4.61		*Hirose* et al. [2001
	(001)	4.49		*Miyazaki* et al. [2001]
	(001)		3.13	*Afanas'ev* et al. [2001]
Si_3N_4	(001)	1.95		*Goodmann* [1968]
	(001)	2.1		*DiMaria & Arnet* [1975]
	(001)	1.78		*Keister* et al. [1999]
	(001)	1.6		*Miyazaki* et al. [2001]
Al_2O_3	(001)		2.13	*Afanas'ev* et al. [2001]
	(001)		2.78	*Ludeke* [2002]
ZrO_2	(001)	3.15		*Miyazaki* et al. [2001]
	(001)		1.98	*Afanas'ev* et al. [2001]
Sm_2O_3	–		1.6	*Rozhkov* et al. [1998]
$SrTiO_3$	(001)	2.12		*Chambers* et al. [2001]

offsets at semiconductor–insulator interfaces. The branch-point energies of the semiconductors are well known and, therefore, the branch-point energies of SiO_2 and Si_3N_4 may be at least estimated from the valence-band offsets shown in Tables 6.4 and 6.5. First of all, however, the SiC/SiO_2 data will be considered in some more detail.

Afanas'ev et al. [1996] and performed internal photoemission yield measurements at *n*-SiC/SiO_2 interfaces and obtained the same barrier height between the SiC valence-band maximum and the SiO_2 conduction-band bottom as the value of 6.0 ± 0.1 eV for the four polytypes 3*C*, 15*R*, 6*H*, and 4*H*. Hence, the conduction-band discontinuities of the four polytypes decrease to the same extent as their band gaps increase, while their valence-band offsets are identical. Using a SiO_2 band gap of 8.9 eV one obtains the valence-band offset $\Delta W_v = 2.95$ eV irrespective of which one of the polytypes is considered. *Afanas'ev* and *Stesmans* [2000] confirmed this conclusion by their additional IPEYS measurements with *p*-type 4*H*- and 6*H*-SiC/SiO_2 interfaces These findings are quite similar to what is observed with Schottky contacts on different SiC polytypes (see Sect. 6.1.6) and it may be explained alike. For one and the same metal on different SiC polytypes doped *n*-type, the barrier heights vary to the same extent as the band-gap widths of the corresponding polytypes differ. Hence, the barrier heights are identical on *p*-type SiC substrates. All these observations are easily explained by identical IFIGS branch-point energies $\Phi_{bp}^{p} = W_{bp} - W_v(\Gamma)$ irrespective of the whether the lattice structure of the specific silicon carbide polytype is cubic or hexagonal. With this in mind, the branch-point energy of SiO_2 may be estimated from the valence-band offsets given in Tables 6.4 and 6.5.

Experimental valence-band offsets are not only available for the four different SiO_2/SiC interfaces mentioned above but also for Si-, ZnS-, and CdTe/SiO_2. The respective data are also given in Tables 6.4 and 6.5. Figure 6.33 displays these experimental data as a function of the branch point energies Φ_{bps}^{p} of the respective semiconductors. Only the SiO_2/CdTe data point deviates from the general trend that the valence-band-offset becomes smaller with increasing branch-point energy of the respective semiconductor. The dashed line is the linear least-squares fit

$$\Delta W_v = (4.88 \pm 0.06) - (1.25 \pm 0.06)\Phi_{bps}^{p} \text{ [eV]} \tag{6.23}$$

to the data points where the CdTe/SiO_2 value is not considered. The sign of the slope parameter indicates that the valence-band maximum of SiO_2 is lower in energy than the valence-band maxima of the semiconductors.

Figure 6.33 also displays the experimental valence-band offsets reported for heterostructures of silicon and gallium nitride with Si_3N_4. The Si/Si_3N_4 data are taken from Table 6.4. For GaN/Si_3N_4 interfaces, *Cook* et al [2003b] determined a valence-band offset of –0.5 eV, where the minus sign indicates that the valence-band maximum of GaN is lower in energy than the valence-band top of Si_3N_4. The dash-dotted line in Fig. 6.33 is the linear least-squares fit

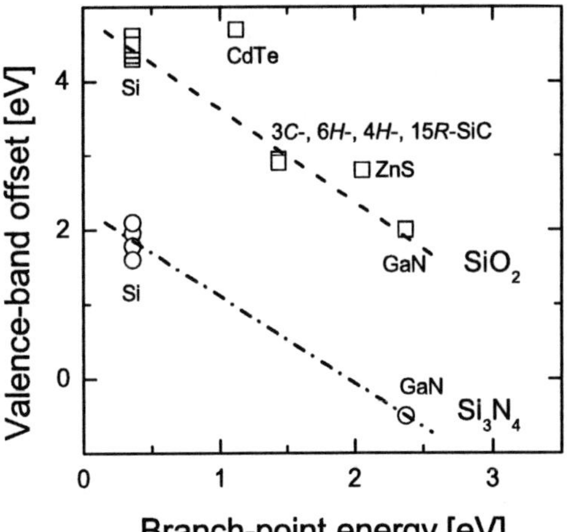

Fig. 6.33. Valence-band offsets at SiO_2 and Si_3N_4 interfaces with various semiconductors versus the branch-point energies of the semiconductors. Data from Tables 6.4 and 6.5. The dashed and dash-dotted lines are linear least-squares fits to the data

Table 6.5. Experimental band offsets at semiconductor–SiO_2 interfaces, in eV

Semiconductor	Orientation	ΔW_v	ΔW_c	Reference
3*C*-SiC	(001)		3.6	*Afanas'ev* et al. [1996]
4*H*-SiC	(0001)		2.7	*Afanas'ev* et al. [1996]
6*H*-SiC	(0001)		2.95	*Afanas'ev* et al. [1996]
15*R*-SiC			3.0	*Afanas'ev* et al. [1996]
6*H*-SiC		2.9		*Mattern* et al. [1998]
4*H*-SiC	(0001)	2.9		*Afanas'ev & Stesmans* [2000]
6*H*-SiC	(0001)	2.9		*Afanas'ev & Stesmans* [2000]
2*H*-GaN	(0001)	2.0		*Cook* et al. [2003a]
ZnS	(111)	2.8		*Ban* et al. [1998]
CdTe	(111)	4.7		*Ban* et al. [1998]

$$\Delta W_v = (2.27 \pm 0.13) - (1.17 \pm 0.12)\Phi^p_{bps} \text{ [eV]} \qquad (6.24)$$

to the data points.

Within the margins of experimental error, the slope parameters of the fits (6.23) and (6.24) are almost equal and close to one. If the dipole terms $D_X(X_{SiO2} - X_s)$ and $D_X(X_{Si3N4} - X_s)$, respectively, in (5.25) are neglected, then the branch-pointpoint energies $\Phi^p_{bp} = W_{bp} - W_v$ of SiO_2 and Si_3N_4 are estimated as approximately 4.9 eV and 2.3 eV, respectively. The band gaps of SiO_2 and Si_3N_4 are approximately 8.9 eV and 5.3 eV, respectively, wide. Obviously, the branch points of both insulators are close to the mid-gap position, as is to be expected for amorphous insulators within the concept of interface-induced gap states.

From the valence-band offsets given in Table 6.4 and the Si branch-point energy, 0.36 eV, the branch-point energies are estimated as approximately 3.4 eV for ZrO_2, and as approximately 2.5 eV for $SrTiO_3$. No branch-point energies will be estimated for the other insulators from the experimental conduction band since the band gaps of the insulator films strongly depend on preparational details and cannot be easily transferred from one study to another.

6.2.8 Langer–Heinrich Rule

The energy levels of *shallow* donors and acceptors in semiconductors are customarily referred to the conduction- and the valence-band edges, respectively, of the host. The 3*d* transition-metal impurities, on the other hand, are *deep* donors and acceptors which act as efficient recombination centers. Some of the early experimental data seemed to suggest that the vacuum level might be an adequate reference for the energy levels of deep impurities [*Ledebo* and *Ridley* 1982]. However, *Vogl* [1985] had already noted that the deep energy levels of 3*d* transition-metal impurities are by no means pinned to the vacuum level. In a heuristic approach, *Langer* and *Heinrich* [1985] assumed the existence of a *common bulk reference*

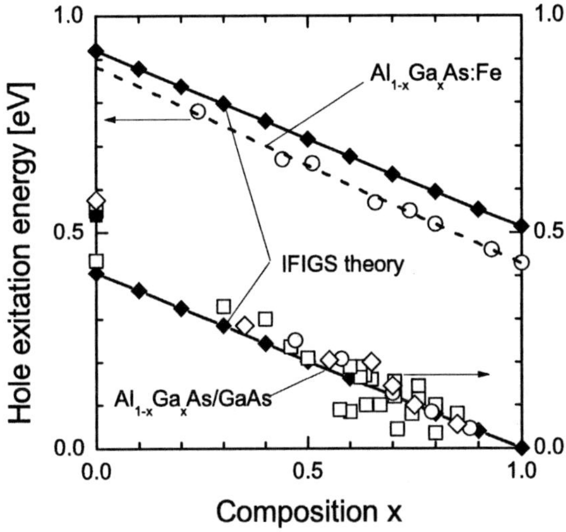

Fig. 6.34. Hole excitation energies of deep Fe impurities in $Al_{1-x}Ga_xAs$ alloys as well as IFIGS branch-point energies, experimental and IFIGS valence-band offsets of $Al_{1-x}Ga_xAs$/GaAs heterostructures and valence-band offsets. Fe impurity data from *Wang* et al. [1984]; all other data the same as in Fig. 6.24

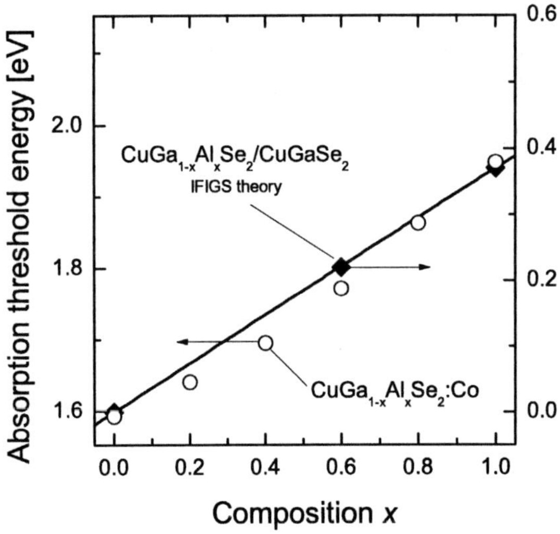

Fig. 6.35. Absorption threshold energies of deep Co impurities in $CuGa_{1-x}Al_xSe_2$ alloys and IFIGS valence-band offsets. Co impurity data from *Jin* et al. [1993]; IFIGS data from Table 5.1 and *Mönch* [2002]

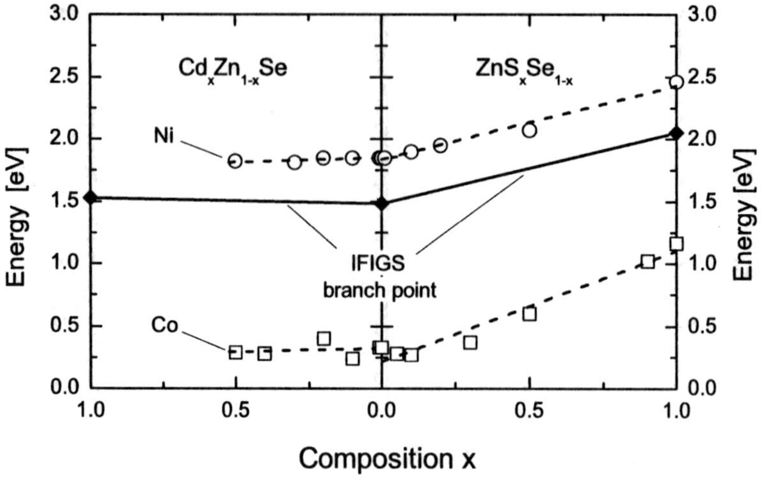

Fig. 6.36. Deep levels of Ni and Co impurities and IFIGS branch-point energies in $Cd_xZn_{1-x}Se$ and ZnS_xSe_{1-x} alloys. Experimental impurity data from *Surkova* et al. [1996, 1997];, IFIGS data from Table 5.1

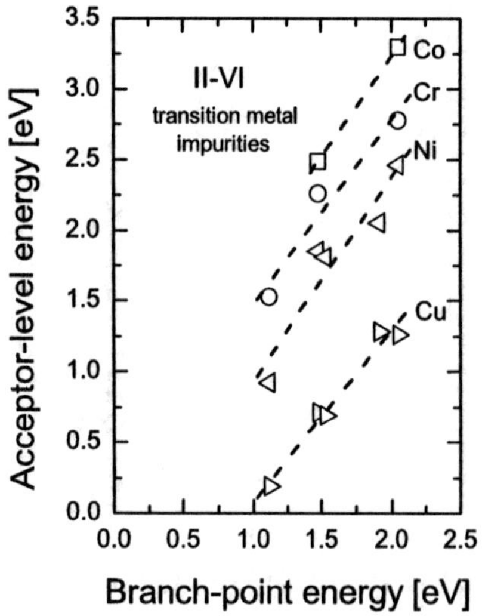

Fig. 6.37. Experimental acceptor-level energies of transition-metal impurities in ZnS, ZnSe, CdS, CdSe, and CdTe as a function of the IFIGS branch-point energies of the semiconductors. Experimental data as compiled by *Langer* and *Heinrich* [1985] and from *Landoldt–Börnstein* [2001]

level for the energy levels of the 3*d* transition-metal impurities in isovalent semiconductors. Consequently, they concluded that *a valence-band-edge discontinuity in heterojunctions is then given by just the difference in the energy-level positions of a transition-metal impurity in the two compounds forming the heterojunction.*

The first and most striking example of this approach was the ^5E acceptor level of Fe impurities in $Al_{1-x}Ga_xAs$ alloys, which were determined by *Wang* et al. [1984] using the photocapacitance technique. Figure 6.34 displays the hole excitation energies of the Fe impurities as a function of the composition of the $Al_{1-x}Ga_xAs$ alloys. The dashed line is a linear least-squares fit to the experimental data and its slope parameter almost fits the compositional variation of the IFIGS branch-point energy calculated in the virtual crystal approximation. Hence, the Fe(^5E) data also reproduce the compositional trend of the experimentally observed valence-band offsets of $Al_{1-x}Ga_xAs$/GaAs heterostructures which are also shown in Fig. 6.34.

Another example is the Co impurity in $CuGa_{1-x}Al_xSe_2$ chalcopyrite alloys, which was investigated by *Jin* et al. [1993] applying optical absorption spectroscopy. The experimental data are displayed in Fig. 6.35. Evidently, they are excellently described by the difference of the IFIGS branch-point energies of the $CuGa_{1-x}Al_xSe_2$ alloys and $CuGaSe_2$. However, it has to be mentioned that the experimentally determined threshold energies are assigned to electron transitions from the valence band into the Co impurity levels [*Mönch* 2002] rather than from the Co levels into the conduction band as inferred earlier by *Chang* et al. [1990].

Further examples are Ni acceptors and Co donors in both $Cd_xZn_{1-x}Se$ and ZnS_xSe_{1-x} alloys, which were studied by *Surkova* et al. [1996, 1997] using optical absorption spectroscopy. Figure 6.36 displays the energy-level positions of Ni and Co impurities $Cd_xZn_{1-x}Se$ and ZnS_xSe_{1-x} alloys in as a function of composition The dashed lines are linear least-squares fits to the experimental data. The experimental data again follow the compositional variation of the IFIGS branch-point energies. However, the slope parameter of the Co donor-levels is approximately 50% larger than the value predicted by the IFIGS theory. These findings clearly illustrate that the 3*d* transition-metal impurity levels are not pinned to a bulk reference level which is the basic assumption of the *Langer–Heinrich* rule. This conclusion is further illustrated in Fig. 6.37 where the available acceptor-level energies of Cr, Co, Ni, and Cu impurities in ZnS, ZnSe, CdS, CdSe, and CdTe are plotted versus the IFIGS branch-point energies of these II–VI compounds. The slope parameters vary between 1.21 and 1.44 instead of 1, the value expected if the levels were pinned to the IFIGS branch-points. Obviously, the *Langer–Heinrich* rule is not a universal one but it holds in favorable cases.

7. First-Principles Calculations of Barrier Heights and Valence-Band Offsets

7.1 Introductory Remarks

The local-density approximation to the density-functional theory or, for short, LDA-DF theory is the most powerful and widely used tool for theoretical studies of the ground-state properties of solids. However, excitation energies such as, for example, the width of the energy gaps between the valence and conduction bands of semiconductors cannot be correctly obtained from such calculations. The fundamental band gaps of the elemental semiconductors C, Si, and Ge as well as of the III–V and II–VI compounds are notoriously underestimated by 25 to 50%. However, it became possible to compute quasi-particle energies and band gaps of semiconductors from first principles by using the so called GW approximation for the electron self-energy which was pioneered by *Hybertsen* and *Louie* [1986] and by *Godby* et al. [1988]. The resulting band-gap energies agree to within 0.1 to 0.3 eV with the experimental values. *Bechstedt* and *DelSole* [1988, 1990], *Zakharov* et al. [1994] and *Grossner* et al. [1999] calculated the quasi-particle shifts of the valence-band maxima of the III–V and the II–VI compound semiconductors in the GW approximation. The quasi-particle corrections move the valence band-maxima to lower energies and, on the average, the GW correction accounts for 63 ± 2 % of the quasi-particle widening of the LDA-DFT band gaps. To mention the most important point first, it is obviously much more demanding to calculate the barrier heights of Schottky contacts than the valence-band offsets of heterostructures even if the most advanced theoretical tools, *ab initio* LDA-DFT plus GW corrections, are employed.

7.2 Schottky Barrier Heights

For some specific metal–semiconductor contacts, the band-structure lineup was also studied by state-of-the-art *ab initio* LDA-DFT calculations. The resulting LDA-DFT barrier heights were then subjected to *a posteriori* corrections which consider quasi-particle effects and, if necessary, spin–orbit interactions and semicore–orbital effects. Table 7.1 contains published first-principles calculations of

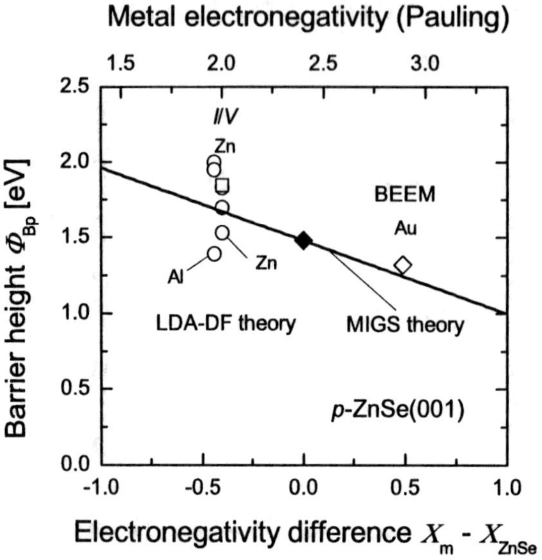

Fig. 7.1. Theoretical and experimental barrier heights of *p*-ZnSe(001) Schottky contacts versus the difference of the *Miedema* electronegativities of the metal and ZnSe. ◇ and □ symbols differentiate experimental *I/V* and BEEM data (see Tables 3.3 and 3.4). ○ data were computed by applying LDA-DF theory [*Rubini* et al. 2001]. The solid MIGS line is drawn with the same parameters as in Fig. 6.6

barrier heights of Schottky contacts. These results will be briefly discussed in the following.

Lazzarino et al. [1998] and *Rubini* et al. [2001] derived the barrier heights of Al- and Zn/*p*-ZnSe contacts, respectively, by using *ab initio* LDA-DF theory and *a posteriori* spin-orbit and quasi-particle corrections. Figure 7.1 displays their results. The three different data points each for Al- and Zn/*p*-ZnSe(001) contacts are for different interface configurations. On the average, these results of *ab initio* LDA-DFT calculations agree with the chemical trend of Schottky barrier-heights predicted by the MIGS-and-electronegativity theory. The experimental Zn and Au data points were obtained by conversion of the *I/V* and BEEM barrier heights of laterally homogeneous Zn- and Au/*n*-ZnSe(001) contacts, respectively, into the corresponding *p*-type values, see Fig. 6.6 and Tables 3.3 and 3.4. Considering the margins of experimental error, these data are excellently explained by the MIGS theory.

Picozzi et al. [2000, 2002] presented barrier heights of Al-, Ag-, and Au/*p*-GaN contacts which they obtained from *ab initio* LDA full-potential linearized augmented plane wave calculations. They neglected quasi-particle effects and spin–orbit coupling. First of all, *Picozzi* et al. [2002] found Al contacts on relaxed surfaces of zincblende- and on [0001] surfaces of wurtzite-structure GaN to differ by 70 to 90 meV only. These results corroborate the earlier calculations of *Stampfl* and *Van der Walle* [1998], who estimated a valence-band offset of only approxi-

Table 7.1. First-principles calculations of Schottky barrier heights

Semiconductor	Metal	Reference
Si(001)	$CoSi_2$	*Stadler* and *Podloucky* [2000]
Si(111)	$CoSi_2$	*Stadler* et al. [1999]
	$NiSi_2$	*Das* et al. [1989]
	$NiSi_2$	*Fujitani* and *Asano* [1990]
3*C*-SiC(001)	Al	*Hoekstra* and *Kohyama* [1998]
	Ti	*Kohyama* and *Hoekstra* [2000]
2*H*-GaN[0001]	Al	*Picozzi* et al. [2000, 2002]
3*C*-GaN[111]		*Picozzi* et al. [2000, 2002]
GaAs(110)	Al	*Zhang* et al. [1986]
	Al	*Charlesworth* et al. [1992]
	Al	*Needs* et al. [1994]
GaAs(001)	Al	*Dandrea and Duke* [1993a, b]
	Al	*Ruini* et al. [1997]
$Al_xGa_{1-x}As$(001)	Al	*Bardi* et al. [1999]
$Al_xGa_{1-x}As$(110)		*Bardi* et al. [1999]
ZnSe(001)	Al	*Lazzarino* et al. [1998]
	Zn	*Rubini* et al. [2001]

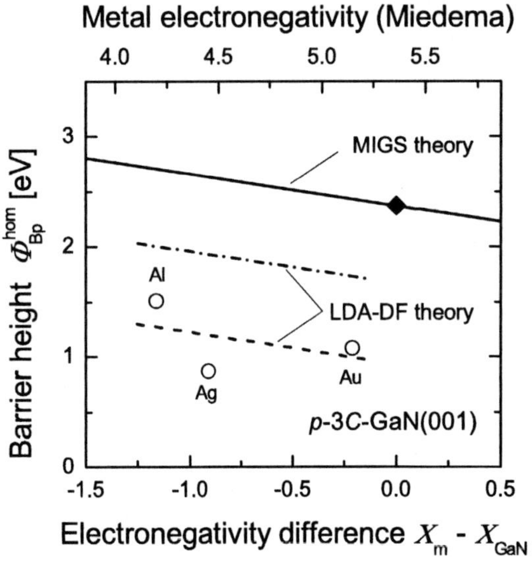

Fig. 7.2. Theoretical barrier heights of *p*-3*C*-GaN Schottky contacts versus the difference of the *Miedema* electronegativities of the metal and GaN. LDA-DFT data from *Picozzi* et al. [2000, 2002]. The solid MIGS line is drawn with the same parameters as in Fig. 6.4

mately 70 meV at zincblende–wurtzite interfaces of GaN. The barrier heights reported by *Picozzi* et al. are plotted in Fig. 7.2 versus the difference of the metal and the GaN electronegativities. The dashed line in the figure is a linear least-squares fit to the data points. Its slope parameter of 0.29 eV/Miedema-unit agrees with the value of the MIGS-and-electronegativity theory; see Sect. 6.4. However, the LDA-DFT barrier heights are on the average 1.34 eV smaller than those predicted by the MIGS-and-electronegativity concept. *Grossner* et al. [1999] calculated a quasi-particle correction of –0.74 eV for the valence-band maximum of 3*C*-GaN. The dashed-dotted line in Fig. 7.2 is shifted by this estimated value towards larger barrier heights. However, even this large correction only partly reduces the difference between the LDA-DFT results and the MIGS line, which, on the other hand, excellently explains the experimental results

Hoekstra and *Kohayama* [1998] and *Kohayama* and *Hoekstra* [2000] applied *ab initio* LDA-DF theory to Al- and Ti/3*C*-SiC(001) interfaces. Their *p*-type barrier heights are plotted in Fig. 7.3 versus the difference of the metal and SiC electronegativities. The two data points for each of the two contacts are for carbon- and silicon-terminated surfaces. These theoretical results strongly deviate from the predictions of the MIGS-and-electronegativity concept which, on the other hand, explain all the experimental data of Schottky contacts not only on 3*C*- but also on

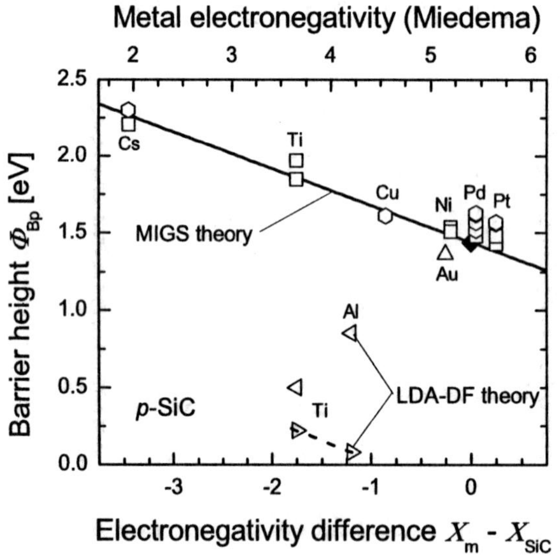

Fig. 7.3. Theoretical barrier heights of *p*-SiC Schottky contacts versus the difference of the *Miedema* electronegativities of the metal and SiC. LDA-DFT data from *Hoekstra* and *Kohayama* [1998] and *Kohayama* and *Hoekstra* [2000]. The solid MIGS line is drawn with the same parameters as in Fig. 6.5. The Δ, □, and O symbols differentiate the converted experimental *n*-type barrier heights of 3*C*-, 4*H*- and 6*H*-SiC, respectively; see Tables 3.3 through 3.6. The solid MIGS line is drawn with the same parameters as in Fig. 6.5

4*H*- and 6*H*-SiC. The LDA-DFT band-gap of 3*C*-SiC differs by 1.08 eV from the experimental value [*Rohlfing* et al. 1993]. By using the rule mentioned above, this value gives an estimated quasi-particle shift of the valence-band maximum of 0.78 eV. Even such a large quasi-particle correction does not remove the discrepancy between the LDA-DFT and the experimental barrier heights of Ti/*p*-4*H*-SiC contacts which, on the other hand, are excellently described by the MIGS-and-electronegativity line.

The results of LDA-DFT calculations for epitaxial $CoSi_2$ and $NiSi_2$ contacts on both *p*-Si(111) and *p*-Si(001) substrates are displayed in Fig. 7.4. The $CoSi_2$ data contain estimated corrections which are based on the quasiparticle concept. The figure also shows the experimental results. The spread of the theoretical data and the deviations from the experimental results clearly reveal the great difficulties encountered in the theoretical approaches. The theoretical results turned out to be extremely sensitive to the geometrical interface structure.

The barrier heights of Schottky contacts and their chemical trends are well explained by the simple MIGS-and-electronegativity theory. The present results of most of the much more elaborate first-principles LDA-DFT calculations, on the other hand, fail to explain the experimental data even if quasi-particle corrections are applied. Up to now, no complete quasi-particle calculations were performed for metal–semiconductor contacts. Furthermore, each of these *ab initio* calculations is a case study in itself, while the MIGS-and-electronegativity theory provides the chemical trends.

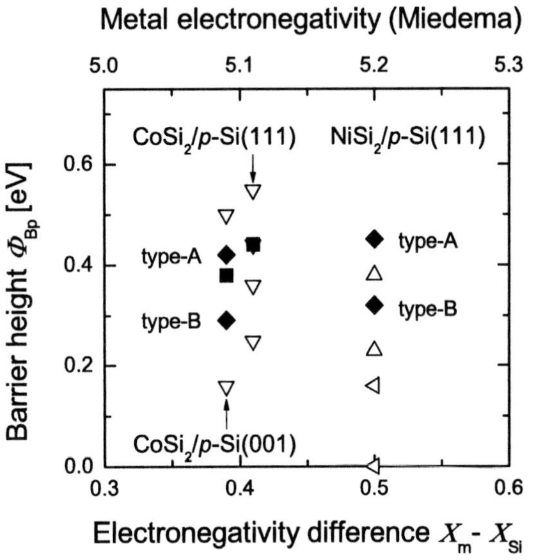

Fig. 7.4. Theoretical and experimental barrier heights of epitaxial $CoSi_2$ and $NiSi_2$ contacts on *p*-Si(001) and *p*-Si(111) substrates. Theoretical ◁ data from *Das* et al. [1989], Δ data from *Fujitani* and *Asano* [1990], and ∇ data from *Stadler* et al. [1999] and *Stadler* and *Podloucki* [2000]. Experimental ◆ and ■ data from Tables 3.3 and 3.4

7.3 Valence-Band Offsets at Heterostructures

7.3.1 Isovalent and Lattice-Matched AlGaAs/GaAs

In a first-principles calculation *Nelson* et al. [1991b] calculated the valence-band offsets of $Al_xGa_{1-x}As/GaAs$ heterostructures. They assumed randomly disordered alloys and employed the virtual-crystal approximation. Their calculations did not employ the GW approximation. Figure 7.5 displays their results. Obviously, the results of this first-principles calculation agree with the experimental results.

Zhang et al. [1990] did not study the complete series of alloys but rather calculated the band-structure lineup at AlAs/GaAs(001) heterostructures only. They used a first-principles quasi-particle theory and obtained a LDA valence-band offset of 0.41 eV, which equals the corresponding result reported by *Nelson* et al. [1991b], and a GW correction of 0.12 eV. The resultant offset of 0.53 eV is in good agreement with the experimental data. In contrast to what is reported for Schottky barrier heights, the valence band offsets of lattice-matched heterostructures as obtained from *ab initio* calculations are in good agreement with the experimental data.

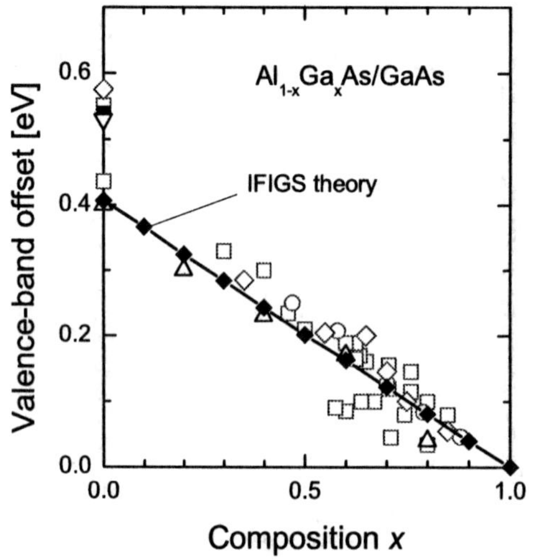

Fig. 7.5. Theoretical valence-band offsets at $Al_{1-x}Ga_xAs/GaAs$ heterostructures: Δ data from *Nelson* et al. [1991b], ∇ data from *Zhang* et al. [1990], and –◆– IFIGS data from *Mönch* [1996b]. The gray symbols reproduce the experimental valence-band offsets shown in Fig. 6.24

7.3.2 SiC Heterostructures

In complete contrast to the calculations of Schottky barrier heights, *ab initio* LDA-DFT calculations yield valence-band offsets of heterostructures which are much closer to both the experimental data and the results of the IFIGS theory. *Städele* et al. [1997], *Majewski* et al. [1997], *Binggeli* et al. [2001], *Agrawal* et al. [2001], and *Laridjani* et al. [2001] calculated valence-band offsets of Al- and GaN/SiC heterostructures. Figure 7.6 displays their data. $C^{III}A^{V}/C^{IV}A^{IV}$ heterostructures are heterovalent and, therefore, 3*C*-(110) interfaces are non-polar, while 3*C*-(111), 3*C*-(001), and 2*H*-(0001) heterostructures will be polar. Most remarkably, the LDA-DFT valence-band offsets at non-polar and pseudomorphic AlN and GaN/3*C*-SiC(110) heterostructures deviate by approximately only 0.1 and –0.4 eV, respectively, from the values predicted by the IFIGS theory. For GaN/3*C*-SiC(110) interfaces, approximately 0.14 eV of this difference might be attributed to the 3% lattice mismatch. C^{III}–C^{IV} and A^{V}–A^{IV} intermixing will avoid the charging of polar interfaces. The correlated electric dipoles, however, will increase and reduce, respectively, the valence-band offsets with regard to the value observed for the non-polar interface. The LDA-DFT calculations considered a 50% intermixing in the atomic layers right at the interface. For one and the same polar interface, the different LDA-DFT valence-band discontinuities are generally quite close. However, the scatter of the GaN/SiC data is somewhat larger. The difference between the C^{III}–C^{IV} and A^{V}–A^{IV} intermixed interfaces amounts to approximately 1 eV irrespective of whether AlN or GaN is considered. A simple point-charge model easily explains this observation by the very slight difference

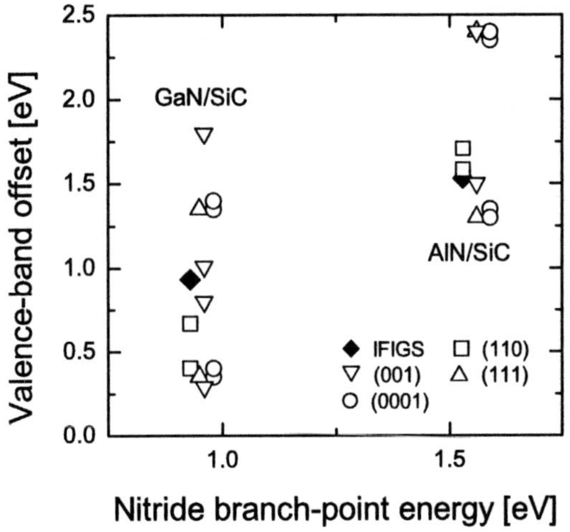

Fig. 7.6. Valence-band offsets of GaN- and AlN/SiC heterostructures calculated using *ab initio* LDA-DF theory [*Städele* et al. 1997, *Majewski* et al. 1997, *Binggeli* et al. 2001, *Agrawal* et al. 2001, *Laridjani* et al. 2001] and predicted by the IFIGS theory

between the electronegativities of Al and Ga. Most importantly, the *ab -initio* valence-band offsets of the 2*H*-(0001) and the 3*C*-(111) interfaces turned out to be equal. However, for a direct comparison of the experimental and the LDA-DFT valence-band offsets a more detailed characterization of the geometrical structure and the chemical composition at the heteropolar III–N/SiC interfaces is highly desirable.

8. Temperature and Pressure Effects

8.1 Temperature Effects

The energies of the optical transitions in the bulk of semiconductors are sensitive to variations in temperature and to the application of external pressure. The effects of the temperature as well as the pressure on the barrier heights of Schottky contacts are well explained by the MIGS model.

The BEEM technique is well-suited for studying the temperature variations of Schottky barrier heights. Figure 8.1 displays the distributions of barrier-heights which *Zhu* et al. [1999] determined from BEEM I_{coll}/V_{tip} Curves recorded with PtSi/*n*-Si(001) contacts in the temperature range between room temperature and 183 K. Within the margins of experimental error the room-temperature data agree with the histogram displayed in Fig. 3.16. The dashed lines are again Gaussian least-squares fits to the data. The energy of their center values is plotted as a function of temperature in Fig. 8.2. As the temperature is reduced the barrier heights increase. This behavior is commonly observed with *n*-type Schottky contacts. The IPEYS barrier heights of $CrSi_2$/*n*-Si(001) diodes, which are displayed in Fig. 8.3a and which were measured by *Aniltürk* and *Turan* [1999] are another example. The barrier heights of *p*-type of Schottky contacts, on the other hand, do not vary as a function of temperature. This is evidenced by the IPEYS data of a $CrSi_2$/*p*-Si(001) contact which are shown in Fig. 8.3a as an example. Similar experimental results were reported by *Newman* et al. [1986b] and *Revva* et al. [1993] for Al- and Cu/GaAs contacts while *Aboelfotoh* et al. [1990] and *Aboelfotoh* [1991] studied the temperature dependence of Cu_3Si- and W/Si(001) diodes, respectively.

Simultaneously with their IPEYS studies *Aniltürk* and *Turan* [1999] also measured the I/V characteristics of their $CrSi_2$ contacts as a function of temperature. By reducing the temperature they again observed the effective barrier height and the ideality factor to decrease and to increase, respectively, for *n*-type contacts but they found both parameters to be insensitive to temperature variations with the *p*-type diodes. For one of their *n*- and *p*-type diodes each, the effective barrier heights are plotted versus the ideality factors in Fig. 8.3b. For the *n*-type diodes the extrapolated barrier height Φ_{Bn}^{nif} equals the IPEYS barrier height measured at 300 K and the barrier heights of the *p*-type contacts are the same irrespective of temperature.

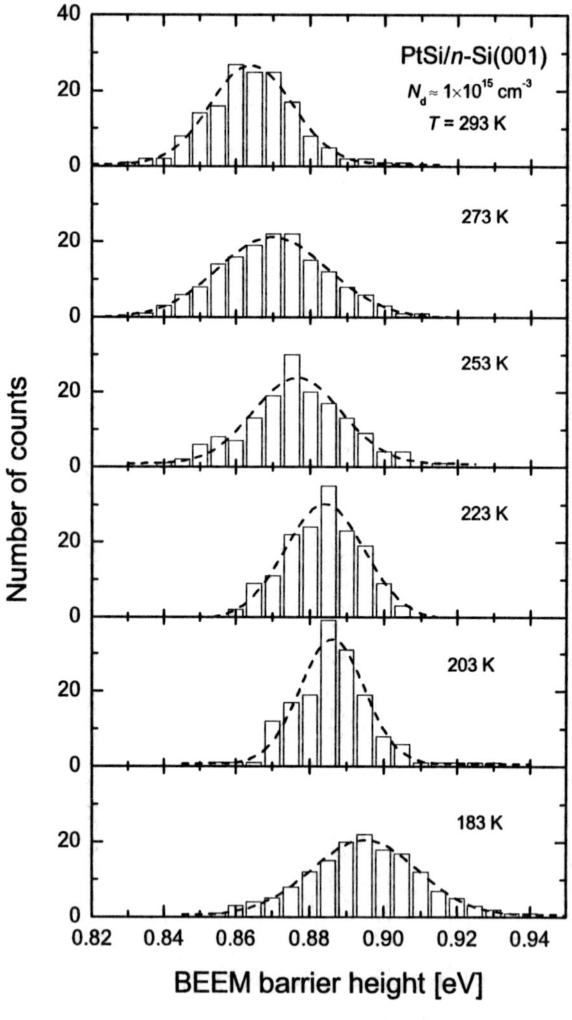

Fig. 8.1. Histograms of BEEM barrier heights on one PtSi/*n*-Si(001) contact at different temperatures. The dashed lines are Gaussian least-squares fits to the data. Data from *Zhu* et al. [1999]

The changes of the barrier heights observed as a function of temperature with *n*-type Schottky contacts are well explained by the temperature dependence of the fundamental bulk band gaps. The semi-empirical relation [*Varshni* 1967]

$$W_g(T) = W_{g0} - \frac{\vartheta T^2}{\Theta + T} \tag{8.1}$$

describes the temperature variation of the bulk band gaps of semiconductors where W_{g0} is the width of the gap at zero temperature and ϑ and Θ are empirical parameters. In Figs. 8.2 and 8.3b, the dashed lines are least-squares fits of this re-

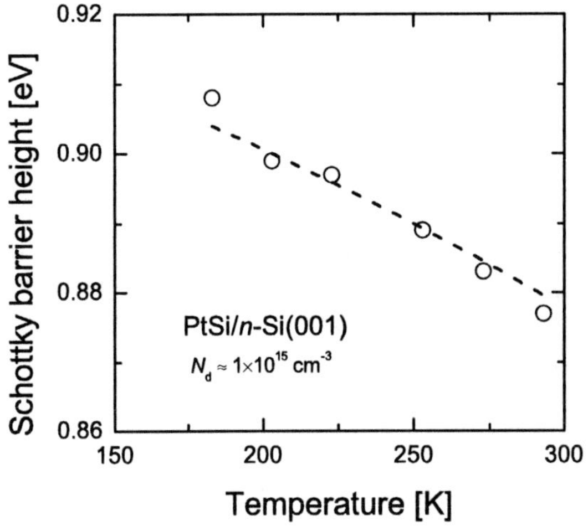

Fig. 8.2. Schottky barrier heights of a PtSi/*n*-Si(001) contact as a function of temperature. The data plotted are the center barrier-heights of the BEEM barrier-height distributions displayed in Fig. 8.1. The dashed curve represents the temperature dependence of the fundamental bulk band gap. Data from *Zhu* et al. [1999]

$$\Phi_{Bn}^{hom}(T) = \Phi_{Bn}^{hom}(0\mathrm{K}) - \frac{\vartheta T^2}{\Theta + T} \tag{8.2}$$

lation to the experimental data, where the empirical parameters $\vartheta_{\mathrm{Si}} = 4.73\times10^{-4}$ eVK^{-1} and $\Theta_{\mathrm{Si}} = 636$ K and of the fundamental, indirect band gap of silicon were used [*Thurmond* 1975]. The fit to the PtSi/*n*-Si(001) data yields a zero-temperature barrier height of 0.923 eV.

The barrier heights of *p*-type Schottky contacts, on the other hand, do not vary as a function of temperature. Obviously, the temperature coefficients of both the branch-point energy $\Phi_{bp}^{p} = W_{bp} - W_{v}(\Gamma)$ and the slope parameter S_X are very small and may be neglected. The slope parameter, for example, is determined by the optical dielectric constant ε_∞; see relations (5.63). The typical temperature coefficient $\mathrm{d}\varepsilon_\infty/\mathrm{d}T \approx 6\times10\text{-}4\ \mathrm{K}^{-1}$ of the optical dielectric constant then yields $\mathrm{d}S_X/\mathrm{d}T \approx -2\times10^{-4}$ eV(Miedema-unit)$^{-1}$K^{-1}, i.e., the temperature variation of the slope parameter may be safely neglected for practical purposes. For one and the same metal–semiconductor contact the *p*- and *n*-type barrier heights add up to the width of the band gap so that the temperature dependence of the *n*-type barrier height may be written as

$$\Phi_{Bn}^{hom}(T) = W_g(T) - \Phi_{bp}^{p} - S_X(X_\mathrm{m} - X_\mathrm{s}) \ . \tag{8.3}$$

The zero-temperature barrier height of *n*-type Schottky contacts then results as

$$\Phi_{Bn}^{hom}(0\,\mathrm{K}) = \Phi_{bp}^{n}(0\,\mathrm{K}) + S_X(X_\mathrm{m} - X_\mathrm{s}) = W_{g0} - \Phi_{bp}^{p} + S_X(X_\mathrm{m} - X_\mathrm{s}) \ . \tag{8.4}$$

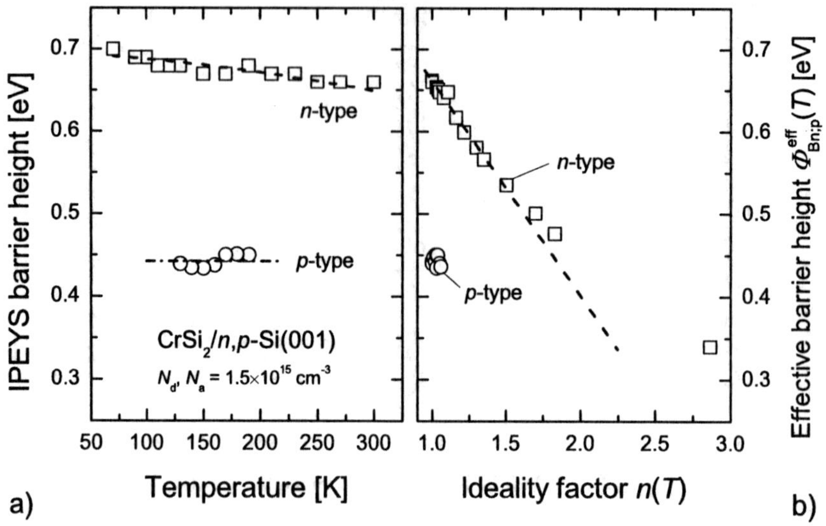

Fig. 8.3. Temperature dependence of IPEYS barrier heights (a) and effective barrier heights versus ideality factors (b) of $CrSi_2$ contacts on one Si(001) substrate doped n- and p-type each. The dashed lines are explained in the text. Data from *Aniltürk* and *Turan* [1999]

With the zero-temperature band gap $W_{g0} = 1.17$ eV, the branch-point energy $\Phi^p_{bp} = 0.36$ eV and the slope parameter $S_X = 0.101$ eV/Miedema-unit of silicon relation (8.4) gives a zero-temperature barrier height $\Phi^{hom}_{Bn}(0\,\text{K}) = 0.906$ eV for ideal PtSi/n-Si contacts. Within the margins of error, this theoretical value of the MIGS-and-electronegativity theory corroborates the experimental value, 0.923 eV, of laterally homogeneous PtSi/n-Si(001) contacts which is obtained from the BEEM data displayed in Figs. 8.1 and 8.2.

The IFIGS branch-point energies are as insensitive to the temperature variations as the ionization energies $I = W_{vac} - W_{vs}$ of semiconductors. Their temperature coefficients were found to range between -6×10^{-5} eV/K for cleaved InP(110) [*Mönch* and *Koenders* 1985] and $+3\times10^{-5}$ eV/K for cleaved Si(111) surfaces [*Bachmann* 1968]. The values of these temperature coefficients are smaller than k_B and their signs vary. This results from a delicate balance between the Debye–Waller and the hydrostatic contributions, which have different signs but are almost equal in absolute values [*Mönch* and *Koenders* 1985].

The valence-band offsets at isovalent and lattice-matched semiconductor heterostructures equal the difference $\Phi^p_{bpl} - \Phi^p_{bpr}$ of the p-type branch-point energies of the two semiconductors in contact. Since the barrier heights of p-type Schottky contacts and, therefore, the p-type branch-point energies of the semiconductors are insensitive to temperature variations, the valence-band discontinuities are expected to behave in the same way. *Von Truchsess* et al. [1996], *Lindell* et al. [1997], and *Zubkov* et al. [1999] indeed found the valence-band offsets of isova-

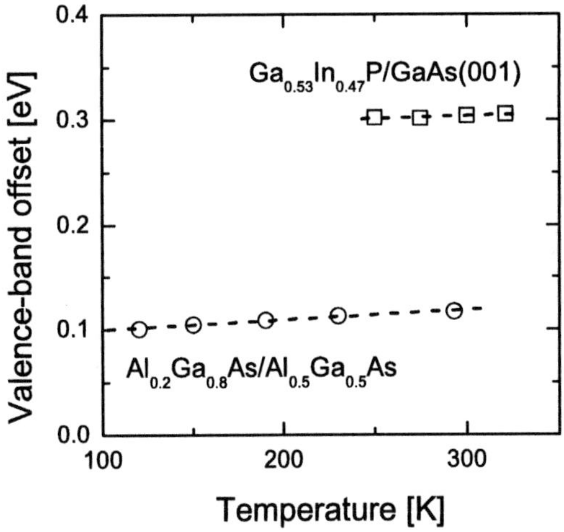

Fig. 8.4. Temperature dependence of valence-band offsets at isovalent and lattice-matched $Ga_{0.53}In_{0.47}P/GaAs(001)$ and $Al_{0.2}Ga_{0.8}As/Al_{0.5}Ga_{0.5}As$ heterostructures. Data from *Lindell* et al. [1997] and *Zubkov* et al. [1999], respectively

lent and lattice-matched $HgTe/Hg_{0.32}Cd_{0.68}Te$, $Ga_{0.53}In_{0.47}P/GaAs$, and $Al_{0.2}Ga_{0.8}As/Al_{0.5}Ga_{0.5}As$ heterostructures, respectively, to be almost independent of temperature. Figure 8.4 displays the experimental data obtained with the $Al_{0.2}Ga_{0.8}As/Al_{0.5}Ga_{0.5}As$ and the $Ga_{0.53}In_{0.47}P/GaAs$ heterostructures. The virtual-crystal approximation (6.21) and the branch-point energies (5.59) give valence-band offsets of 0.12 and 0.32 eV for $Al_{0.2}Ga_{0.8}As/Al_{0.5}Ga_{0.5}As$ and $Ga_{0.53}In_{0.47}P/GaAs$ heterostructures, respectively. These theoretical values agree well with the discontinuities of 0.117 and 0.304 eV, respectively, found experimentally at room temperature. The experimentally observed temperature coefficients amount to -5×10^{-5} and -9×10^{-5} eV/K, respectively, and are again close to k_B.

8.2 Pressure Effects

8.2.1 Schottky Barrier Heights

The response of the Schottky barrier heights to externally applied hydrostatic pressure is much more varied than their temperature behavior. Figure 8.5 displays the variations of silicon, GaAs, and $Al_xGa_{1-x}As$ Schottky diodes as a function of external pressure applied at room temperature [*Shan* et al. 1988, *Werner* 1989, *Langer* et al. 1995, *Phatak* et al. 1995]. Again, the barrier heights of the contacts prepared on *p*-type semiconductors remain almost unchanged as a function of applied pressure. The behavior is more complicated when substrates doped *n*-type

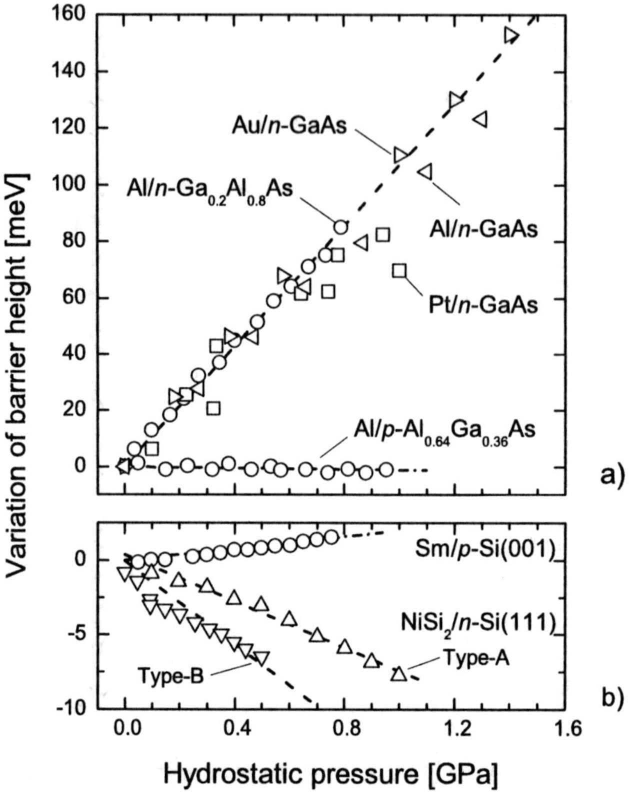

Fig. 8.5. Variation of barrier heights of Al-, Au-, and Pt/n-GaAs, Al/n-$Ga_{0.2}Al_{0.8}As$, Al/p-$Al_{0.64}Ga_{0.36}As$, type-A and type-B $NiSi_2$/n-Si(111), and Sm/p-Si(001) contacts as a function of the externally applied pressure. The dashed lines give the pressure changes of the respective bulk band gaps. □ data from *Shan* et al. [1988], Δ and ∇ data from *Werner* [1989], O data from *Langer* et al. [1995], ▷ and ◁ data from *Phatak* et al. [1995]

are considered. The pressure coefficients of the barrier heights have different signs when n-type Si and $Al_xGa_{1-x}As$ substrates are used. They are negative for n-Si but positive for n-$Al_xGa_{1-x}As$ Schottky diodes.

The pressure coefficients of the fundamental gaps were found to be typically +100 meV/GPa for direct transitions in the middle of the Brillouin zone but as −10 meV/GPa for indirect and then phonon-assisted transitions from the valence-band maximum at Γ into conduction-band states near to the X point at the Brillouin zone boundary. The different magnitude and the different signs of these bulk pressure coefficients immediately explain the increase and the reduction of the barrier heights at n-$Al_{1-x}Ga_xAs$ and n-Si Schottky contacts, respectively, as a function of externally applied pressure. The dashed line of Fig. 8.5b represents the pressure variation of the band gap of GaAs. The pressure dependence observed for the barrier heights in n-$Al_{1-x}Ga_xAs$ Schottky diodes is excellently described by the GaAs band-gap data.

The experimental results plotted in Fig. 8.5a give pressure coefficients of approximately –8 meV/GPa and –9 meV/GPa for the barrier heights at type-A and type-B $NiSi_2$/n-Si(111) interfaces, respectively. They have the same sign but are slightly smaller than the respective coefficient, –14 meV/GPa, of the indirect bulk band gap in silicon. For other n-Si Schottky contacts investigated, the pressure coefficients of their barrier heights were found to vary between –2.2 and –1.1 meV/GPa [*Werner* and *Güttler* 1993].

Externally applied hydrostatic pressure causes almost no changes of p-type Schottky barrier heights. In this respect p-type contacts behave the same as if the temperature is varied. Again, the pressure coefficients of both the branch-point energy $\Phi^p_{bp} = W_{bp} - W_v(\Gamma)$ and the slope parameter S_X are very small and may be neglected. The slope parameter, for example, is determined by the optical dielectric constant ε_∞; see relation (5.63). The pressure coefficient of the optical dielectric constant typically amounts to $d\varepsilon_\infty/dP \approx 9\times10^{-2}\,GPa^{-1}$, which value yields $dS_X/dT \approx -2\times10^{-2}\,eV(Miedema\text{-}unit)^{-1}GPa^{-1}$, i.e., the variation of the slope parameter caused by hydrostatic pressure may be neglected for practical purposes. This means that the n-type Schottky barrier height (8.3) may be rewritten as

$$\Phi^{\text{hom}}_{Bn}(T,P) = W_g(T,P) - \Phi^p_{bp} + S_X(X_\text{m} - X_\text{s}) \tag{8.5}$$

in excellent agreement with the experimental findings.

The experimental data plotted in Fig. 8.5b give pressure coefficients of the barrier heights of – 8 meV/GPa for type-A and of – 9 meV/GPa for type-B $NiSi_2$/n-Si(111) diodes. In comparison with type-B interfaces, type-A contacts have barrier heights which are lower by 0.13 eV. This difference is attributed to extrinsic, structure-induced interface dipoles at type-A contacts, see Sect. 10.2.2. The respective pressure coefficient of the type-A barrier heights thus contains two contributions. One contribution, $d\Phi_{Bn}/dP$, originates from the intrinsic MIG states and is also present at type-B interfaces, while the other, $d(\delta\Phi_{Bn})/dP$, is assigned to the extrinsic interface dipoles. This extrinsic contribution will have a positive sign since the dipole length will be reduced as pressure is applied externally, and its magnitude is estimated as

$$d(\delta\Phi_{Bn})/dP = +\delta\Phi_{Bn}\cdot\kappa_i/3 \quad . \tag{8.6}$$

Taking the experimental difference $\delta\Phi_{Bn} = 130$ meV of the barrier heights and equating the interface compressibility κ_i with the inverse of the bulk modulus $B_b = 97.9$ GPa of silicon, one obtains the extrinsic contribution to the pressure coefficient of the barrier heights of type-A $NiSi_2$/n-Si(111) contacts as

$$d(\delta\Phi_{Bn})/dP = +0.5 \text{ meV/GPa} \quad .$$

First of all, the sign of the difference of the pressure coefficients obtained from the experimental data plotted in Fig. 8.5b agrees with what relation (8.6) predicts. Secondly, the estimate of the extrinsic contribution to the barrier-height pressure coefficients is surprisingly close to the difference of 1 meV/GPa which is evaluated from the data plotted in Fig. 8.5b. This finding strongly supports the concept developed in Sect. 10.2.2 that structure-induced interface dipoles are responsible for the lowering of the barrier heights at type-A interfaces relative to type-B $NiSi_2$/*n*-Si(111) contacts.

For $ErSi_2$/Si(111) contacts quite the opposite temperature behavior was reported [*Duboz* et al. 1989a, 1989b]. As a function of temperature, the effective barrier heights remain unchanged on *n*-type substrates but vary as the width of the bulk band gap on *p*-type samples. The effective room-temperature barrier height of $ErSi_2$/*n*-Si(111) contacts was evaluated as 0.25 eV and is much smaller than the value predicted by the MIGS-and-electronegativity model. This lowering of the barrier height is attributed to a very high density of interface defects of donor type. The temperature effects might then be caused by a pinning of the Fermi level at the energy position of these interface donors.

8.2.2 Valence-Band Offsets

The effect of hydrostatic pressure on the barrier heights of Schottky contacts depends on the type of doping of the semiconductors. While for *n*-type doping the barrier heights vary like the bulk band gaps with increasing pressure, there is no variation or a very small variation when the semiconductor substrates are doped *p*-type. This experimental finding indicates that the energy positions $W_{bp} - W_v(T)$ of the branch points with regard to the valence-band maximum are almost insensitive to external pressure. Therefore, it might be expected that the valence-band offsets of semiconductor heterostructures will show, if at all, slight changes under hydrostatic pressure only.

Studies on the hydrostatic-pressure dependence of valence-band offsets should be carried out with isovalent and lattice-matched heterostructures since they are initially unstrained. An appropriate example is the $In_{0.52}Al_{0.48}As$/InP structure. Their band lineup is such that the conduction- and the valence-band edges of the $In_{0.52}Al_{0.48}As$ are both higher in energy than those of the InP. *Yeh* et al. [1995] observed three photoluminescence lines with such heterostructures. They attributed them to transitions of photo-injected carriers in the bulk of the semiconductors and from the $In_{0.52}Al_{0.48}As$ conduction to the InP valence band right at the interface. The peak photon energy of the latter photoluminescence signal showed the same pressure dependence as the fundamental band gap of InP. Consequently, the variation of the $In_{0.52}Al_{0.48}As$/InP valence-band offset is, within the margins of experimental error, insensitive to the application of hydrostatic pressure. Figure 8.6 displays the corresponding experimental data.

In the virtual crystal approximation (6.21), the IFIGS concept predicts a valence-band offset of

$$\begin{aligned}\Delta W_v(\mathrm{In}_{0.52}\mathrm{Al}_{0.48}\mathrm{As/InP}) &= \Phi_{bp}^{p}(\mathrm{InP}) - [0.52 \cdot \Phi_{bp}^{p}(\mathrm{InAs}) + 0.48 \cdot \Phi_{bp}^{p}(\mathrm{AlAs})] \\ &= 0.158\ \mathrm{eV}\end{aligned}$$

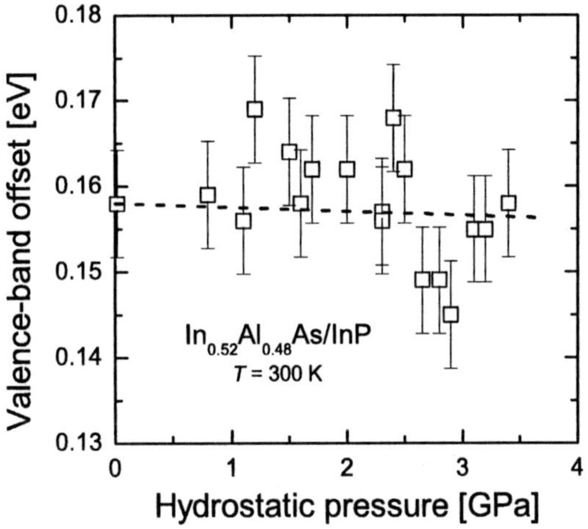

Fig. 8.6. Valence-band offset of a lattice-matched $In_{0.52}Al_{0.48}As/InP$ heterostructure as a function of hydrostatic pressure at room temperature. The dashed line is a linear least-squares fit to the data. Data from *Yeh* at al. [1995]

for $In_{0.52}Al_{0.48}As/InP$ heterostructures. The experimental value at zero pressure agrees with this theoretical result.

8.3 Pseudomorphic Heterostructures

Pseudomorphic growth occurs for slightly different lattice parameters of the semiconductors in contact. Such overlayers are tetragonally distorted and their lattice parameters $a_{l\|}$ and $a_{l\perp}$ parallel and normal to the interface, respectively, are related as [*Ke* et al. 1994]

$$a_{l\perp} = a_0(\text{sub}) - 2[a_{l\|} - a_0(\text{sub})]\frac{C_{12}}{C_{11}} \, , \tag{8.7}$$

where a_0(sub) is the lattice parameter of the substrate and C_{11} and C_{12} are the elastic moduli of the pseudomorphic film. With increasing difference of the lattice parameters the overlayers become more and more strained. Above a certain limit misfit dislocations accommodate the lattice mismatch. Such growth is said to be metamorphic.

The valence-band offsets of lattice-matched semiconductor heterostructures do not change under hydrostatic pressure. However, quite a different behavior is observed when uniaxial strain is applied. Even lattice-matched heterostructures are naturally strained. This phenomenon is due to the generally different covalent radii of the constituent atoms of the two semiconductors in contact. It is illustrated schematically in Fig. 8.7. As an example, the case of a lattice-matched $InP/Ga_{0.47}In_{0.53}As/InP(001)$ sandwich structure will be briefly considered. From the

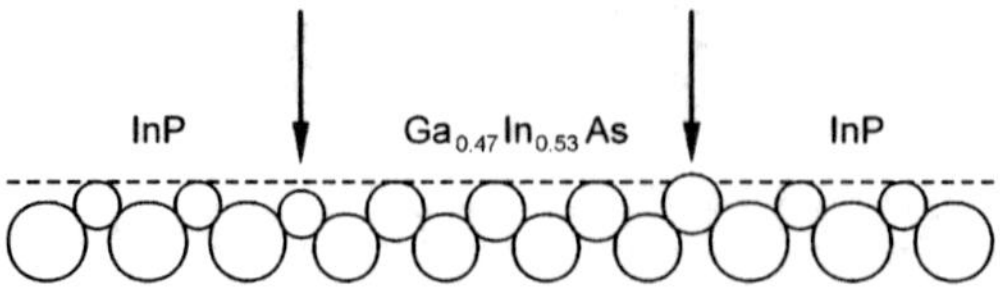

Fig. 8.7. Schematic illustration of the lattice distortions in the (001) direction of a ... $InP/Ga_{0.47}In_{0.53}As/InP$ ... sequence as an example of two lattice-matched zincblende-structure compounds

data of Table A.3 in the Appendix one obtains $r_{cov}(\text{In}) > \hat{r}_{cov}(\text{Ga}_{0.47}\text{In}_{0.53})$ and $r_{cov}(P) < r_{cov}(\text{As})$, where $\hat{r}_{cov}(\text{Ga}_{0.47}\text{In}_{0.53}) = 0.47 \cdot r_{cov}(\text{Ga}) + 0.53 \cdot r_{cov}(\text{In})$ is the covalent radius of virtual $Ga_{0.47}In_{0.53}$ cations. Hence, ... In–P–$Ga_{0.47}In_{0.53}$–As ... and ... $Ga_{0.47}In_{0.53}$–As–In–P ... interfaces are under compressive and tensile stress, respectively. The interfaces will respond by tetragonal distortions. However, in most lattice-matched systems this effect will be small and may be neglected. *Vandenberg* et al. [1988, 1990] performed high-resolution X-ray diffraction of lattice-matched $InP/Ga_{0.47}In_{0.53}As/InP$ superlattices. Their results clearly demonstrated the presence of strained regions on the scale of about one interfacial atomic layer at each interface.

Ohler et al. [1995] systematically investigated the influence of uniaxial lattice strain on the valence-band discontinuities of InAs/AlAs, InAs/GaAs, and AlAs/GaAs heterostructures. They strain varied within these heterostructures by pseudomorphic growth on thick and then fully relaxed $Ga_{1-x}In_xAs$ buffer layers of different compositions. These layers were used as *virtual* substrates. The pseudomorphic layers are tetragonally distorted. Their lattice parameters $a_{/\|}$ and $a_{/\perp}$ parallel and normal to the interface, respectively, are related as indicated by relation (8.7). The valence-band offsets were determined by using XPS as described in Sect. 3.12.2. Figure 8.8 displays the experimental valence-band offsets of strained InAs/AlAs, InAs/GaAs, and AlAs/GaAs heterostructures as a function of the lattice constant of the virtual $Ga_{1-x}In_xAs$ substrates. The discontinuities become smaller with increasing strain for InAs/AlAs and InAs/GaAs but remain constant for AlAs/GaAs heterostructures.

Ke et al. [1994] calculated the influence of uniaxial strain on the branch point of the virtual gap states of the complex band structure self-consistently for the group-III phosphides and arsenides. They applied the so-called average-bond-energy model developed by *Wang* et al. [1992]. They used a linearized muffin-tin-orbital method in the atomic spheres approximation and calculated the densities of states $D_v^{mv}(W)$ of the valence- and $D_c^{mv}(W)$ of the conduction bands at the mean-value point $\underline{k}_{mv}$ of the Brillouin zone. They then defined the bonding and the anti-bonding energies

$$W_b(\underline{k}_{mv}) = \frac{1}{4}\int W \cdot D_v^{mv}(W)\mathrm{d}W \qquad (8.8)$$

and

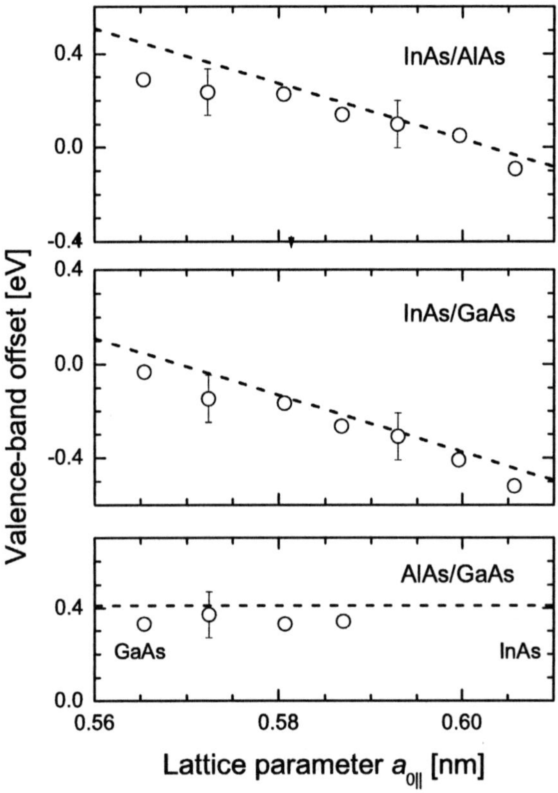

Fig. 8.8. Valence-band offsets of strained InAs/AlAs, InAs/GaAs, and AlAs/GaAs heterostructures as a function of the lattice constant parallel to the interface. Experimental data from *Ohler* et al. [1995] and theoretical results (dashed lines) from *Ke* et al. [1994] (Fig. 8.9)

$$W_{ab}(\underline{k}_{mv}) = \frac{1}{4}\int W \cdot D_c^{mv}(W)\mathrm{d}W \quad , \tag{8.9}$$

respectively. The factor 1/4 in front of the integrals accounts for the number of bands considered. The arithmetic mean

$$W_m = \frac{W_b(\underline{k}_{mv}) + W_{ab}(\underline{k}_{mv})}{2} \tag{8.10}$$

of these bonding and antibonding energies was named the average-bond energy. This model takes the average-bond energies (8.10) as reference levels for the band lineup at the heterostructures or, in other words, it replaces the branch-point energies Φ_{bp}^{p} of the IFIGS theory by the average-bond energies and completely neglects the existence of the intrinsic interface-induced gap states.

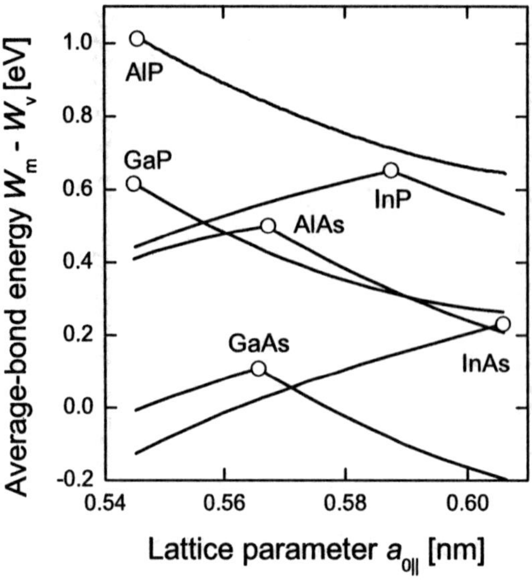

Fig. 8.9. Average-bond energies (7.10) of tetragonally strained group-III phosphides and arsenides with respect to the valence-band maximum as a function of the lattice parameter $a_{0\|}$. Circles indicate data of unstrained compounds. After *Ke* et al. [1994]

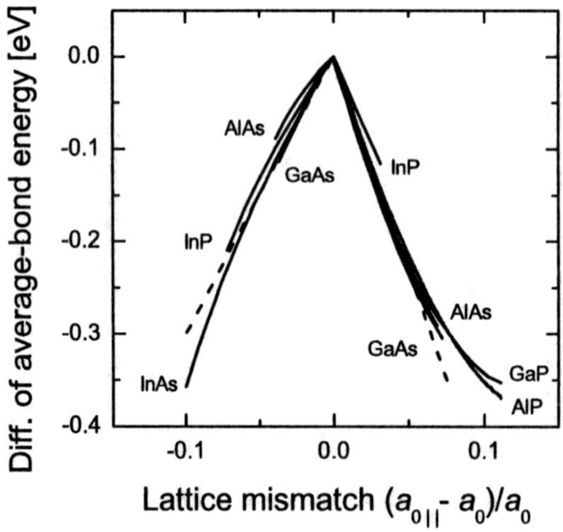

Fig. 8.10. Difference $W_m(a_{0\|}) - W_m(a_0)$ of the average-bond energies of the tetragonally strained group-III phosphides and arsenides with respect to the valence-band maximum as a function of the lattice mismatch $(a_{0\|} - a_0)/a_0$. The dashed lines indicate mean values. Same data as in Fig. 8.9

Figure 8.9 displays the average-bond energies calculated by *Ke* et al. [1994] for strained group-III phosphides and arsenides as a function of the lattice parameter $a_{0\|}$ parallel to the interface. The circles mark the data of the unstrained semiconductors. The average-bond energy or, what is the same, the branch-point energy decreases as a function of uniaxial lattice strain. Most remarkably, the theoretical data indicate a universal behavior, at least for the group-III phosphides and arsenides. Figure 8.10 displays the difference $W_m(a_{l\|}) - W_m(a_0)$ of the average-bond energies of the tetragonally strained and the unstrained semiconductors as a function of the lattice mismatch $(a_{l\|} - a_0)/a_0$. The dashed straight lines are interpolations and have slope parameters of +3 eV and –4 eV for compressive and tensile strain, respectively.

The valence-band offsets of pseudomorphic and then strained heterostructures of group-III phosphides and arsenides may be directly read from Fig. 8.9. The dashed lines in Fig. 8.8 give the corresponding data for the respective heterostructures. Obviously, the experimental data are excellently described by the theoretical results obtained from the average-bond model.

9. Barrier Heights and Extrinsic Interface Defects

9.1 Defect-Induced Changes of Barrier Heights

Adatoms at semiconductor surfaces may induce surface states within the band gap of the semiconductor. These energy levels are discrete as long as the adatoms are sparsely distributed so that they are non-interacting. This behavior is quite different from what occurs on metal surfaces. The interaction of the adatoms with the continuum of conduction-band states of the metal, which reaches up to the Fermi level, broadens the sharp atomic levels into wide resonances [*Gurney* 1935]. This is schematically explained in Fig. 9.1. For the same reason, foreign atoms or, more generally speaking, defects at metal–semiconductor interfaces will also show broad interface states.

Following *Gadzuk* [1967], the local density of states of such adatom-induced resonance levels on metal surfaces is customarily assumed to be of Lorentzian shape

$$D_{id}(W)\mathrm{d}W = \frac{N_{idef}\Gamma_{m-v}}{\pi[W - W_{idef})^2 + \Gamma_{m-v}^2]}\mathrm{d}W \quad , \tag{9.1}$$

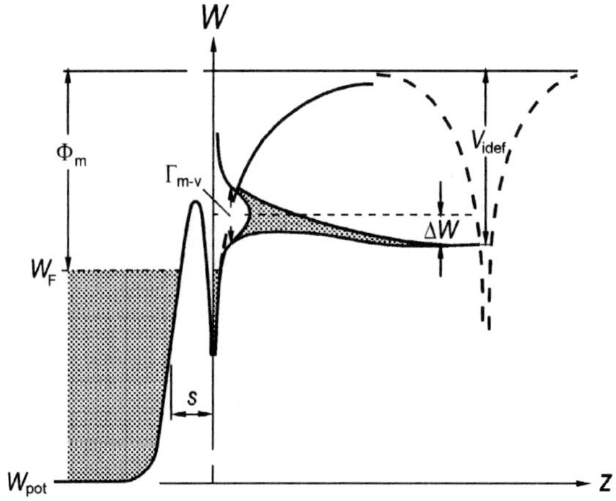

Fig. 9.1. Broadening and energy shift of adatom energy levels on metal surfaces. *After* Gurney [1935] and *Gadzuk* [1967]

where N_{idef} is the total number of surface defects per unit area. The energy level W_{idef} of the adatoms is shifted with respect to its energy in the free atom. By using Fermi's Golden Rule, one obtains the line width

$$\Gamma_{m-v} \propto |\langle a | V_{idef} | m \rangle|^2 \quad , \tag{9.2}$$

where $|a\rangle$ and $|m\rangle$ are the unperturbed states of the free adatom and the uncovered metal, respectively. The interaction matrix element is a Coulomb potential

$$V_{idef} = -e_0^2 / 4\pi\varepsilon_i\varepsilon_0 r \quad . \tag{9.3}$$

It is the dielectric function ε_i in (9.3) where metal surfaces or, more generally speaking, metal–vacuum interfaces and metal–semiconductor surfaces differ. For metal surfaces, the dielectric function is $\varepsilon_{m\text{-}v} = 1$ while for metal–semiconductor interfaces one obtains an interface dielectric constant $\varepsilon_{m\text{-}s} \approx 4$ on the average; see Sect. 5.6.

Combining (9.2) and (9.3) gives a line width

$$\Gamma_{m-s} = \Gamma_{m-v}(\varepsilon_{m-v)} / \varepsilon_{m-s})^2 \tag{9.4}$$

for defect-induced interface states within the semiconductor band-gap at metal–semiconductor interfaces [*Ludeke* et al. 1988a, b]. *Gadzuk* [1969] estimated the resonance levels of metal-adatoms on metal surfaces to be typically broadened by $\Gamma_{m\text{-}v} \approx 1.5$ eV, so that one obtains $\Gamma_{m\text{-}s} \approx 0.06$ eV on the average for metal defects at metal–semiconductor interfaces [*Ludeke* 1989]. Defects in Schottky contacts may thus be assumed to induce sharp levels.

Figure 9.2 illustrates the influence of donor-type interface defects on the barrier heights of Schottky contacts. In analogy to Sect. 5.2, the density of states of the intrinsic MIGS is again taken to be constant across the whole energy gap and extrinsic interface-donors with discrete levels are assumed to exist above the branch point of the intrinsic MIGS continuum. As long as enough negative charge is transferred into the MIGS so as to keep the Fermi level well above the defect levels, the interface donors will all be neutral. With decreasing the negative charge in the MIGS continuum, the Fermi level moves to lower energies, i.e., further away from the conduction-band minimum and approaches the interface-defect level W_{idef}. As a result, the donors gradually become charged positive and *the Fermi level gets pinned* at the energy position of the interface-defect levels. When all interface donors are eventually charged, the MIGS continuum will again take up additional charge and will again determine the position of the Fermi level in the band gap as a function of the charge on the semiconductor side of the metal–semiconductor interface. In Fig. 9.2b, the solid line runs parallel to the dashed one which would be followed at the defect-free interface. Figure 9.2c is a re-plot of Fig. 92 b and demonstrates that interface donors lower the barrier heights of *n*-type Schottky contacts with respect to the case when no interface defects are present.

Extrinsic defects at semiconductor interfaces have to be considered in the condition of interface charge neutrality. They will be present in addition to the intrin-

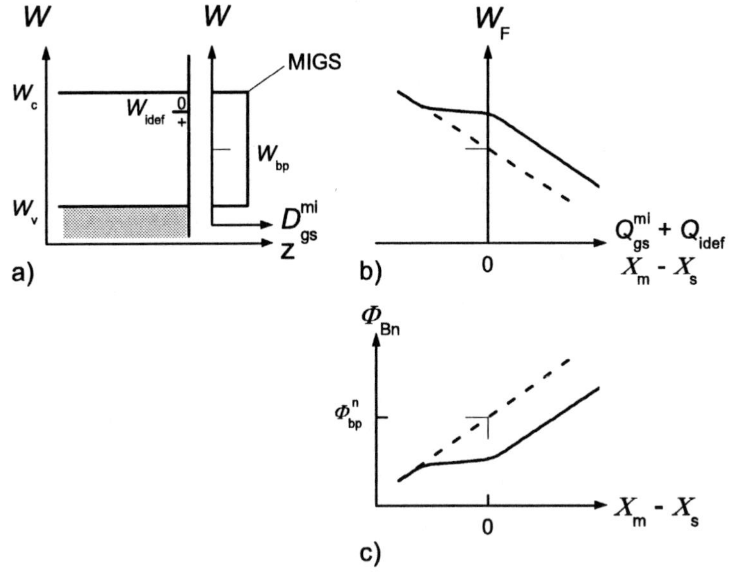

Fig. 9.2. Band diagram (a), charge transfer (b), and barrier height (c) at metal–semiconductor contacts containing a continuum of intrinsic MIGS states with constant density of states across the band gap and extrinsic interface donors with energy levels above the branch-point energy of the MIGS (schematic)

sic metal-induced gap states and the space charge so that condition (5.9) has to be replaced by

$$Q_{\mathrm{m}} + Q_{\mathrm{s}} = Q_{\mathrm{m}} + (Q_{gs}^{mi} + Q_{idef} + Q_{sc}) = 0 \quad . \tag{9.5}$$

Here, Q_{idef} means the net charge per unit area in the interface defects. The charge density Q_{m} on the metal side is now balanced by the net charge density $Q_{gs}^{mi} + Q_{idef} + Q_{sc}$ in the MIGS continuum, the interface defects, and the space-charge layer on the semiconductor side.

Interface defects of donor character and with discrete energy levels contribute a net charge per unit area

$$Q_{idef} = +e_0 N_{idef}[1 - f_o(W_{idef} - W_F)] \quad , \tag{9.6}$$

where $f_0(W_{idef} - W_F)$ is the Fermi–Dirac function. Combining relations (2.16), (5.7), (5.11), and (9.6) with the condition of interface charge neutrality (9.5) yields the barrier height

$$\Phi_{Bn} = \Phi_{bp}^{n} + S_X(X_{\mathrm{m}} - X_{\mathrm{s}}) - (1 - S_X / \mathrm{A}_X)(1 - f_0) N_{idef} / D_{gs}^{mi} \quad . \tag{9.7}$$

Comparison of (9.7) with (5.22), which is obtained with no interface defects present, shows that positively charged interface defects of donor character lower the barrier height of metal contacts on n-type semiconductors. With no interface defects present at all, relation (5.22) is retrieved.

A similar analysis may be carried out for interface acceptors. With respect to what is expected for defect-free interfaces, interface acceptors increase the barrier heights.

For a given density of defects, the maximum defect-induced variation of the barrier height is achieved when all defects are charged. For donor-type interface defects and *n*-type semiconductors relation (9.7) yields

$$\delta\Phi_{Bn}^{\max} = (1 - S_X / A_X) N_{idef} / D_{gs}^{mi} \quad . \tag{9.8}$$

The density of states of the intrinsic MIGS amounts to some 10^{14} states per cm^2 and eV; see Table 5.1. Therefore, small densities of interface states will not significantly alter the barrier heights with respect to defect-free Schottky contacts.

Zur et al. [1983], *Duke* and *Mailhiot* [1985] and *Zhang* et al. [1985] modeled such interface doping in somewhat more refined theoretical approaches. In agreement with the much simpler approach presented above, they found that approximately 10^{14} defects per cm^2 to be sufficient to *pin* the Fermi-level at the energy position of the defect levels and that interface donors and acceptors will reduce and increase, respectively, the barrier heights of *n*-type, and vice versa for *p*-type Schottky contacts.

9.2 Application to *p*-GaP(110) Schottky Contacts

The barrier heights of intimate, abrupt, clean, and laterally homogeneous Schottky contacts on *p*-GaP(110) surfaces are, on the average, approximately 130 meV larger compared to what is predicted by the MIGS-and-electronegativity theory;, see Fig. 6.4. This observation may be attributed to extrinsic interface defects which are positively charged. Relation (9.8) then yields a corresponding density of approximately 4.5×10^{13} extrinsic interface defects per cm^2.

Defects on semiconductor surfaces were studied by using scanning tunneling microscopy.[1] In their STM investigations of cleaved GaP- and InP(110) surfaces, *Ebert* and *Urban* [1993] and *Ebert* et al. [1994] observed positively charged P vacancies on *p*-type doped surfaces. These surface vacancies originate from low-temperature Langmuir evaporation of P atoms after cleavage even at room temperature. Their density grows as the substrate temperature is increased. Correlated with this, *Ebert* et al. [2000] observed the surface band-bending to increase. For this purpose they studied the energy shift of the In(4*d*) core levels by using angle-resolved photoemission spectroscopy (PES) with 50-eV photons. The density of surface vacancies determined from the STM images and the surface band-bending obtained from the PES measurements give the 0/+ charge-transition level of P surface-vacancies on *p*-InP(110) surfaces at 0.75 ± 0.1 eV above the valence-band maximum.

---

[1] For a recent review on defects in III–V semiconductor surfaces see, for example, a review article by *Ebert* [2002a].

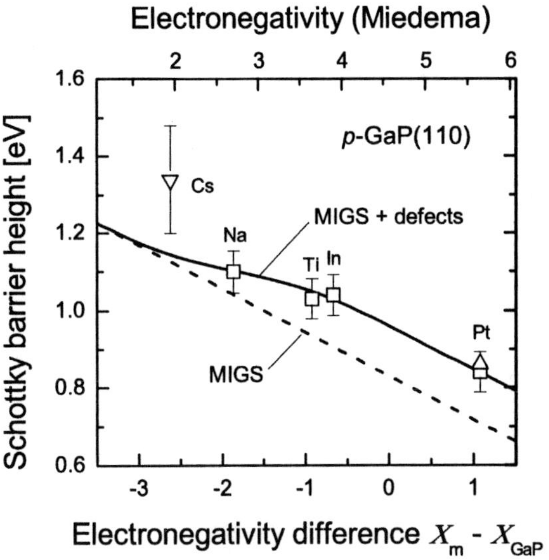

Fig. 9.3. Barrier heights of laterally homogeneous *p*-GaP(110) Schottky contacts versus the difference of the *Miedema* electronegativities of the metal and gallium phosphide. The experimental data and the prediction of the MIGS-and-electronegativity theory (dashed line) are the same as in Fig. 6.4. The solid line is a plot of relation (7.7) assuming donor-type defects at 1.1 eV above the valence-band maximum and a maximum increase of the barrier heights by 130 meV

Unfortunately, no such detailed experimental data are available for P surface-vacancies on *p*-GaP(110) surfaces. However, based on DFT calculations by *Schwarz* [2002] the energy position of the charge-transition level of P surface vacancies on *p*-GaP(110) surfaces above the valence-band maximum was estimated between 0.75 and 0.95 eV [*Ebert* 2002b].

The observation of positively charged surface vacancies on *p*-GaP(110) surfaces suggests that the larger barrier heights of *p*-GaP(110) Schottky contacts with regard to those predicted by the MIGS-and-electronegativity theory can be explained by interfacial P vacancies. The defect density estimated above from the observed increase of the barrier heights would correspond to approximately 8% of the P sites in a GaP(110) layer. The interfacial P vacancies would form during the preparation of the contacts, i.e., they were fabrication-induced. The full line in Fig. 9.3 shows the dependence (9.7) of the barrier height of *p*-GaP Schottky contacts with interfacial defects of donor-type as a function of the electronegativity difference $X_{\mathrm{m}} - X_{\mathrm{GaP}}$. Their 0/+ charge-transition level of the interface vacancies was assumed to be 1.1 eV above the valence-band maximum. An interface-induced shift of 0.2 eV with regard to the energy position at the free surface appears to be plausible; see Fig. 9.1. The calculated line excellently describes the experimental data with the exception of the Cs data point, which anyway has a very large margin of experimental error.

9.3 Unified Defect Model

Spicer's *Unified defect model* assumed a pinning of the Fermi level at metal–semiconductor interfaces by metal-induced but native interface defects of the semiconductor [*Spicer* et al. 1979]. As a consequence of this, the barrier heights of Schottky contacts on one and the same semiconductor should be equal irrespective of the metal used. However, this assumption is not justified by the experimental data of well-prepared and laterally uniform Schottky contacts. As Figs. 6.1 through 6.8 reveal, the experimental barrier heights vary as a function of the difference of the metal and the semiconductor electronegativities. The chemical trends are well explained by the metal-induced gap states rather than by native interface defects of the semiconductors.

10. Extrinsic Interface Dipoles

10.1 Interface Doping of Schottky Contacts

The deliberate manipulation of barrier heights at metal–semiconductor contacts by doping of the interface with foreign atoms has turned out to be a difficult task. Predeposited atoms were found to form compounds with subsequently evaporated metal atoms, to segregate at the surface of the growing metal film, or to desorb during metal evaporation. Hydrogen preadsorbed on diamond surfaces, on the other hand, happens to be stable against subsequent deposition of most metals.

Waldrop [1985a, b] systematically studied Schottky diodes prepared on GaAs(001) surfaces precovered with chalcogen atoms. Interlayers of S, Se, and Te increase the barrier heights of Ag-, Au-, and Pd/*n*-GaAs(001) diodes while the metals Ti, Mn, and Al show quite the opposite behavior. The two groups of metals differ strongly in their reactivities with chalcogens. For titanium, manganese, and, above all, aluminum chalcogenides, the heats of formation are larger by a factor of up to ten than for respective silver, gold, and palladium compounds.

The influence of monolayers of antimony and bismuth deposited on cleaved GaAs(110) surfaces prior to the evaporation of Ag and Mn films was studied by *Zahn* et al. [1992]. Such monolayers are well-ordered and ideally terminate all dangling bonds.[1] The barrier heights turned out to be the same irrespective of whether the GaAs surfaces were initially clean or covered with a monolayer of Sb or Bi. Photoelectron emission from Sb(4*d*) and Bi(5*d*) core levels revealed that both group-V elements segregate at the surface of the evaporated metal films. *Troost* et al. [1994] obtained similar results with halogen-covered GaAs(110) surfaces. Again, the halogens either strongly react with silver evaporated at room temperature or segregate on top of the growing Ag film.

Silicon surfaces precovered with hydrogen were found to behave quite interestingly during metal evaporation. Hydrogen-terminated silicon surfaces are easily obtained by, for example, *HF dips* of oxidized silicon samples [*Ubara* et al. 1984, *Yablonowitch* et al. 1986, *Grundner* and *Jacob* 1986]. This process consists of growing thick SiO_2 layers by either conventional thermal or chemical oxidation and their stripping in low-concentration (5%), high-purity aqueous HF solutions. The lowest amounts of residual carbon and oxygen are obtained if the stripping

[1] See *Mönch* [2001], Sects. 16.3 and 16.4.

solution is continuously diluted first by NH_4F-buffered HF and eventually by deionized water. After such wet chemical etching, the silicon surfaces are found to be hydrophobic and low-energy electron diffraction (LEED) detects 1×1 patterns with both {111}- and {100}-oriented Si samples. Such HF dips of silicon leave all dangling silicon bonds saturated with hydrogen. Both Auger electron (AES) and X-ray photoemission spectroscopy revealed only very minor traces of fluorine, if at all, to be present. High-resolution low-energy electron energy-loss (HREELS) and infrared absorption spectra, on the other hand, show lines which are characteristic of ≡Si–H and $= \mathrm{Si} <^{\mathrm{H}}_{\mathrm{H}}$ configurations on {111}- and {100}-oriented silicon samples, respectively. Hence, Si(111):H-1×1 and Si(100):H-1×1 surfaces are obviously terminated by bulk-like Si planes and all dangling bonds are saturated by hydrogen atoms.

The HF dips may be optimized by using basic solutions of pH ≈ 7.8 to 9 [*Higashi* et al. 1990, *Dumas* et al. 1990, *Jakob* et al. 1993]. However, prolonged etching in the NH_4F-buffered HF solutions roughens silicon surfaces. With {111}-oriented silicon samples, a minimum rms roughness of 0.14 ± 0.03 nm is obtained after an etching duration of 90 s only [*Hsiao* et al. 1993].

Under Pb films evaporated onto H-terminated Si(111):H-1×1 surfaces at room temperature the hydrogen persists but evaporated silver atoms remove the preadsorbed H atoms. *Tsunoda* et al. [1994] and *Fukutani* et al. [1997], respectively, studied this different behavior of the preadsorbed hydrogen by the resonant nuclear reaction $^{1}H(^{15}N,\alpha\gamma)^{12}C$. Naturally, the electronic properties of these contacts differ.

Kampen et al. [1995] evaporated silver onto Si(111):H-1×1 surfaces at room temperature or onto clean Si(111)-7×7 surfaces followed by an annealing treatment to produce $(1\times1)^{i}$-unreconstructed interfaces and found the Ag/*n*-Si(111) diodes to be rectifying. The barrier heights determined from their *I*/*V* curves were the same irrespective of the preparatory details. Lead contacts produced on H-terminated *n*-Si(111):H-1×1 surfaces, on the other hand, have ohmic *I*/*V* characteristics [*Kampen* and *Mönch* 1995] in contrast to Pb/*n*-Si(111)-$(1\times1)^{i}$ contacts that are rectifying [*Schmitsdorf* and *Mönch* 1999]. Figure. 4.10 shows Φ^{eff}_{Bn} versus n data for such diodes. Their linear extrapolation to n_{if} yields $\Phi^{nif}_{Bn} = 0.72$ eV. Pb/H/*p*-Si(111) contacts, on the other hand, are rectifying [*Kampen* and *Mönch* 1995]. The linear extrapolation of their Φ^{eff}_{Bp} versus n data to n_{if} gives Φ^{nif}_{Bp} = 0.71 eV. This value converts to a barrier height Φ^{nif}_{Bn} of approximately 0.4 eV for laterally homogeneous Pb/H/*n*-Si(111) contacts. Corresponding simulations using the *I*/*V* relationship (3.38) and a reasonable value of the series resistance R_s indeed produces ohmic behavior as observed experimentally. The interface doping with hydrogen thus lowers the barrier heights of Pb contacts on *n*-type by 0.35 eV but increases it on *p*-type Si(111) substrates by the same amount.

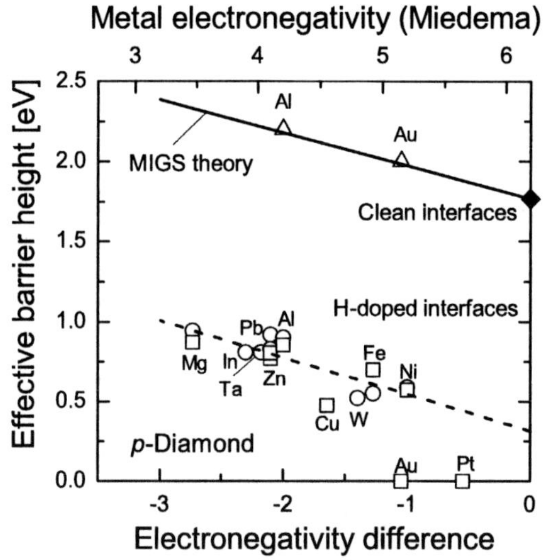

Fig. 10.1. Barrier heights of metal/*p*-diamond contacts versus the differences of the Miedema electronegativities of the metal and carbon. Δ data from *Mead* and *McGill* [1976]; O and □ data from *Kawarada* et al. [1994]. The dashed line is a linear least-squares fit to the data of the H-doped interfaces. The solid MIGS line is drawn with Φ^{p}_{bp} = 1.77 eV [*Mönch* 1996] and S_X= 0.205 eV/Miedema-unit (see text)

Hydrogen-terminated diamond surfaces behave somewhat differently. Figure 10.1 displays barrier heights determined by *Mead* and *McGill* [1976] from IPEYS measurements and by *Kawarada* et al. [1994] from *I*/*V* characteristics of metal contacts on clean and on hydrogen-covered *p*-diamond surfaces, respectively. The metal contacts produced on H-terminated surfaces have low barrier heights. The ideality factors range between 1.1 for Pb and 1.6 for Fe diodes. The gold and platinum contacts exhibit ohmic *I*/*V* characteristics. Most probably, the latter two contacts have finite but small barrier heights so that their *I*/*V* curves only appear to be ohmic. The dashed line in Fig. 10.1 is the linear least- squares fit

$$\Phi^{eff}_{Bp} = 0.314 - 0.231(X_m - X_C) \ \text{[eV]} \tag{10.1}$$

to the effective barrier heights of the Schottky contacts fabricated on the H-terminated diamond substrates. The MIGS line

$$\Phi^{MIGS}_{Bp} = 1.77 - 0.205(X_m - X_C) \ \text{[eV]} \tag{10.2}$$

excellently describes the trend of the IPEYS barrier heights that were reported for Schottky diodes on clean diamond surfaces. The slopes of the MIGS line and of the linear least-squares fit to the experimental data are almost equal. This observation suggests that the metal evaporation does not remove the H-termination of the diamond surfaces and, consequently, the hydrogen doping reduces the barrier heights of metal/H/*p*-diamond contacts by approximately 1.4 eV.

The changes of the barrier heights caused by interfacial hydrogen in Pb/H/silicon and metal/H/diamond thus have opposite sign. This different behavior is easily explained if nearest-neighbor interactions between interlayer (ia) and substrate atoms (s) only are considered. By analogy to the "surface molecule" approach for adsorbates, chemical bonds between monovalent interlayer atoms, such as hydrogen, and substrate atoms may be described as diatomic "interface molecules" [*Mönch* 1994b]. By applying *Pauling*'s concept, the ionicities of the interface molecules are modeled by the difference of the corresponding atomic electronegativities. The electronegativity of hydrogen, $X_{\mathrm{H}} = 2.2$ Pauling-units, is intermediate between the values for carbon, $X_{\mathrm{C}} = 2.55$ Pauling-units, and silicon, $X_{\mathrm{Si}} = 1.9$ Pauling-units. The corresponding interface dipoles $\mathrm{H}^{+\Delta q}$-$\mathrm{C}^{-\Delta q}$ and $\mathrm{H}^{-\Delta q}$-$\mathrm{Si}^{+\Delta q}$ are thus oppositely oriented. This finding immediately explains the experimentally observed reduction of the barrier heights by interfacial hydrogen at *n*-silicon but at *p*-diamond contacts.

Interfaces are electrically neutral. Therefore, extrinsic interface dipoles (eid) must be considered in the condition of interface charge neutrality, (1.4) and (5.9), in addition to the intrinsic MIGS dipole and the space charge. Extrinsic interface dipoles add a charge density Q_{eid} to the charge density on the semiconductor side, i.e.,

$$Q_s = Q_{gs}^{mi} + Q_{eid} + Q_{sc} \quad . \tag{10.3}$$

For one and the same metal deposited, the charge density Q_{s} will be the same at clean and at interface-doped contacts so that one obtains from relations (5.7), (5.9), and (10.3)

$$Q_{eid} = e_0 D_{gs}^{mi} \cdot (\Phi_{Bn}^{clean} - \Phi_{Bn}^{eid}) = e_0 D_{gs}^{mi} \cdot (\Phi_{Bp}^{eid} - \Phi_{Bp}^{clean}) \quad . \tag{10.4}$$

Interfacial hydrogen induces extrinsic charge densities Q_{eid} of positive and negative sign at silicon and diamond interfaces, respectively. Correspondingly, relation (10.4) predicts that, for example, hydrogen-doping will reduce the barrier heights of Schottky contacts on *n*-type silicon and on *p*-type diamond in complete agreement with the experimental findings.

By analogy to the adatom-induced surface dipoles that are discussed in Sect. 3.11, *Mönch* [1994b] described the extrinsic interface dipoles by an electric double layer. The corresponding variation of the barrier height due to extrinsic interface dipoles then equals the potential drop across the double layer, i.e.,

$$\delta\Phi_B^{eid} = \Phi_{Bn}^{clean} - \Phi_{Bn}^{eid} = \Phi_{Bp}^{eid} - \Phi_{Bp}^{clean} = \pm\frac{e_0}{\varepsilon_i \varepsilon_0} p_{eid\perp} N_{eid} \quad , \tag{10.5}$$

where ε_i and N_{eid} are the dielectric constant and the number of extrinsic dipoles per unit area at the interface, respectively. The signs of Q_{eid} and $\delta\Phi_B^{eid}$ are identical. The simple point-charge model of the electronegativity concept approximates the component of the dipole moment normal to the interface by

$$p_{eid\perp} = e_0[0.16\,|X_{\mathrm{ia}} - X_{\mathrm{s}}| + 0.035(X_{\mathrm{ia}} - X_{\mathrm{s}})^2] \cdot (r_{\mathrm{cov}}^{\mathrm{ia}} + r_{\mathrm{cov}}^{\mathrm{s}}) \quad . \tag{10.6}$$

Mutual interactions between the interface dipoles that will lower their dipole moments are neglected. This effect will become important only for atoms having large covalent radii and, therefore, also large polarizibilities. Thus, it may be safely neglected for interfacial hydrogen.

For H-modified diamond and silicon Schottky contacts relations (10.5) and (10.6) give [*Mönch* 1994b]

$$\delta\Phi_B^{\text{H-C}} = -1.86/\varepsilon_i \text{ eV} \quad \text{and} \quad \delta\Phi_B^{\text{H-Si}} = +1.00/\varepsilon_i \text{ eV} ,$$

respectively. The interface dielectric constants were estimated as $\varepsilon_i(\text{C}) \approx 2$ for diamond and $\varepsilon_i(\text{Si}) \approx 4.6$ for silicon; see Sect. 5.6. One then obtains $\delta\Phi_B^{\text{H-C}} \approx -0.9$ eV and $\delta\Phi_B^{\text{H-Si}} \approx +0.2$ eV, respectively. These estimates have the correct sign but are slightly smaller than the values experimentally observed, -1.4 and $+0.35$ eV, respectively. The condition of charge neutrality at interface-modified metal–semiconductor contacts and the simple point-charge model of the electronegativity concept thus provide not only the correct sign of H-induced variations of the barrier heights but also provide quantitative predictions that are close to the experimentally observed values.

Cantile et al. [1994] studied the influence of Si interlayers on the barrier heights of Al/*n*-GaAs(001) contacts. They followed the variations of the energy position $W_{ci} - W_F$ of the Fermi level at the interface by X-ray photoemission spectroscopy of the Ga(3*d*) and the As(3*d*) core levels. Figure 10.2. displays their data.

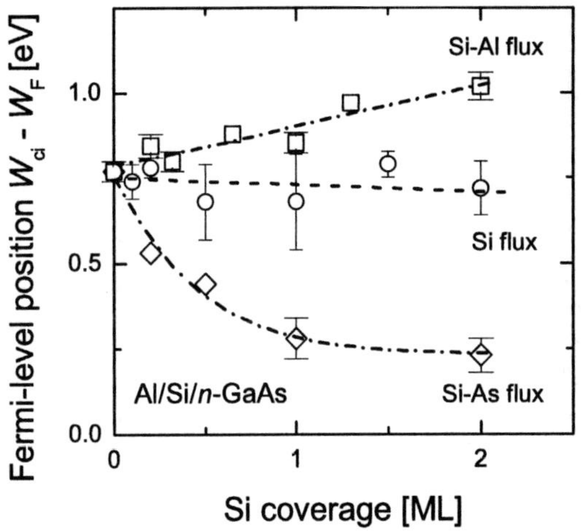

Fig. 10.2. XPS barrier heights of Si-doped Al/*n*-GaAs(001) contacts as a function of the interfacial Si coverage. ○, □, and ◇ data points are for interfacial doping achieved with pure Si, Si–Al, and Si–As flux, respectively. The dashed line is a linear least-squares fit to the □ data, while the dash-dotted lines are only meant to guide the eye. Data from *Cantile* et al. [1994]

When they deposited the Si atoms under ultrahigh-vacuum conditions then the barrier height remained almost constant. [2] This finding is plausible since the electronegativities of silicon and gallium arsenide are almost equal so that the dipole moment of Si-induced interface dipoles will be small. However, *Cantile* et al. were able to tune the barrier heights of Al/Si/*n*-GaAs(001) contacts between 0.2 and 1 eV if they additionally applied sufficiently high As or Al fluxes during the growth of the Si interlayer. The sign of the observed variations are again explained by the simple dipole considerations outlined above. Virtual $Al_{1-y}Si_y$ and $As_{1-y}Si_y$ adatoms are slightly less and somewhat more electronegative than GaAs. Hence, Al–Si and As–Si interlayers are expected to increase and to reduce, respectively, the barrier heights of metal contacts on *n*-GaAs, in complete agreement with the experimental observations of *Cantile* et al., which are displayed in Fig. 10.2. However, the interface compositions of the Al–Si and As–Si interlayers are not known, so that the corresponding increase and reduction of the barrier height cannot be estimated.

Berthod et al. [1996, 2000] performed *ab initio* LDA-DFT calculations to simulate the interface doping of Al/*n*-GaAs(001) contacts with a silicon interlayer. They assumed As- and Ga-terminated GaAs substates and modeled the interfacial Si doping by $As_{1-x}Si_x$ and $Ga_{1-x}Si_x$ pseudo-ions. Their calculations revealed that increasing Si coverage moves the Fermi level towards the valence-band maximum at Ga-terminated and towards the conduction-band minimum at As-terminated Al/Si/GaAs(001) interfaces.

10.2 Interface Structure

10.2.1 Metal/Si(111)-$(7\times7)^i$ Contacts

The interface structure has a pronounced influence on the barrier heights of silicon Schottky contacts. The barrier heights of laterally homogeneous diodes with $(7\times7)^i$-reconstructed and $(1\times1)^i$-unreconstructed interfaces differ by 76 meV. The differences are even larger with type-A and type-B interfaces of epitaxial $NiSi_2$/*n*-Si(111) and $CoSi_2$/*n*-Si(001) contacts. Interface doping of diamond and silicon Schottky contacts with, for example, hydrogen also alters the barrier heights. The latter observations are explained by chemical interface dipoles; see Sect. 5.1. A characteristic of surface and interface reconstructions is that the surface and interface atoms of the substrate are displaced in comparison to the atomic positions in equivalent bulk planes. Such structural rearrangements are connected with redistributions of the valence charge. The bonds in perfectly ordered bulk silicon are purely covalent and, therefore, reconstructions are accompanied by $Si^{-\Delta q}$–$Si^{+\Delta q}$ dipoles. The shifted surface components of the Si(2*p*) core levels demonstrate the

---

[2] The barrier heights of Al/*n*-GaAs(001) diodes obtained by *Cantile* et al. [1994] from their XPS measurements are approximately 0.12 eV smaller than the value determined from Φ_{Bn}^{eff} versus n curves for laterally homogeneous contacts; see Table. 3.3.

existence of such reconstruction-induced charge transfer.[3] In a simple point-charge model, reconstruction-displaced and then charged silicon interface atoms may be treated in the same way as foreign atoms at interfaces.

The geometrical structure of buried Ag- and Pb/Si(111) interfaces was determined by applying grazing-incidence X-ray diffraction. After evaporation of 26 nm of silver or 100 nm of Pb onto clean Si(111)-7×7 surfaces at room temperature the 7×7 *periodicity* was found to persist at the Ag- [*Hong* et al. 1992] and the Pb/Si(111) interfaces [*Grey* et al. 1989, *Feidenhans'l* et al. 1990, *Howes* et al. 1995]. As Fig. 3.4 shows, the clean-surface 7×7 unit mesh contains 12 adatoms and a stacking fault in one of its triangular subunits that is surrounded by nine dimers and a corner hole. Best fits to the diffraction intensities observed with both the Ag- and the Pb/Si(111)-$(7\times7)^i$ interface reconstructions were obtained with the assumption that the stacking fault as well as the associated corner hole and dimers still exist but no silicon adatoms are present.

Annealing of such Ag- and Pb/Si(111)-$(7\times7)^i$ contacts at temperatures above 250°C converts the $(7\times7)^i$ interface structure into a $(1\times1)^i$ structure, or, in other words, removes the stacking fault as well as the dimers and corner holes associated with them [*Hong* et al. 1992, *Lucas* and *Loretto* 1995]. Surprisingly, a $(\sqrt{3}\times\sqrt{3})R30°$ structure, which is obtained by, for example, depositing one monolayer of Ag at 500 K[4], does not form. Quite on the contrary, the Si(111):Ag-$(\sqrt{3}\times\sqrt{3})R30°$ structure is destroyed by further deposition of Ag even at room temperature and the interface structure becomes $(1\times1)^i$-unreconstructed. The same happens to the Si(111):Pb-$(\sqrt{3}\times\sqrt{3})R30°$ surface reconstruction [*Howes* et al. 1995]. This behavior seems more plausible since only the Pb adatoms are arranged in a $(\sqrt{3}\times\sqrt{3})R30°$ lattice while the silicon surface atoms underneath still pertain a bulk-like 1×1 arrangement [*Grey* et al. 1989]. With further Pb evaporation, the now buried Pb adatoms obviously assume bulk positions. The 7×7 structure has the lowest surface free-energy of all clean-surface Si(111) structures. Obviously, the stacking fault and the dimers characteristic of the 7×7 periodicity of Si(111) surfaces are destabilized by the silver and lead overlayers, since on clean Si(111) surfaces the then reversible 7×7 $\leftrightarrows$ 1×1 structural phase transition occurs in the range between 1040 and 1126 K.[5]

Chou et al. [1985] studied the electronic structure of stacking faults in bulk silicon using *ab initio* density-functional theory within the local-density approximation. Both extrinsic and intrinsic stacking faults were considered. They are distinguished by removing or adding of a double layer, i.e., the ideal stacking sequence ... AA'BB'CC'AA'BB'CC' ... is changed to ... AA'BB'CC'BB'CC' ... and ... AA'BB'AA'CC'AA'BB'CC' ..., respectively. The electronic density near both types of stacking faults in the bulk slightly deviates from the undisturbed silicon crystal. Figure 10.3 shows the difference of the integrated charge density over the

[3] See *Mönch* [2001], Sect. 9.4.
[4] See *Mönch* [2001], Sect 14.4.
[5] See *Mönch* [2001], Sect. 12.1.

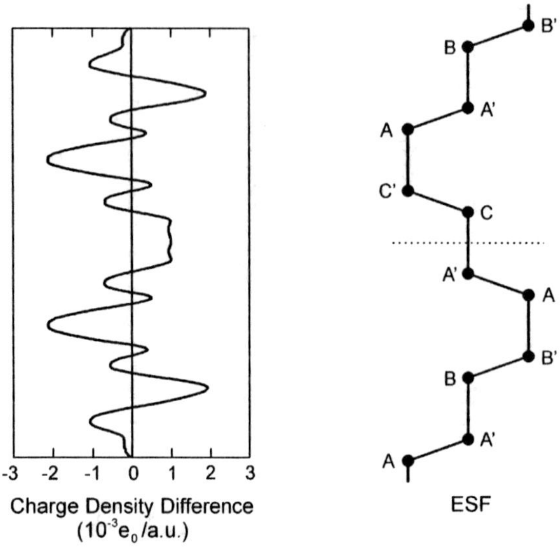

Fig. 10.3. Integrated difference of the charge density over (111) planes between a silicon crystal with an extrinsic stacking fault (ESF) and a perfect silicon crystal, and atomic positions in the (110) plane of a silicon crystal with an extrinsic stacking fault (1 a.u. = a_B^3 = 1.48×10^{-4} nm^3, where a_B is the Bohr radius). After *Chou* et al. [1985]

(111) plane between a sample with and without an extrinsic stacking fault. It may be described by two electric double layers that are symmetrically arranged.

The *surface* stacking fault of the Si(111)-7×7 structure equals one half of an extrinsic stacking fault in the bulk; see Figs. 3.4 and 10.3. The charge-density difference shown in the left-hand panel of Fig. 10.3 indicates that the stacking fault of the Si(111)-7×7 structure indeed represents an electric double layer; therefore, the same also holds for the Si(111)-$(7\times7)^i$ interface structure. The negatively charged sheet of the structure-induced electric double layer is on the vacuum side at clean Si(111)-7×7 surfaces and, consequently, on the metal side of metal/Si(111)-$(7\times7)^i$ interfaces, which are of interest here. In both cases, the charge density Q_{eid} on the silicon side has a positive sign and, as a result of this, relation (10.4) predicts the stacking fault to lower the barrier heights of metal/*n*-Si(111)-$(7\times7)^i$ contacts with respect to unreconstructed metal/*n*-Si(111)-$(1\times1)^i$ interfaces. This finding agrees with what was observed experimentally. The magnitude of the charge transfer within the interfacial stacking fault at metal/Si(111)-$(7\times7)^i$ interfaces may be evaluated from the theoretical charge distribution plotted in Fig. 10.3.

In a more general approach, electric dipole moments are defined by

$$\underline{p} = \int \underline{r} \cdot \rho(\underline{r}) \mathrm{d}^3\underline{r} \ . \tag{10.7}$$

Provided that the space-charge density $\rho(x,y,z)$ varies along the z-direction only, then (10.7) may be rewritten as

$$p_z = \iint \mathrm{d}x\mathrm{d}y \cdot \int z \cdot \rho(z)\mathrm{d}z \quad . \tag{10.8}$$

The second term on the right-hand side of (10.8) is a dipole moment N_p per unit area. Using this quantity, relation (10.5) may be rewritten as

$$\delta\Phi_B^{eid} = \pm \frac{e_0}{\varepsilon_i \varepsilon_0} N_p \quad . \tag{10.9}$$

A numerical integration of the charge density of one half of the extrinsic Si(111) stacking fault, which is displayed in Fig. 10.3, yields

$$N_p^{sf} = 4.93 \times 10^7\, e_0\, \mathrm{m/m^2} \, .$$

For metal/*n*-Si(111)-(7×7)^{i} contacts one then obtains a lowering of the barrier height

$$\delta\Phi_{Bn}^{(7\times7)i} = \Phi_{Bn}^{(7\times7)i} - \Phi_{Bn}^{(1\times1)i} = -\frac{e_0}{\varepsilon_i \varepsilon_0} \frac{N_p^{sf}}{2} = -\frac{446}{\varepsilon_i} \quad \mathrm{meV} \quad .$$

The factor ½ accounts for the fact that only half of the (7×7)^{i} unit mesh contains a stacking fault. For silicon Schottky contacts the interface dielectric constant is estimated as $\varepsilon_i \approx 4.6$, see Sect. 5.6. With this value the dipole model predicts the barrier heights of metal/*n*-Si(111)-(7×7)^{i} contacts to be lower by

$$\delta\Phi_{Bn}^{(7\times7)i} = \Phi_{Bn}^{(7\times7)i} - \Phi_{Bn}^{(1\times1)i} \approx -97 \quad \mathrm{meV}$$

than the values observed for corresponding (1×1)^{i}-unreconstructed contacts. This value is remarkably close to the difference of –76 meV that resulted from the experimental data displayed in Fig. 6.1. It is also noteworthy that in the framework of the MIGS-and-electronegativity concept the same interface dielectric constant yielded a slope parameter $S_X = \partial\Phi_{Bn}/\partial X_{\mathrm{m}} = 0.101$ eV/Miedema-unit. This finding corroborates the value deduced from the barrier heights experimentally observed with laterally homogeneous silicon Schottky contacts; see (6.1) to (6.4) in Sect. 6.1.

10.2.2 Epitaxial Silicide/Silicon Interfaces

Epitaxial metal–semiconductor interfaces are important for an understanding of the mechanisms which determine the barrier heights of such contacts. The disilicides $NiSi_2$ and $CoSi_2$ have the cubic CaF_2 structure and their lattice constants match that of silicon to within 0.44 and 1.2%, respectively, at room temperature. Epitaxial layers of both silicides were fabricated on {100}- as well as on {111}-oriented silicon substrates by using either solid-phase epitaxy (SPE) or molecular beam epitaxy (MBE).

Single-crystalline $NiSi_2$ films of high perfection can be grown on Si(111) surfaces. *Cherns* et al. [1982] and *Gibson* et al. [1982] identified two different interface structures by application of high-resolution transmission electron microscopy (HRTEM). Figure 10.4 shows lattice images of such contacts, which were obtained by using this technique. They reveal the interfaces to be abrupt. The corre-

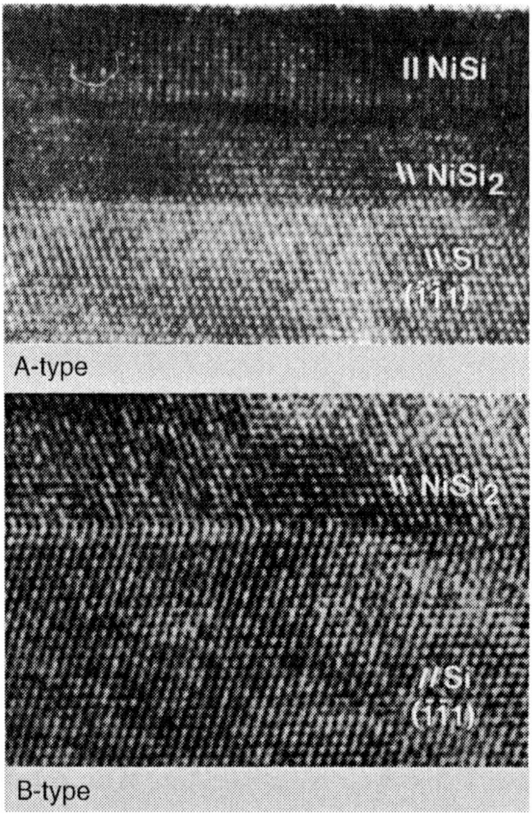

Fig. 10.4. Cross-section HRTEM micrographs of type-A and type-B $NiSi_2$/Si(111) interfaces. From *Ospelt* et al. [1988]

sponding epitaxial structures are illustrated in Fig. 10.5. In type-A interfaces, the lattices are identically aligned on both sides of the interface while they are rotated by 180° around the interface normal for the case of type-B contacts. The atomic arrangement at the latter interfaces is similar to a stacking fault. At the interface, the Ni atoms are sevenfold coordinated [*Cherns* et al. 1982, *Föll* 1982, *van Loenen* et al. 1985, *Vlieg* et al. 1986, *Robinson* et al. 1988, *Zegenhagen* et al. 1989].

By proper choice of the growth conditions, one obtains one of the two interface structures otherwise a mixture of both modifications forms. *Tung* et al. [1983b] found films grown by codeposition of Ni and Si or MBE at approximately 600°C to be of type-B. *Tung* et al. [1983a, b] also demonstrated that layers of both types may be prepared by using the template technique. Evaporation of different amounts of Ni at room temperature and subsequent annealing at temperatures between 450 to 600°C yield either pure type-A or type-B interfaces by solid-phase epitaxy. Figure 10.6 shows the respective experimental data. A conclusive explanation of these observations is still lacking. In any case, growth studies suggest

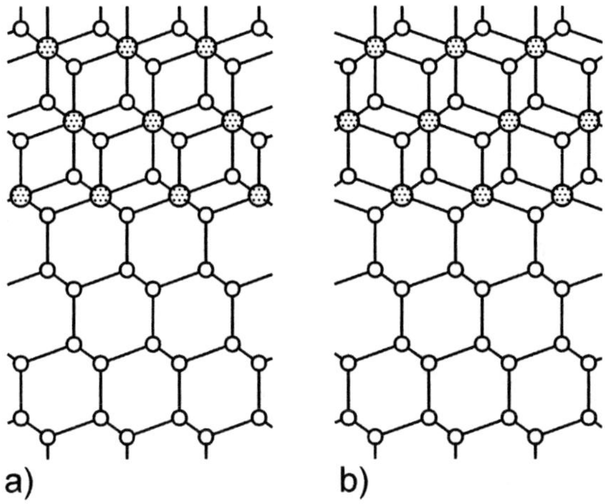

Fig. 10.5. Atomic structure of type-A and type-B $NiSi_2$/Si(111) interfaces. Shaded symbols represent Ni atoms

that the interface free energy is lower for type-B than for type-A interfaces. *Fujitani* [1998] confirmed this experimental observation with results of his *ab initio* calculations using the full-potential linear augmented plane-wave method (FLAPW) in the local-density approximation. The FLAPW method currently gives the most reliable total energies, provided that the number of layers per $NiSi_2$/Si(111)/$NiSi_2$ slab is sufficiently large.

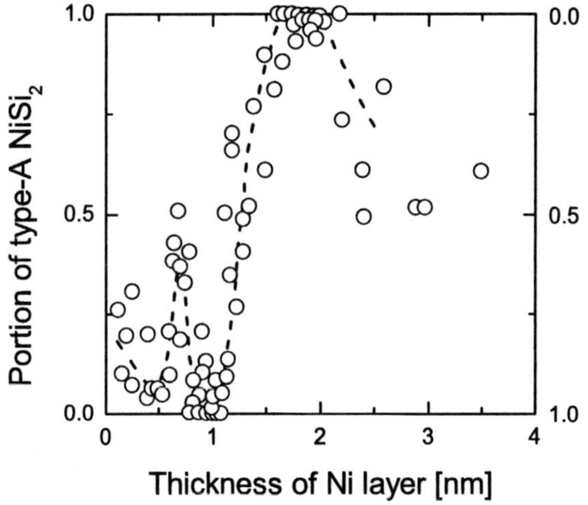

Fig. 10.6. Orientation of thin $NiSi_2$ layers grown on Si(111) surfaces by Ni evaporation at room temperature and subsequent annealing at approximately 450°C. After *Tung* [1989]

Single-crystal $CoSi_2$ layers were grown on Si(111) surfaces by using SPE [*Tung* et al. 1982] as well as MBE [*Henz* et al. 1987]. Both growth techniques resulted in type-B interfaces only, as was established by using high-resolution transmission electron microscopy (HRTEM) [*Gibson* et al. 1982], X-ray standing waves (XSW) [*Fischer* et al. 1987, *Zegenhagen* et al. 1987], and medium-energy ion scattering (MEIS) [*Fischer* et al. 1988]. The atomic arrangements are quite different for $NiSi_2$/Si(111) and $CoSi_2$(111) interfaces. While the Ni interface-atoms are sevenfold coordinated, *de Jong* and *Bulle-Lieuwma* [1990] detected an eightfold coordination of the Co atoms at type-B $CoSi_2$/Si(111) interfaces in their HRTM studies.

Werner et al. [1993] produced buried layers of epitaxial $CoSi_2$ by Co^+-ion implantation into {001}-oriented *n*-silicon substrates and subsequent rapid annealing at temperatures below 1150°C. They used quantitative high-resolution transmission electron microscopy to investigate the atomic structure of the two $CoSi_2$/Si(001) interfaces. Their atomic arrangements differ and the interfacial Co atoms are four- and six-fold coordinated (type-A) at the outer interface and six- and eight-fold coordinated (type-B) at the inner interface. Experimental effective barrier heights and ideality factors of the two types of interfaces are displayed in Fig. 3.9. The Schottky barrier heights of the laterally homogeneous contacts differ by 130 meV. Figure 6.2 reveals that the data point of type-B interfaces matches the predictions of the MIGS-and-electronegativity theory for ideal silicon Schottky contacts. The value characteristic of the type-A contacts, on the other hand, is set off towards lower barrier heights.

The barrier heights of the two types of $NiSi_2$/Si(111) contacts differ by 0.13 eV. They measure 0.66 and 0.45 eV for type-A and 0.79 and 0.32 eV for type-B interfaces prepared on substrates doped *n*-and *p*-type, respectively [*Tung* 1984a, b, 1989; *Tung* et al. 1986]. These data were obtained by extrapolation of $\Phi^{eff}_{Bn,p}$ versus *n* plots to n_{if}. In both cases Φ^{hom}_{Bn} and Φ^{hom}_{Bp} add up to the silicon band-gap at room temperature. Figure 6.2 reveals that the data point for type-B interfaces is close to the predictions of the MIGS-and-electronegativity theory for ideal silicon Schottky contacts while the barrier height of type-A contacts deviates towards lower values. In this regard the two types of $CoSi_2$/*n*-Si(001) and of $NiSi_2$/*n*-Si(111) contacts behave alike. The barrier height difference of the two types of silicide/silicon interfaces was first attributed to fabrication-induced interface defects. Another explanation is the presence of interface-structure-induced dipoles at the type-A interfaces in addition to the intrinsic MIGS that obviously describe the barrier heights of the type-B contacts quite well.

Based on measured capacitance in excess of the space-charge capacitance of the depletion layer, *Ho* et al. [1986] initially assigned the lower barrier heights of type-A $NiSi_2$/*n*-Si(111) contacts to interface defects. Later on, *Werner* et al. [1988] attributed this additional capacitance to an injection of holes into the neutral silicon outside the depletion layer at the $NiSi_2$/*n*-Si(111) interface and at least a high density of interface defects was excluded. *Vrijmoeth* et al. [1990] applied medium-energy ion scattering to such interfaces and their results further supported this conclusion. The channeling method makes this experimental technique quite sensitive to defects in epitaxial structures. Such measurements gave upper limits of 1×10^{12} and 3×10^{13} Si atoms per cm^2 being displaced from lattice sites in type-A and type-B $NiSi_2/Si(111)$ contacts, respectively. The corresponding lowering of the barrier height may be estimated using relation (7.8). With the slope parameter $S_X = 0.101$ eV/Miedema-unit and the MIGS density of states $D_{gs}^{mi} \approx 4.25\times10^{14}\ cm^{-2}eV^{-1}$ characteristic of silicon, the above defect densities give defect-induced reductions $\delta\Phi_{Bn}^{idef}$ of the barrier height by only 2 and 60 meV, respectively, at maximum. Thus, the low densities of interface defects detected by MEIS in the epitaxial $NiSi_2/Si(111)$ contacts cannot account for the observed difference of 130 meV in barrier heights.

As outlined Sect. 10.2.1, extrinsic interface dipoles correlated with specific interface structures explain deviations from the barrier heights predicted by the MIGS-and-electronegativity theory for ideal and then unreconstructed Schottky contacts. This concept works quite well for metal/Si(111)-$(7\times7)^i$ interfaces. Since the observed defect densities cannot explain the deviations of the Schottky barrier heights of both type-A $NiSi_2/n$-Si(111) and type-A $CoSi_2/n$-Si(001) contacts from the linear least-squares fit to the Schottky barrier heights of laterally homogeneous $(1\times1)^i$-unreconstructed interfaces towards lower values extrinsic interface-structure-induced dipoles in addition to the intrinsic MIGS dipoles are a most probable explanation. According to relation (10.4), the reduction of the barrier heights indicates that the extrinsic interface dipoles will be oriented such that their positive and negative charges are on the silicon and the silicide side of the contact, respectively. The corresponding extra charge-transfer across the interface may be estimated from relations (10.5) and (10.6). For the epitaxial silicide/silicon interfaces, the dipole length is approximated by the Si bond length, the area density of dipoles is taken as 7.84×10^{14} per cm^2, the atomic density in bulk Si{111} planes, and the interface dielectric constant ε_i is again assumed as 4.6. The barrier height reduction of 0.13 eV then corresponds to an extra charge of

$$\Delta q_{eid}^{A} \approx 0.02\cdot e_0$$

per interface atom due to extrinsic, structure-induced dipoles at type-A $NiSi_2/Si(111)$ interfaces. The extra valence-charge at interface atoms, which is attributed to structural differences, is quite small.

The electronic properties of type-A and type-B $NiSi_2/Si(111)$ contacts have been calculated using the linear muffin-tin orbitals method in the atomic-sphere approximation (LMTO-ASA) [*Das* et al. 1989, *Fujitani* and *Asano* 1990, *Ossicini* et al. 1990]. This method works well in close-packed structures and it is fast and reliable for calculating energy bands and total energies. However, both the CaF_2 and the diamond structure are relatively open and, therefore, the interstitial space is filled with "empty spheres". These fictitious atoms have zero atomic number

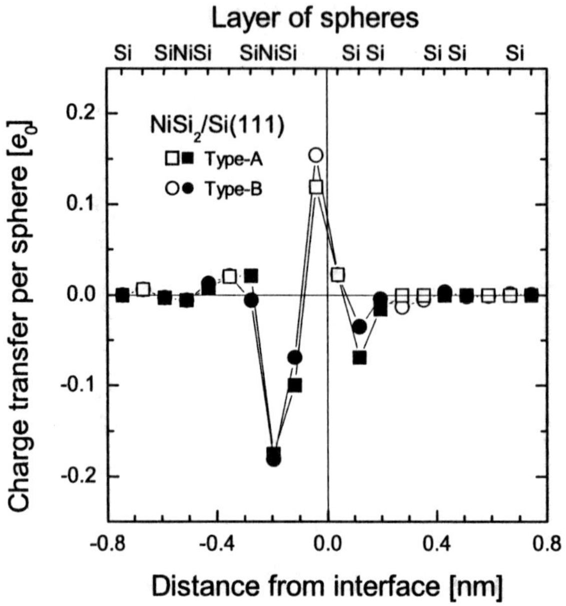

Fig. 10.7. Calculated charge transfer normal to type-A and type-B $NiSi_2$/*Si*(111) interfaces. ■ and □ data for atomic and empty spheres, respectively, at type-A interfaces; ● and ○ data for atomic and empty spheres, respectively, at type-B interfaces. After *Das* et al. [1989]

and, consequently, the potential at these sites is repulsive. The interstitial lattice of silicon consists of one "empty" site next to each silicon atom while there is one empty sphere per molecular unit in $NiSi_2$. The atomic plus the empty spheres thus form body-centered cubic lattices for both silicon and nickel disilicide. Figure 10.7 displays the charge transfer across type-A and type-B $NiSi_2$/Si(111) interfaces calculated by *Das* et al. [1989]. The number of electrons assigned to the atomic and empty spheres in the bulk are taken as the reference. The electronegativity of Ni is larger than that of silicon. Chemical intuition thus expects that electronic charge is transferred from the silicon to the silicide side of the contact. The data of the LMTO-ASA calculations concur with this simple chemical argument. However, there exists an additional and even larger shift of charge within the silicide towards the interface.

The negative sign of the charge transfer on the silicide side translates into positively charged MIGS. Figure 10.8 displays the charge distribution on the silicon side or, in other words, in the MIGS as calculated by *Fujitani* and *Asano* [1990]. Least-squares fits to the data produce decay lengths of 0.30 ± 0.05 nm. This value agrees excellently with the result of APW calculations; see Table 5.1. Furthermore, the data clearly show that the extrinsic dipole induced by the atomic type-A arrangement increases the positive charge in the MIGS. The data shown in Figs. 10.7 and 10.8 yield MIGS charges that are larger by approximately $0.01 \cdot e_0$

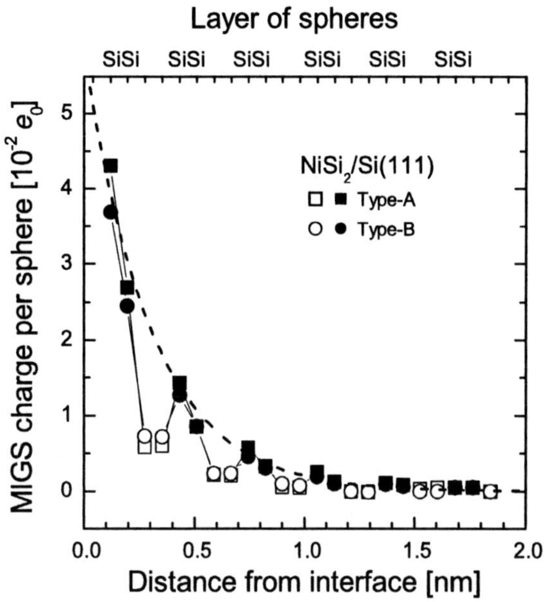

Fig. 10.8. Charge in atomic and empty Si spheres normal to $NiSi_2$/Si(111) interfaces obtained by integrating the electron density of metal-induced gap states in the Si gap. ■ and □ data for atomic and empty spheres, respectively, at type-A interfaces; ● and O data for atomic and empty spheres, respectively, at type-B interfaces. After *Fujitani* et al. [1990]

and $0.028 \cdot e_0$ per interfacial Si atom, respectively, at type-A than at type-B interfaces. The average of these theoretical values equals the extra charge of $0.02 \cdot e_0$ per Si interface atom that was estimated above from the barrier-height difference of 130 meV by using the simple nearest-neighbor dipole relation (10.5). This small charge difference explains the great difficulties encountered in theoretical computations of the type-A and type-B barrier heights even if identical approaches such as the LMTO–ASA or the FLAPW methods are employed. The barrier height of type-A interfaces turned out to be especially sensitive to variations of the interface geometry.

Tung et al. [1991] obtained extremely low barrier heights of 0.4 eV with uniform epitaxial layers of $NiSi_2$ grown on {100}-oriented *n*-Si substrates. Such uniform junctions were prepared by co-deposition of $NiSi_2$ followed by an annealing treatment above 700°C. The barrier heights of these $NiSi_2$/*n*-Si(001) interfaces are 0.4 eV smaller than those measured with type-B $NiSi_2$/n-Si(111)contacts. Attributing the difference again to interface-structure-induced dipoles, relations (10.5) and (10.6) yield an extra dipole charge of

$$\Delta q_{eid}^{001} \approx 0.1 \cdot e_0 \quad ,$$

a value that is a factor of 5 larger than that estimated above for type-A interfaces on {111}-oriented Si substrates. In both cases, $NiSi_2$/*n*-Si(111) interfaces of type-B are taken as the reference since they have the lowest interface free-energy and their barrier heights are closest to the MIGS line of silicon in Fig. 6.2. The atomic arrangements at {100}- and {111}-oriented $NiSi_2$/Si interfaces differ in that the Ni atoms are six- and sevenfold coordinated, respectively [*Gibson* et al. 1983, *Cherns* et al. 1984].

10.2.3 Polar Heterostructures

The atomic arrangement at polar interfaces of real heterostructures between heterovalent semiconductors will differ from the bulk-like termination since anion–cation bonds of donor- and acceptor-type should be present in equal numbers to keep the interface electrically neutral. At Ge/GaAs (001) interfaces, the example chosen in Fig. 6.21, this requirement may be met by an atomic intermixing so that the actual interface layer consists of 50% Ge and As each. Then, Ga^- and As^+ species exist in adjacent planes parallel to the interface and the corresponding electric dipoles form an electric dipole double layer at the actual interface.

The voltage drop across such interfacial dipole layers contributes to the band-edge offsets by

$$\Delta W_{id} = \frac{e_0}{\varepsilon_i \varepsilon_0} Q_{id} d_{id} \quad . \tag{10.10}$$

For (001) interfaces, to name an example, the spacing d_{id} of the interfacial double layer equals the distance $d_{100} = a_0/4$ between two {100} lattice planes. The interface charge per unit area may be written as

$$Q_{id} = 2e_0 \,|\, (\nu_{cl} + \nu_{ar}) - 2 \,|\, \frac{\sigma_{100}}{2} = 2e_0 \,|\, (\nu_{rl} + \nu_{al}) - 2 \,|\, \frac{\sigma_{100}}{2} \quad , \tag{10.11}$$

where $\sigma_{100} = (a_0/\sqrt{2})^{-2}$ is the density of sites in a bulk {100} plane. The subscripts a and c denote anions and cations, respectively, of the semiconductors on the left-hand, l, and right-hand side, r, of the interface. For the presently discussed Ge/GaAs(001) heterostructures, one has $\nu_{cl} = \nu_{al} = \nu_4 = 4/4$, $\nu_{cr} = 3/4$, and $\nu_{ar} = 5/4$. The interface dielectric constant ε_i may be approximated by the average $(\varepsilon_{bl} + \varepsilon_{br})/2$ of the bulk dielectric constants of the two semiconductors in contact. For 50% intermixing in one plane right at the Ge/GaAs(001) interface, one then obtains an extrinsic dipole contribution to the valence-band offset of

$$\Delta W_{id} \approx 0.28 \text{ eV} \quad .$$

The simple electrostatic model thus predicts the valence-band offsets at polar Ge/GaAs(001) interfaces with one mixed Ge:Ga = 1:1 or Ge:As = 1:1 plane to be larger and smaller, respectively, by approximately 0.3 eV relative to non-polar Ge/GaAs(110) interfaces.

Relation (5.27) and the ETB branch-point energies listed in Table 5.1 yield a valence-band discontinuity of 0.34 eV for non-polar Ge/GaAs(110) interfaces. The above simple estimate of the dipole contribution then predicts valence-band offsets of 0.62 and 0.06 eV for Ga- and As-intermixed Ge/GaAs(001) interfaces, respectively. A self-consistent calculation that used a first-principles pseudopotential method and treated exchange and correlation within the local-density approximation [*Dandrea* et al. 1990] obtained valence-band discontinuities of 0.48 eV for non-polar Ge/GaAS(110) and of 0.7 and 0.1 eV for Ga- and As-intermixed Ge/GaAs(001) interfaces, respectively. The band offsets that resulted from the ETB branch-point energies and the simple electrostatic model agree with the value obtained by using the much more refined first-principles calculations. Unfortunately, no studies have been reported where the chemical composition at Ge/GaAs(001) interfaces was systematically varied. However, the band alignment across polar GaAs/ZnSe(001) interfaces was investigated both experimentally as well as by applying state-of-the-art theoretical calculations.

Epitaxial ZnSe films were grown on GaAs(001) surfaces by MBE and their composition was monitored *in situ* by recording the intensity of the Zn(3*d*) and Se(3*d*) XPS lines [*Nicolini* et al. 1994]. Near to the interfaces the composition was found to vary from Se- to Zn-rich when the ratio of the Zn and Se beam fluxes was systematically increased from 0.1 to 10. The valence-band offsets were evaluated from the relative shifts of the Ga(3*d*) and Zn(3*d*) core levels as observed with XPS and found to increase from 0.58 eV to 1.2 eV as the initial composition varied from Se- to Zn-rich. This correlation is remarkable although the true interface composition could not be established from these data. The trend of the valence-band discontinuity as a function of the interface composition will be again analyzed by applying the electrostatic model outlined above.

At polar GaAs/ZnSe(001) interfaces, Ga–Se donor- and As–Zn acceptor-bonds exist. A charge-compensated interface structure is again obtained when, for example, one intermixed Zn:Ga = 1:1 or one intermixed As:Se = 1:1 interface layer exists. This interface structure is illustrated schematically in Fig. 6.21b. For such ZnSe/GaAs(001) interfaces, relations (10.10) and (10.11) then produce an extrinsic dipole contribution

$$\Delta W_{id} \approx 0.38 \text{ eV}$$

to the valence-band discontinuity. The relation (5.27) and the ETB branch-point energies listed in Table 5.1 give a valence-band offset of 0.96 eV for non-polar ZnSe/GaAs heterostructures. This value is close to the valence-band offset of 1.05 eV which was experimentally determined from core-level data recorded with non-polar ZnSe/GaAs(110) heterostructures [*Kowalczyk* et al. 1982a], see Fig. 6.22.

The simple electrostatic model predicts valence-band discontinuities of 1.34 eV and 0.58 eV for polar ZnSe/GaAs(001) heterostructures with Zn:Ga = 1:1 and Se:As = 1:1 interface-layer composition, respectively. First-principles self-consistent-field calculations within the local-density approximation to density-functional theory, on the other hand, produced valence-band offsets of 1.59 and 0.62 eV for polar ZnSe/GaAs(001) heterostructures with Zn:Ga = 1:1 and Se:As = 1:1 interface-layer composition, respectively [*Nicolini* et al. 1994]. These results agree with the data from ETB branch-point energies together with the simple electrostatic model. Furthermore, the theoretical results explain the trend as well as the magnitude of the experimental valence-band offsets of the ZnSe/GaAS(001) heterostructures as a function of relative interface composition.

11. Ohmic Contacts

Metal contacts on moderately doped semiconductor are generally rectifying. This behavior is due to the fact that the branch points of almost all semiconductors are well within their fundamental band gaps. For sufficiently large barrier heights the current transport is then dominated by thermionic emission over the barrier. However, the *I*/*V* characteristics will become apparently ohmic if the barrier height drops to below approximately 0.3 eV. The *I*/*V* characteristics displayed in Fig. 11.1a,b illustrate this behavior and also indicate the influence of the series resistance. When the doping level of the semiconductor is increased and, as a conse-

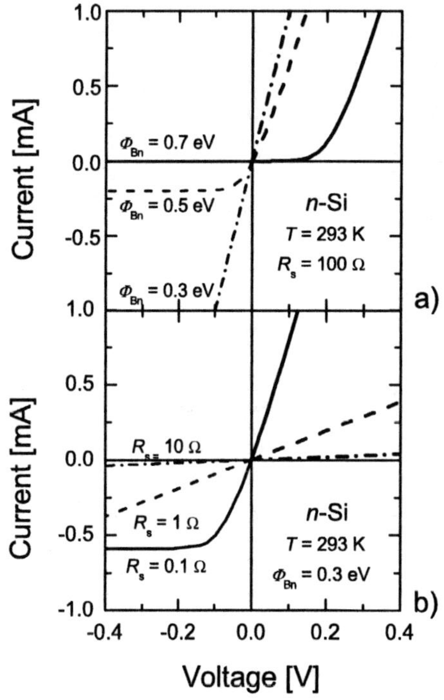

Fig. 11.1. Current–voltage characteristics for thermal emission over the barrier of *Schottky* contacts with different barrier heights (a) and different series resistance (b)

quence of this, the depletion layer narrows, tunneling or field emission through the barrier becomes increasingly important. As a consequence of this, the I/V characteristics of metal contacts on highly doped semiconductors show ohmic behavior. There are, however, also metal–semiconductor contacts that are *naturally* ohmic.

Semiconductors with branch points that are close to one of the band edges or even resonant with one of the bulk bands will exhibit metal contacts that are *naturally* ohmic, provided that the type of doping of the semiconductor substrate is appropriately chosen. The theoretical data displayed in Table 5.1 indicate that the branch points are close to the valence-band maximum for GaSb and InSb and even above the conduction-band minimum for InN and InAs. This means that metal contacts are naturally ohmic on p-type GaSb and InSb and on n-type InN and InAs.

The most important property of ohmic metal–semiconductor contacts is their so-called contact resistance. It is defined as

$$R_c \equiv \left(\frac{\partial j}{\partial V_c}\right)^{-1}_{V_c=0} . \tag{11.1}$$

As an example, Fig. 11.2 displays experimental R_c values of PtSi/n-Si(001) contacts [*Chang* et al. 1971]. The data clearly demonstrate that the contact resistance is large and strongly depends on temperature for low donor densities where the current transport is expected to be dominated by thermal emission over the barrier. On the other hand, the contact resistance becomes small and insensitive to temperature variations in the high-doping range where tunneling through the barrier should prevail.

From the standard I/V characteristic (3.38) one obtains the contact resistance

$$R_c^{te} = \frac{n k_B^2}{e_0 A_R^{**} k_B T} \exp\left(\frac{\Phi_B^{eff}}{k_B T}\right) \tag{11.2}$$

for thermionic emission over the barrier. The contact resistance strongly depends on both temperature and barrier height. It decreases exponentially with increasing temperature and with decreasing barrier height. The dash-dot-dotted lines labeled "*te*" in Fig. 11.2 are theoretical R_c^{te} values (11.2) calculated for metal/n-Si(111) contacts with an effective barrier height Φ_{Bn}^{eff} of 0.85 eV, which value is characteristic of the corresponding PtSi/n-Si(001) contacts [*Chang* et al. 1971], and at the temperatures indicated in the subscripts. The theoretical results give excellent fits to the experimental data in the low-doping range.

In metal contacts on highly doped semiconductors, on the other hand, the current transport occurs by tunneling through the correspondingly thin barriers. As a very crude approximation, the shape of the depletion barrier may be thought of as triangular. The consequently constant electric-field strength is assumed to equal

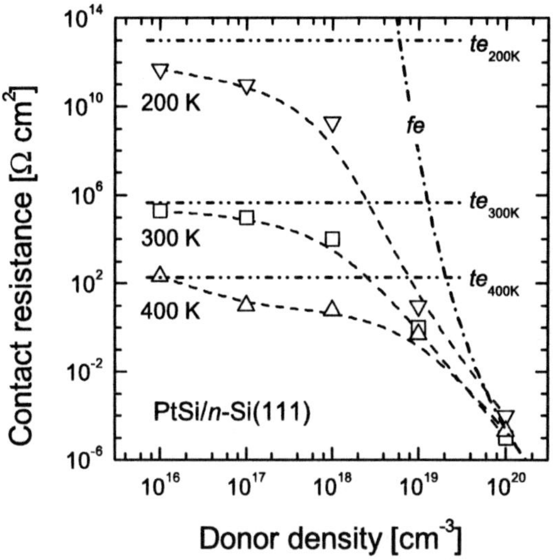

Fig. 11.2. Contact resistance of PtSi/*n*-Si(001) contacts as a function of the donor density of the silicon substrate. The dashed lines are meant to guide the eye of the reader only. The dash-dot-dotted and the dash-dotted lines were calculated using (11.2) and (11.6) for thermal emission (*te*) over and tunneling or field emission (*fe*) through the barrier, respectively. The experimental barrier height of the contacts was 0.85 eV. Data from *Chang* et al. [1971]

the maximum value of the real but quadratic barrier. The transmission probability of electrons at the bottom of the conduction band, i.e., right at the depletion-layer edge, is then found to vary as

$$P_{fe}^{tr} = \exp\left[\frac{2}{3}\frac{e_0(V_i - V_c)}{E_{00}}\right] \tag{11.3}$$

as a function of the applied voltage. The parameter

$$E_{00} = e_0 \frac{\hbar}{2}\left(\frac{N_{d,a}}{m_{n,p}^{*}\varepsilon_b\varepsilon_0}\right)^{1/2}, \tag{11.4}$$

which characterizes the tunneling of electrons through the depletion layers [*Padovani* and *Stratton* 1966], is determined by the dopant density but does not depend on the temperature at all. Figure 11.3 displays the tunneling parameter E_{00} as a function of the doping density with three values of $m^*\varepsilon_b/m_0$. Tunneling or field emission through and thermionic emission over the barrier are expected to dominate the current transport in metal–semiconductor contacts if $k_BT \ll E_{00}$ and $k_BT \gg E_{00}$, respectively.

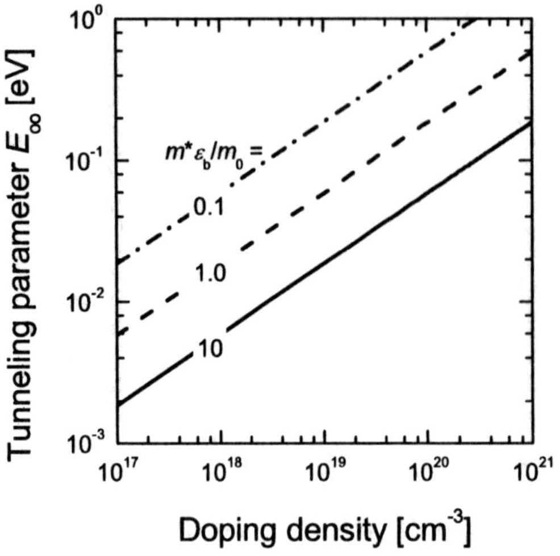

Fig. 11.3. Tunneling parameter E_{00} as a function of doping density for three different values of $m^* \varepsilon_b/m_0$

The tunneling current of forward-biased metal–semiconductor contacts can be expressed as [*Padovani* and *Stratton* 1966]

$$I_{fe} \cong \frac{4\pi A A_R^{**} E_{00}^2}{k_B^2} \exp\left(-\frac{\Phi_{Bn}^{eff}}{E_{00}}\right) \cdot \exp\left(\frac{e_0 V_c}{n E_{00}}\right) . \tag{11.5}$$

As in the I/V relationship (3.38) for thermal emission over the barrier, the effective barrier height Φ_B^{eff} and the ideality factor n again account for the lateral inhomogeneities of the barrier height. The respective contact resistance then results as

$$R_c^{fe} \approx \frac{n k_B^2}{4\pi e_0 A_R^{**} E_{00}} \exp\left(\frac{\Phi_B^{eff}}{E_{00}}\right) . \tag{11.6}$$

It does not depend on temperature but is rather determined by the dopant density. The dash-dotted line labeled "*fe*" in Fig. 11.2 was calculated using relation (11.6) for tunneling contacts with an effective barrier height Φ_{Bn}^{eff} of 0.85 eV on *n*-Si(111) substrates. The theoretical line excellently reproduces the experimental data obtained for Si samples with donor densities of 10^{20} cm^{-3}. The transition range between predominantly thermionic emission over or mainly field emission through the barrier requires a more elaborate theoretical approach [*Padovani* and *Stratton* 1966, *Crowell* and *Rideout* 1969] that goes beyond the objectives of the present discussions.

Appendix

Table A.1 Bulk properties of semiconductors at room temperature: lattice parameter a_0, in nm; widths W_{gd} and W_{gi} of direct and indirect gand gaps, respectively, in eV; static and optical dielectric constant ε_b and ε_∞, respectively; plasmon energy $\hbar\omega_p$ of the bulk valence electrons [*Landoldt-Börnstein* 2001]

Semiconductor	a_0	W_{gd}	W_{gi}	ε_b	ε_∞	$\hbar\omega_p$
C(diamond)	0.357	6.5	5.47	5.7		
Si	0.543	3.37	1.12	11.9		16.9
Ge	0.566	0.81	0.66	16.2		16.0
3*C*-SiC	0.436	6.00	2.4	9.52	6.38	22.5
4*H*-SiC	hexag.		3.26			
6*H*-SiC	hexag.		3.02			
BN	0.362	14.5	6.4	7.1	4.5	
2*H*-AlN	wurtzite	6.2		8.5	4.84	
AlP	0.546	3.62	2.45	9.80	7.54	
AlAs	0.566	3.03	2.15	10.06	8.16	
AlSb	0.614	2.21	1.62	12.04	10.24	14.8
3*C*-GaN	0.452	3.22		9.87	5.8	
2*H*-GaN	wurtzite	3.39		9.5	5.35	
GaP	0.545	2.78	2.27	11.11	9.11	16.5
GaAs	0.565	1.42	1.81	12.85	10.9	15.7
GaSb	0.610	0.25	0.80	15.69	14.44	
3*C*-InN	0.498					
2*H*-InN	wurtzite	0.7			9.3	
InP	0.587	1.34	1.84	12.56	9.61	13.8
InAs	0.606	0.35	1.21	15.15	12.25	13.8
InSb	0.648	0.18	0.62	17.88	15.68	12.8
ZnS	0.541	3.84	4.38	8.32	5.14	
ZnSe	0.567	2.7	3.47	9.2	5.7	
ZnTe	0.610	2.26		9.67	7.28	
CdS	0.582	2.50	4.30	8.42	5.27	
CdSe	0.605	1.75	3.4	9.3	6.1	
CdTe	0.648	1.49	3.	10.76	7.21	
γ-CuCl	0.541	3.17	(7.8)	7.9	3.61	
γ-CuBr	0.569	2.91		7.9	4.06	

Table A.1 continued

Semiconductor	a_0	W_{gd}	W_{gi}	ε_b	ε_∞	$\hbar\omega_p$
$CuGaS_2$	0.536	2.53			6.15	
$CuInS_2$	0.552	1.53			6.3*	
$CuAlSe_2$	0.561	2.72			6.3*	
$CuGaSe_2$	0.561	1.68			7.3*	
$CuInSe_2$	0.578	1.04			9.0	
$CuGaTe_2$	0.599	1.23			8.0*	
$CuInTe_2$	0.617	1.06			9.2	
$AgGaSe_2$	0.599	1.81			6.8	
$AgInSe_2$	0.609	1.24			7.2	

* $\varepsilon_\infty = n^2$

Table A.2 Hartree- Fock atomic term values $E_s = (W_s - W_{vac})$ and $E_p = (W_p - W_{vac})$, in eV, of outer *s*- and *p*-electrons, respectively, and covalent radii $r_{cov,}$ in nm, of some of the elements. Data from *Fischer* [1972] and *Sargent–Welch Table of Periodic Properties of the Elements* [1980]

	H 1									He 2
$-E_s$	13.6									24.97
r_{cov}	0.032									0.093
	Li 3	Be 4			B 5	C 6	N 7	O 8	F 9	Ne 10
$-E_s$	5.34	8.41			13.46	19.37	26.22	34.02	42.78	52.51
$-E_p$		5.79			8.43	11.07	13.84	16.72	19.86	23.13
r_{co}	0.123	0.106			0.088	0.077	0.070	0.073	0.064	
	Na 11	Mg 12			Al 13	Si 14	P 15	S 16	Cl 17	Ar 18
$-E_s$	4.95	6.88			10.70	14.79	19.22	24.01	29.19	34.75
$-E_p$		3.84			5.71	7.58	9.54	11.60	13.78	16.08
r_{cov}	0.154	0.140			0.126	0.117	0.110	0.104	0.099	
	K 19	Ca 20	Cu 29	Zn 30	Ga 31	Ge 32	As 33	Se 34	Br 35	Kr 36
$-E_s$	4.01	5.32	7.72	7.96	11.55	15.15	18.91	22.86	27.00	31.37
$-E_p$			2.37	4.02	5.67	7.33	8.98	10.68	12.43	14.26
r_{cov}	0.203	0.174	0.135	0.131	0.126	0.122	0.118	0.114	0.111	
	Rb 37	Sr 38	Ag 47	Cd 48	In 49	Sn 50	Sb 51	Te 52	I 53	Xe 54
$-E_s$	3.75	4.85	7.06	7.21	10.14	13.04	16.02	19.12	22.34	25.69
$-E_p$			2.61	3.99	5.37	6.76	8.14	9.54	10.97	12.44
r_{cov}	0.216	0.191	0.152	0.148	0.144	0.140	0.136	0.132	0.138	
	Cs 55	Ba 56	Au 79	Hg 80	Tl 81	Pb 82	Bi 83	Po 84	At 85	Rn 86
$-E_s$	3.36	4.29	6.98	7.10	9.82	12.48	15.19	17.96	20.82	23.78
$-E_p$			2.67	3.95	5.23	6.53	7.79	9.05	10.33	11.64
r_{cov}	0.235	0.198	0.134	0.148	0.148	0.147	0.146	0.146	0.145	

Table A.3 Table of atomic electronegativities after *Pauling* [1939/1960] and *Miedma* et al. [1973, 1980]

H 1 2.20																	He 2
Li 3 0.98 2.85	Be 4 1.57 4.20											B 5 2.04 4.75	C 6 2.55 6.20	N 7 3.04 7.00	O 8 3.44	F 9 3.98	Ne 10
Na 11 0.93 2.70	Mg 12 1.31 3.45											Al 13 1.61 4.20	Si 14 1.90 4.70	P 15 2.19 5.10	S 16 2.58	Cl 17 3.16	Ar 18
K 19 0.82 2.25	Ca 20 1.00 2.55	Sc 21 1.36 3.25	Ti 22 1.54 3.65	V 23 1.63 4.25	Cr 24 1.66 4.65	Mn 25 1.55 4.45	Fe 26 1.83 4.93	Co 27 1.88 5.10	Ni 28 1.91 5.20	Cu 29 1.90 4.55	Zn 65 4.10	Ga 31 1.81 4.10	Ge 32 2.01 4.55	As 33 2.18 4.80	Se 34 2.55	Br 35 2.96	Kr 36
Rb 37 0.82 2.10	Sr 38 0.95 2.40	Y 39 1.22 3.20	Zr 40 1.33 3.40	Nb 41 1.60 4.00	Mo 42 2.16 4.65	Tc 43 1.90 5.30	Ru 44 2.20 5.40	Rh 45 2.28 5.40	Pd 46 2.20 5.45	Ag 47 1.93 4.45	Cd 48 1.69 4.05	In 49 1.78 3.90	Sn 50 1.96 4.15	Sb 51 2.05 4.40	Te 52 2.10	I 53 2.66	Xe 54
Cs 55 0.79 1.95	Ba 56 0.89 2.32	La 57 1.10 3.05	Hf 72 1.30 3.55	Ta 73 1.50 4.05	W 74 2.36 4.80	Re 75 1.90 5.40	Os 76 2.20 5.40	Ir 77 2.20 5.55	Pt 78 2.28 5.65	Au 79 2.54 5.15	Hg 80 2.00 4.20	Tl 81 2.04 3.90	Pb 82 2.33 4.10	Bi 83 2.02 4.15	Po 84 2.00	At 85 2.20	Rn 86
Fr 87 0.7	Ra .90	Ac 89 1.10															

Ce 58 1.12	Pr 59 1.13	Nd 60 1.14	Pm 61 1.13	Sm 62 1.17	Eu 63 1.20	Gd 64 1.20	Tb 65 1.20	Dy 66 1.22	Ho 67 1.23	Er 68 1.24	Tm 69 1.25	Yb 70 1.10	Lu 71 1.27
Th 90 1.30 3.30	Pa 91 1.50	U 92 1.38 4.05											

References

Abay, B., Çanakaya, G., Güder, H.S., Efeoğlu, H., Yoğurtç, Y.K. (2003): Semicond. Sci. Technol. **18**, 75. Sect. 6.18, Table 3.5

Aboelfotoh, M.O. (1988): J. Appl. Phys. **64**, 4046. Sect. 6.1.2, Table 3.3

Aboelfotoh, M.O. (1991): Solid-State Electron. **34**, 51. Sect. 8.1

Aboelfotoh, M.O., Tu, K.N. (1986): Phys. Rev. B **34**, 2311. Sect. 6.1.2, Table 3.3

Aboelfotoh, M.O., Cros, A., Svensson, B.G., Tu, K.N. (1990): Phys. Rev. B **41**, 9819. Sect. 8.1

Aboelfotoh, M.O., Fröjdh, C., Peterson, C.S. (2003): Phys. Rev. B **67**, 075312. Sect. 6.16, Table 3.3

Adams, W.G., Day, R.E. (1876): Proc. Roy. Soc. **25**, 113. Sect. 1.1.1

Afanas'ev, V.V., Stesmans, A. (2000): Appl. Phys. Lett. **77**, 2024. Sects. 3.12.1, 6.2.7; Table 6.5

Afanas'ev, V.V., Bassler, M., Pensl, G., Schulz, M.J., Stein von Kamienski, E. (1996): J. Appl. Phys. **79**, 3108. Sects. 3.12.1, 6.2.7; Tables 6.4, 6.5

Afanas'ev, V.V., Houssa, M., Stesmans, A. , Heyns, M.M. (2001): Appl. Phys. Lett. **78**, 3073. Sect. 6.2.7, Table 6.4

Agrawal, B.K., Agrawal, S., Srivastava, R., Srivastava, P. (2001): Physica E **11**, 27. Sect. 7.3.2

Alay, J.L., Hirose, M. (1997): J. Appl. Phys. **81**, 1606. Table 6.4

Allongue, P., Souteyrand, E. (1987): J. Vac. Sci. Technol. B **5**, 1644. Sect. 1.1.3

Almeida, J., dell'Orto, T., Coluzza, C., Margaritondo, G., Bergossi, O., Spajer, M., Courjon, D. (1996): Appl. Phys. Lett. **69**, 2361. Sects. 3.8, 6.1.4; Table 3.5

An, K.S., Park, R.J., Kim, J.S., Park, C.Y., Kim, C.Y., Chung, J.W., Abukawa, T., Kono, S., Kinoshita, T., Kakizaki, A., Ishii, T. (1995): Surf. Sci. **337**, L789. Sect. 3.11, Table 3.7

Andersen, J.N., Wigren, C., Karlsson, U.O. (1991): J. Vac. Sci. Technol. B **9**, 2384. Sect. 3.11, Table 3.7

Anderson, R.L. (1962): Solid State Electron. **5**, 341. Sect. 1.2

Anderson, C.L., Crowell, C.R., Kao, T.W. (1975): Solid-State Electron. **18**, 705. Sect. 3.8

Andres, P.L. de, Garcia-Vidal, F.J., Reuter, K., Flores, F. (2001): Prog. Surf. Sci. **66**, 3. Sect. 3.6

Andrews, J.M., Lepselter, M.P. (1970): Solid-State Electron. **13**, 1011. Sect. 3.3

Andrews, J.M., Phillips, J.C. (1975): CRC Cri. Rev. Solid State Sci. **5**, 405. Sect. 1.1.4

Aniltürk, Ö.S., Turan, R. (1999): Semicond. Sci. Technol. **14**, 1060. Sects. 6.1.2, 8.1; Tables 3.3, 3.5

Archer, R.J., Yep, T.O. (1970): J. Appl. Phys. **41**, 303. Sect. 6.1.2, Table 3.5

Arnaud D'Avitaya, F., Badoz, P.-A., Campidelli, Y., Chroboczek, J.A., Duboz, J.-Y., Peruo, A. (1990): Thin Solid Films **184**, 283. Sect. 6.1.2, Table 3.5

Arnold, D., Ketterson, A., Henderson, T., Klem, J., Morkoç (1984): Appl. Phys. Lett. **45**, 1237. Sect. 1.2

Arulkumaran, S., Arokiaraj, J., Dharmarasu, N., Kumar, J. (1996): Nucl. Instr. and Meth. B **116**, 519. Sect. 6.1.3, Table 3.3

Asai, M., Ueba, H., Tatsuyama, C. (1985): J. Appl. Phys. **58**, 2577. Sect. 3.9.2

Bachmann, R. (1968): Phys. kondens. Materie **8**, 31. Sect. 8.1

Bai, S., Ke, Y., Shishkin, Y., Shigiltchoff, O., Devaty, R.P., Choyke, W.J., Strauch, D., Stojetz, B., Dorner, B., Hobgood, D., Serrano, J., Cardona, M., Nagasawa, H., Kimoto, T., Porter, L.M. (2003): Mat. Res. Soc. Symp. Proc. **742**, K 3.1. Sect. 6.1.6, Tables 3.3, 3.5

Baldereschi, A. (1973): Phys. Rev. B **7**, 5212. Sects. 1.1.5, 5.5.1

Ban, D., Yang, F., Fang, R., Xu, S., Xu, P. (1996): J. Electron. Spectrosc. Relat. Phenom. **80**, 197, Sect. 6.2.6

Ban, D., Xue, J., Fang, R. (1998): J. Vac. Sci. Technol. B **16**, 989. Sects. 6.2.2, 6.2.6; Table 6.5

Bardeen, J. (1947): Phys. Rev. **71**, 717. Sects. 1.1.3, 5.2

Bardi, J., Binggeli, N., Baldereschi, A. (1999): Phys. Rev. B **59**, 8054. Table 7.1

Barnard, W.O., Myburg, G., Auret, F.D., Goodman, S.A., Meyer, W.E. (1996): J. Electron. Mater. **25**, 1695. Sect. 6.1.3, Table 3.3

Bastys, A.J., Bikbaev, V.B., Vaitkus, J.J., Karpinskas, S.C. (1988): Litov. Fiz. Sb. **28**, 191. Sect. 4.1

Batra, I.P., Ciraci, S. (1986): Phys. Rev. B **33**, 4312. Sect. 3.11

Bauer, E. (1958): Z. Kristallogr. **110**, 423. Sect. 3.9.2

Bauknecht, A., Blieske, U., Kampschulte, T., Albert, J., Sehnert, H., Lux-Steiner, M.C., Klein, A., Jaegermann, W. (1999): Appl. Phys. Lett. **74**, 1099. Sect. 6.2.4, Table 6.3

Bäumer, A., Jäger, D. (1976): Thesis Universität Münster (unpublished). Sect. 6.1.2, Table 3.3

Bayard, R.T., Alpert, D. (1950): Rev. Sci. Instrum. **21**, 571. Sect. 1.1.3

Bechstedt, F., DelSole, R. (1988): Phys. Rev. B 38, 7710. Sect. 7.1

Bechstedt, F., DelSole, R. (1990): Solid State Commun. **74**, 41. Sect. 7.1

Bell. L.D., Kaiser, W.J. (1988): Phys. Rev. Lett. **61**, 2368. Sect. 3.6

Bell, L.D., Smith, R.P., McDermott, B.T., Gertner, E.R., Pittman, R., Pierson, R.L., Sullivan, G.J. (1998): Appl. Phys. Lett. **72**, 1590. Sect. 6.1.5, Table 3.4

Berthod, C., Binggeli, N., Baldereschi, A. (1996): Europhys. Lett. **36**, 67. Sect. 10.1, Table 5.1

Berthod, C., Binggeli, N., Baldereschi, A. (1998): Phys. Rev. B **57**, 9757. Sect. 5.3.1

Berthod, C., Binggeli, N., Baldereschi, A. (2000): J. Vac. Sci. Technol. B **18**, 2114. Sect. 10.1

Best, J.S. (1979): Appl. Phys. Lett. **34**, 522. Sect. 6.1.10

Bethe, H.A. (1942): MIT Radiation Lab. Rep. 43-12. Sects. 1.1.2, 1.1.5, 3.1, 4.5.1

Bhuiyan, A.S., Martinez, A., Esteve, D. (1988): Thin Solid Films **161**, 93. Sect. 6.1.3, Table 3.3

Binggeli, N., Ferrara, P., Baldereschi, A. (2001): Phys. Rev. B **63**, 245306. Sects. 6.1.6, 7.3.2; Table 6.1

Binnig, G., Rohrer, H., Gerber, Ch., Weibel, E. (1982a): Appl. Phys. Lett. **40**, 178. Sect. 3.6

Binnig, G., Rohrer, H., Gerber, Ch., Weibel, E. (1982b): Phys. Rev. Lett. **49**, 57. Sect. 3.6

Binnig, G., Rohrer, H., Gerber, Ch., Weibel, E. (1983): Phys. Rev. Lett. **50**, 120. Sect. 3.6

Blauärmel, A., Brauer, M., Hoffmann, V., Schmidt, M. (2000): Appl. Surf. Sci. **166**, 108. Sect. 6.1.2, Table 3.5

Blöchl, P., Das, G.P., Fischmeister, H.F., Schönberger, U. (1990): in *Metal–Ceramic Interfaces*, ed. by M. Rühle, A.G. Evans, M.F. Ashby, and J.P. Hirth (Pergamon Press, Oxford), p. 9. Sects. 5.3.1, 6.1.12

Böhrer, J., Krost, A., Bimberg, D.B. (1993a): Appl. Phys. Lett. **63**, 1918. Sect. 6.2.3

Böhrer, J., Krost, A., Wolf, T., Bimberg, D.B. (1993b): Phys. Rev. B **47**, 6439. Sect. 6.2.3

Bondarenko, V.B., Kudinov, Yu.A., Ershov, S.G., Korablev, V.V. (1998): Semiconductors **32**, 495. Sect. 4.5.1

Bortz, M.L., Ohuchi, F.S., Parkinson, B.A. (1989): Surf. Sci. **223**, 285. Sect. 6.2.5

Bose, J.C. (1904): U.S. Patent No. 775,840. Sect. 1.1.1

Bose, D.N., Pal, S. (1997): Phil. Mag. B **75**, 311. Sect. 6.1.8

Boudville, W.J., McGill, T.C. (1985): J. Vac. Sci. Technol. B **3**, 1192. Sect. 4.5.1

Brattain, W.H. (1940): see Bardeen, J. (1947). Sect. 1.1.2

Brattain, W.H., Bardeen, J. (1953): Bell System Tech. J. **32**, 1. Sect. 1.1.3

Braun, F. (1874): Pogg. Ann. Physik und Chemie **153**, 556. Sect. 1.1.1

Braun, F. (1906): Elektrotechn. Zeitschr. **52**, 1199. Sect. 1.1.1

Brillson, L.J. (1978): Phys. Rev. Lett. **40**, 260. Sect. 1.1.4

Bringans, R.D., Olmstead, M.A., Uhrberg, R.I.G., Bachrach, R.Z. (1987): Phys. Rev. B **36**, 9569. Sect. 3.11, Table 3.7

Bucher, E. (1992): in *Photoelectrochemistry and Photovoltaics of Layered Semiconductors*, ed. by A. Aruchamy (Kluwer Academic, Dordrecht), p.1. Sect. 6.1.8

Bulle-Lieuwma, C.W.T., Jong, A.F. de, Vandenhoudt, D.E.W. (1991): Phil. Mag. A **64**, 255. Sect. 4.5.2

Burns, O. (1985): *Solid State Physics* (Academic Press, Orlando). Sect. 5.5.1

Cantile, M., Sorba, L., Faraci, P., Yildirim, S., Biasol, G., Bratina, G., Franciosi, A., Miller, T.J., Nathan, M.I., Tapfer, L. (1994): J. Vac. Sci. Technol. B **12**., 2653. Sect. 10.1

Cardona, M., Christensen, N.E. (1987): Phys. Rev. B **35**, 6182. Sect. 1.1.5, 5.5.1

Carlisle, J.A., Miller, T., Chiang, T.-C. (1993): Phys. Rev. B **47**, 3790. Sect. 3.11, Table 3.7

Chadi, D.J., Cohen, M.L. (1973): Phys. Rev. B **8**, 5747. Sect. 5.5.1

Chadi, D.J., Cohen, M.L. (1975): phys. stat. sol. (b) **68**, 405. Sect. 5.5.2

Chambers, S.A., Liang, Y., Yu, Z., Droopard, R., Ramdani, J. (2001): J. Vac. Sci. Technol. A **19**, 934. Table 6.4

Chan, C.L., Shih, I. (1990): J. Appl. Phys. **68**, 156. Table 6.2

Chan, Y.-J., Wu, C.-S., Chen, C.-H., Shieh, J.-L., Chyi, J.-I. (1997): IEEE Trans. Electron Devices **44**, 708. Sect. 6.2.3

Chand, S., Kumar, J. (1996): Appl. Phys. A **63**, 171. Sect. 6.1.2, Table 3.3

Chang, C.Y., Fang, Y.K., Sze, S.M. (1971): Solid-State Electron. **14**, 541. Sect. 11

Chang, E.Y., Lai, Y.-L., Lin, K.-C., Chang, C.-Y. (1993): J. Appl. Phys. **74**, 5622. Sect. 6.1.10

Chang, S.K., Park, H.L., Kim, H.K., Hwang, J.S., Chung, C.H., Kim, W.T. (1990): phys. stat. sol. (b) **158**, K115. Sect. 6.2.8

Chao, Y.-C., Johansson, L.S.O., Karlsson, C.J., Landemark, E., Uhrberg, R.I.G. (1995): Phys. Rev. B **52**, 2579. Sect. 3.11, Table 3.7

Chao, Y.-C., Johansson, L.S.O., Uhrberg, R.I.G. (1996): Phys. Rev. B **54**, 5901. Sect. 3.11, Table 3.7
Chao, Y.-C., Johansson, L.S.O., Uhrberg, R.I.G. (1997): Phys. Rev. B **55**, 7198. Sect. 3.11, Table 3.7
Chao, Y.-C., Johansson, L.S.O., Uhrberg, R.I.G. (1998): Surf. Sci. **123**, 76. Sect. 3.11, Table 3.7
Charlesworth, J.P.A., Godby, R.W., Needs, R.J., Sham, L.J. (1992): Mater. Sci. Eng. B **14**, 262. Table 7.1
Chen, T.P., Lee, T.C., Ling, C.C., Beling, C.D., Fung, S. (1993): Solid-State Electron. **36**, 949. Sects. 3.5, 6.1.2; Table 3.3
Cherns, D., Anstis, G.R., Hutchinson, J.L., Spence, J.C.H. (1982): Phil. Mag. A **46**, 849. Sect. 10.2.2
Cherns, D., Hetherington, C.J.D., Humphreys, C.J. (1984): Phil. Mag. A **49**, 165. Sect. 10.2.2
Chin, V.W.L., Storey, J.W.V., Green, M.A. (1989): Solid-State Electron. **32**, 475. Sects. 3.5, 6.1.2; Table 3.3
Chin, V.W.L., Storey, J.W.V., Green, M.A. (1990): J. Appl. Phys. **68**, 4127. Sect. 6.1.2, Table 3.3
Cho, Y.-H., Kim, K.-S., Ryu, S.-W., Kim, S.-K., Choe, B.-D., Lim, H. (1995): Appl. Phys. Lett. **66**, 1785. Sect. 6.2.3
Cho, Y.-H., Choe, B.-D., Lim, H. (1996): Appl. Phys. Lett. **69**, 3740. Sect. 6.2.3
Chou, M.Y., Cohen, M.L., Louie, S.G. (1985): Phys. Rev. B **32**, 7979. Sect. 10.2.1
Christensen, N.E. (1988): Phys. Rev. B **37**, 4528. Sect. 5.1.2
Chu, P., Lin, C.L., Wieder, H.H. (1988): Appl. Phys. Lett. **53**, 2423. Sect. 6.1.10
Chyi, J.-I., Shieh, J.-L., Lin, R.-J., Pan; J.-W., Lin, R.-M. (1995): J. Appl. Phys. **77**, 1813. Sect. 6.1.10
Clark, S.A., Wilks, S.P., Morris, J.I., Woolf, D.A., Williams, R.H. (1994): J. Appl. Phys. **75**, 2481. Sect. 6.2.3
Cohen, J., Vilms, J., Archer, R.J. (1969): Hewlett-Packard R&D Report AFCRL-69-0287. Sects. 3.8, 6.1.2; Table 3.5
Constantinidis, G., Kuzmic, J., Michelakis, K., Tsagaraki, K. (1998): Solid-State Electron. **42**, 253. Sect. 6.1.6, Table 3.3
Cook, T.E.jr., Fulton, C.C., Mecouch, W.J., Tracy, K.M., Davis, R.F., Hurt, E.H., Lucovsky, G., Nemanich, R.J. (2003a): J. Appl. Phys. **93**, 3995. Sect. 6.2.7, Table 6.5
Cook, T.E.Jr., Fulton, C.C., Mecouch, W.J., Davis, R.F., Lucovsky, G., Nemanich, R.J. (2003b): J. Appl. Phys. **94**, 3949. Sect. 6.2.7
Copel, M., Reuter, M.C., Kaxiras, E., Tromp, R.M. (1989): Phys. Rev. Lett. **63**, 632. Sect. 3.9.2
Cowley, A.M., Sze, S.M. (1965): J. Appl. Phys. **36**, 3212. Sects. 1.1.3, 5.2
Crowell, C.R. (1965): Solid-State Electron. **8**, 395. Sect. 3.3
Crowell; C.R., Beguwala, M (1971): Solid-State Electron. **14**, 1149. Sect. 3.1
Crowell, C.R., Rideout, V.L. (1969): Solid-State Electron. **12**, 89. Sect. 11
Crowell, C.R., Sze, S.M. (1966): Solid-State Electron. **9**, 1035. Sects. 3.1, 3.3
Crowell, C.R., Spitzer, W.G., Howarth, L.E., LaBate, E.E. (1962): Phys. Rev. **127**, 2006. Sect. 6.1.2, Table 3.5
Crowell, C.R., Sze, S.M., Spitzer, W.G. (1964): Appl. Phys. Lett. **4**, 91. Sects.3.8, 6.1.2; Table 3.5

Crowell, C.R., Kao, T.W., Anderson, C.L., Rideout, V.L. (1972): Surf. Sci. **32**, 591. Sect. 3.8

Cuberes, M.T., Bauer, A., Wen, H.J., Vandré, D., Prietsch, M., Kaindl, G. (1994): J. Vac. Sci. Technol. B **12**, 2422. Sect. 3.6

Dandrea, R.G., Duke, C.B. (1993a): J. Vac. Sci. Technol. B **11**, 1553. Table 7.1

Dandrea, R.G., Duke, C.B. (1993b): J. Vac. Sci. Technol. A **11**, 848. Table 7.1

Dandrea, R.G., Froyen, S., Zunger, A. (1990): Phys. Rev. B **42**, 3213. Sect. 10.2.3

Das, G.P., Blöchl, P., Andersen, O.K., Christensen, N.E., Gunnarsson, O. (1989): Phys. Rev. Lett. **63**, 1168. Sects. 1.1.5, 10.2.2, 7.2; Table 7.1

Davies, A., Craighead, H.G. (1994): Appl. Phys. Lett. **64**, 2833. Sect. 2.3

Deal, B.E., Snow, E.H., Mead, C.A. (1966): J. Phys. Chem. Solids **27**, 1873. Sect. 6.1.11

Defives, D., Noblanc, O., Dua, C., Brylinski, C., Barthula, M., Meyer, F. (1999): Mater. Sci. Eng. B **61**, 395. Sect. 6.1.6, Table 3.3

Dharmadasa, I.M., McLean, A.B., Patterson, M.H., Williams, R.H. (1987): Semicond. Sci. Technol. **2**, 404. Table 3.3

Dharmadasa, I.M., Blomfield, C.J., Scott, C.G., Coratger, R., Ajustron, F., Beauvillain, J. (1998): Solid-State Electron. **42**, 595. Table 3.4

Dharmarasu, N., Arulkumaran, S., Sumathi, R.R., Jayavel, P., Kumar, J., Magudapathy, P., Nair, K.G.M. (1998): Nucl. Instr. and Meth. B **140**, 119. Sect. 6.1.3, Table 3.3

Di Dio, M., Cola, A., Lupo, M.G., Vasanelli, L. (1995): Solid-State Electron. **38**, 1923. Sect. 6.1.3, Table 3.3

Di Maria, D.J., Arnett, P.C. (1975): Appl. Phys. Lett. **26**, 711. Table 6.4

Ding, S.A., Barman, S.R., Horn, K., Yang, H., Yang, B., Brandt, O., Ploog, K. (1997): Appl. Phys. Lett. **70**, 2407. Sects. 6.2.2, 6.2.6

Donoval, D., Barus, M., Zdimal, M. (1991): Solid-State Electron. **34**, 1365. Sect. 6.1.2, Table 3.3

Donoval, D., Drobny, V., Luza, M. (1998): Solid-State Electron. **42**, 235. Sect. 6.1.2, Table 3.3

Duboz, J.Y., Badoz, P.A., d'Avitaya, F.A., Rosencher, E. (1989a): Phys. Rev.. B **40**, 10607. Sect. 8.2.1

Duboz, J.Y., Badoz, P.A., Arnaud d'Avitaya, F., Chroboczek, J.A. (1989b): Appl. Phys. Lett. **55**, 84. Sect. 8.2.1

Duc, R.M., Hsu, C., Faurie, J.P. (1987): Phys. Rev. Lett. **58**, 1127. Sects. 6.2.2, 6.2.6

Duggan, G. (1985): J. Vac. Sci. Technol. B **3**, 1225. Sects. 6.2.2, 6.2.3

Duke, C.B., Mailhiot, C. (1985): J. Vac. Sci. Technol. B **3**, 1170. Sect. 9.1

Dumas, P., Chabal, Y.J., Higashi, G.S. (1990): Phys. Rev. Lett. **65**, 1124. Sect. 10.1

Ebert, P. (2002a): Appl. Phys. A **75**, 101. Sect. 9.2

Ebert, P. (2002b): private communication. Sect. 9.2

Ebert, P., Urban, K. (1993): Ultramicroscopy **49**, 344. Sect. 9.2

Ebert, P., Urban, K., Lagally, M.G. (1994): Phys. Rev. Lett. **72**, 840. Sect. 9.2

Ebert, P., Urban, K., Aballe, L., Chen, C.H., Horn, K., Schwarz, G., Neugebauer, J., Scheffler, M. (2000): Phys. Rev. Lett. **84**, 58.16. Sect. 9.2

Eftekhari, G. (1991): Thin Solid Films **199**, 23. Sect. 6.1.2, Table 3.3

Elsbergen, V. van (1998): *Untersuchungen an reinen und Adsorbat-bedeckten SiC-Oberflächen und Charakterisierung von Metall-SiC-Grenzflächen*; Dissertation, Universität Duisburg. Sects. 3.9.2, 6.1.6; Tab.3.6

Elsbergen, V. van, Kampen, T.U., Mönch, W. (1996): J. Appl. Phys. **79**, 316. Sects. 1.1.4, 3.9.2, 6.1.6; Table 3.6
Emde, M. von der (1991): Diplomarbeit, TU Berlin. Sect. 6.1.4, Table 3.3
Emde, M. von der, Zahn, D.R.T., Schultz, C., Evans, D.A., Horn, K. (1992): J. Appl. Phys. **72**, 4486. Sect. 6.1.4, Table 3.3
Escher, J.S., James, L.W., Sankaran, R., Antypas, G.A., Moon, R.L., Bell, R.L. (1976): J. Vac. Sci. Technol. **13**, 874. Sect. 6.2.3
Evans, D.A., Chen, T.P., Chasseé, T., Horn, K., Emde, M. von der, Zahn, D.R.T. (1992): Surf. Sci. **269/270**, 979. Sect. 6.1.4, Table 3.3
Eyckeler, M., Mönch, W., Kampen, T.U., Dimitrov, R., Ambacher, O., Stutzmann, M. (1998): J. Vac. Sci. Technol. B **16**, 2224. Sects. 1.1.4, 3.9.2, 6.1.5; Table 3.6

Feidenhans'l, R., Grey, F., Nielsen, M., Johnson, R.L. (1990): in *Kinetics of Ordering and Growth at Surfaces*, ed. by M.G. Lagally (Plenum Press, New York), p. 189. Sect. 10.2.1
Fernandez, A., Hallen, H.D., Huang, T., Buhrman, R.A., Silcox, J. (1991): J. Vac. Sci. Technol. B **9**, 590. Sect. 6.1.2, Table 3.4
First, P.N., Stroscio, J.A., Dragoset, R.A., Pierce, D.T., Celotta, R.J. (1989): Phys. Rev. Lett. **63**, L55. Sect. 6.1.12
Fischer, C. (1972): Atomic Data **4**, 301. Sect. 5.5.2
Fischer, A.E.M.J., Vlieg, E., Veen, J.F. van der, Clausnitzer, M., Materlick, G. (1987): Phys. Rev. B **36**, 4769. Sect. 10.2.2
Fischer, A.E.M.J., Gustavson, T., Veen, J.F. van der (1988): Phys. Rev. B **37**, 6305. Sect. 10.2.2
Föll, H. (1982): phys. stat. sol. (a) **69**, 779. Sect. 10.2.2
Forchhammer, T., Veje, E., Tidemand-Petersson, P. (1995): Phys. Rev. B **52**, 14693. Sect. 6.2.3
Forrest, S.R., Schmidt, P.H., Wilson, R.B., Kaplan, M.L. (1984): Appl. Phys. Lett. **45**, 1199. Sect. 6.2.3
Fowell, A.E., Williams, R.H., Richardson, B.E., Shen, T.-H. (1990): Semicond. Sci. Technol. **5**, 348. Sects. 1.1.5, 4.1; Table 3.4
Fowler, R.H. (1931): Phys. Rev. **38**, 45. Sect. 3.8
Frank, F.C., Merve, J.H. van der (1949): Proc. Roy. Soc. London Ser. A **198**, 216. Sect. 3.9.2
Freeouf, J.L., Woodall, J.M. (1981): Appl. Phys. Lett. **39**, 727. Sects. 1.1.4, 4.5.2
Freeouf, J.L., Jackson, T.N., Laux, S.E., Woodall, J.M. (1982a): Appl. Phys. Lett. **40**, 634. Sects. 1.1.5, 4.1, 4.2, 4.3
Freeouf, J.L., Jackson, T.N., Laux, S.E., Woodall, J.M. (1982b): J. Vac. Sci. Technol. B **21**, 570. Sects. 1.1.5, 4.1, 4.2., 4.3
Fritsch, O. (1935): Ann. Physik **22**, 375. Sect. 1.1.1
Fritts, C.E. (1883): Amer. J. Sci. **26**, 465. Sect. 1.1.1
Fujitani, H. (1998): Phys. Rev. B **57**, 8801. Sect. 10.2.2
Fujitani, H., Asano, S. (1990): Phys. Rev. B **42**,1696. Sects. 1.1.5, 10.2.2, 7.2; Table 7.1
Fujitani, H., Asano, S. (1994): Phys. Rev. B **50**, 8681. Sect. 6.1.2
Fukutani, K., Iwai, H., Yamashita, H., Murata, Y., Hatori, S., Kobayashi, K. (1997): Surf. Sci. **377**, 1010. Sects. 3.5, 10.1

Gadzuk, J.W. (1967): Surf. Sci. **6**, 133. Sect. 9.1
Gadzuk, J.W. (1969): in *Structure and Chemistry of Solid Surfaces,* ed. by G.A. Somorjai (Wiley, New York), p. 43-1. Sect. 9.1
Garcia-Moliner, F., Flores, F. (1979): *Introduction to the Theory of Solid Surfaces* (Cambridge Univ. Press, Cambridge). Sect. 5.4
Gibson, J.M., Benn, J.C., Poate, J.M., Tung, R.T. (1982): Appl. Phys. Lett. **41**, 818. Sect. 10.2.2
Gibson, J.M., Tung, R.T., Poate, J.M. (1983): Mater. Res. Soc. Symp. Proc. **14**, 395. Sect. 10.2.2
Gilleo, M.A. (1953): Phys. Rev. **91**, 534. Sect. 3.8
Godby, R.W., Schlüter, M., Sham, L.J. (1988): Phys. Rev. B 37, 10159. Sect. 7.1
Goodman, A.M. (1968): Appl. Phys. Lett. **13**, 275. Table 6.4
Goodwin, E.T. (1939a): Proc. Cambridge Philos. Soc. **35**, 205. Sects. 1.1.3, 5.3.1, 5.4
Goodwin, E.T. (1939b): Proc. Cambridge Philos. Soc. **35**, 221. Sect. 1.1.3
Goodwin, E.T. (1939c): Proc. Cambridge Philos. Soc. **35**, 232. Sects. 1.1.3
Gordy, W., Thomas, W.J.O. (1956): J. Chem. Phys. **24**, 439. Sects. 1.1.4, 5.2
Gossick, B.R. (1963): Solid-State Electron. **6**, 445. Sect. 3.1
Göthelid, M., Hammar, M., Björkqvist, M., Karlsson, U.O., Flodström, S.A., Wigren, C., LeLay, G. (1994): Phys. Rev. B **50**, 4470. Sect. 3.11, Table 3.7
Grant, R.W., Waldrop, J.R., Kraut, E.A. (1978): Phys. Rev. Lett. **40**, 656. Sect. 3.12.2
Grehk, T.M., Larsson, C.U.S., Prince, N.P., Flodström, S.A. (1993): Surf. Sci. **284**, L384. Sect. 3.11, Tab: 3.7
Grehk, T.M., Johansson, L.S.O., Gray, S.M., Johansson, M., Flodström, S.A. (1995): Phys. Rev. B **52**, 16593. Sect. 3.11, Table 3.7
Grey, F., Feidenhans'l, R., Nielsen, M., Johnson, R.L. (1989): J. Phys. Coll. **C7**, 181. Sect. 10.2.1
Grondahl, L.O. (1925): U.S. Patent No. 1,640,335. Sect. 1.1.1
Grondahl, L.O. (1926): Phys. Rev. **26**, 813. Sect. 1.1.1
Grondahl, L.O. (1933): Rev. Mod. Phys. **5**, 141. Sect. 1.1.1
Grossner, U., Furthmüller, J., Bechstedt, F. (1999): phys. stat. sol. (b) **216**, 675. Sects. 7.1, 7.2
Grundner, M., Jacob, H. (1986): Appl. Phys. A **39**, 73. Sect. 10.1
Grunthaner, F.J., Grunthaner, P.J. (1986): Mater. Sci. Rep. **1**, 65. Table 6.4
Grunwald, F. (1987): Diploma Thesis, Universität Duisburg (unpublished). Sects. 1.1.5, 3.9.2, 6.1.3; Table 3.6
Grupp, C., Taleb-Ibrahimi, A. (1998): Surf. Sci. **408**, 160. Sect. 3.11, Table 3.7
Gurney, R.W. (1935): Phys. Rev. **47**, 479. Sect. 9.1
Gyulai, Z. (1926): Z. Physik **35**, 411. Sect. 3.8

Hackam, R., Harrop, P. (1972): IEEE Trans. Electron. Devices **ED-19**, 1231. Sect. 6.1.3, Table 3.3
Hanney, N.B., Smith, C.P. (1946): J. Am. Chem. Soc. **68**, 171. Sect. 3.11,
Hardikar, S., Hudait, M.K., Modak, P., Krupanidhi, S.B., Padha, N. (1999): Appl. Phys. A **68**, 49. Sect. 6.1.3, Table 3.3
Harrison, W.A. (1973): Phys. Rev. B **8**, 4487. Sect. 5.5.2
Harrison, W.A. (1977): J. Vac. Sci. Technol. **14**, 1016. Sect. 1.2
Harrison, W.A. (1980): *Electronic Structure and the Properties of Solids* (Freeman, San Francisco, Ca.). Sect. 5.5.2

Harrison, W.A. (1981): Phys. Rev. B **24**, 5835. Sect. 5.5.2

Harrison, W.A., Kraut, E.A., Waldrop, J.R., Grant, R.W. (1978): Phys. Rev. B **18**, 4402. Sect. 6.2.1

Hasegawa, H., Ohno, H. (1986): J. Vac. Sci. Technol. B 4, 1130. Sect. 1.1.4

Heine, V. (1965): Phys. Rev. **138**, A1689. Sects. 1.1.3, 5.3.1, 5.3.2, 5.4

Henisch, H.K. (1957): *Rectifying Semi-Conductor Contacts* (Clarendon Press, Oxford). Sect. 1.1.1

Henz, J., Ospelt, M., Känel, H. von (1987): Solid State Commun. **63**, 445. Sect. 10.2.2

Heske, C., Eich, D., Fink, R., Umbach, E., Buuren, T.van, Bostedt, C., Terminello, L.J., Kakar, S., Grush, M.M., Callcott, T.A., Himpsel, F.J., Ederer, D.L., Perera, R.C.C., Riedl, W., Karg, F. (1999): Appl. Phys. Lett. **74**, 1451. Sect. 6.2.4

Heslinga, D.R., Weitering, H.H., Werf, D.P. van der, Klapwijk, T.M., Hibma, T. (1990): Phys. Rev. Lett. **64**, 1589. Sect. 6.1.2, Table 3.3

Higashi, G.S., Chabal, Y.J., Trucks, G.W., Raghavachari, K. (1990): Appl. Phys. Lett. **56**, 656. Sect. 10.1

Higashiyama, K., Kono, S., Kinoshita, T., Miyahara, T., Kato, H., Ohsawa, H., Enta, Y., Maeda, F., Yaegashi, Y. (1987): Surf. Sci. **1**, L568. Sect. 3.11, Table 3.7

Himpsel, F.J., McFeely, F.R., Taleb-Ibrahimi, A., Yarmoff, J.A., Hollinger, G. (1988): Phys. Rev. B **38**, 6084. Table 6.4

Hirose, K., Sakano, K., Nohira, H., Hattori, T. (2001): Phys. Rev. B **64**, 155325. Table 6.4

Hiyamizu, S., Fujii, T., Muto, S., Inata, T., Nakata, Y., Sugiyama, Y., Sasa, S. (1987): J. Crystal Growth **81**, 349. Sect. 6.2.3

Ho, P.S., Yang, E.S., Evans, H.L., Wu, X. (1986): Phys. Rev. Lett. **56**, 177. Sect. 10.2.2

Hoekstra, J., Kohyama, M. (1998): Phys. Rev. B **57**, 2334. Sects. 5.3.1, 7.2; Table 7.1

Holloway, P.H., Kim, T.-J., Trexler, J.T., Miller, S., Fijol, J.J., Lampert, W.V., Haas, T.W. (1997): Appl. Surf. Sci. **117**, 362. Sect. 6.1.10

Hong, H., Aburano, R.D., Lin, D.-S., Chen, H., Chiang, T.-C. (1992): Phys. Rev. Lett. **68**, 507. Sects. 1.1.4, 3.4, 3.5, 10.2.1

Horváth, Zs.J. (1999): private communication. Sect. 6.1.2, Table 3.3

Horváth, Zs.J., Molnár, G., Kovács, B., Petö, G., Andrási, M., Szentpáli, B. (1993): J. Crystal Growth **126**, 163. Sect. 6.1.2, Table 3.3

Howes, P.B., Edwards, K.A., Hughes, D.J., Macdonald, J.E., Hibma, T., Bootsma, T., James, M.A. (1995): Phys. Rev. B **51**, 17740. Sects. 1.1.4, 3.5, 10.2.1

Hsiao, G.S., Virtanen, J.A., Penner, R.M. (1993): Appl. Phys. Lett. **63**, 1119. Sect. 10.1

Hübers, H.-W., Röser, H.P. (1998): J. Appl. Phys. **84**, 5326. Sect. 6.1.3, Table 3.3

Hybertsen, M.S., Louie, S.G. (1986): Phys. Rev. B **34**, 5390. Sects. 5.5.1, 7.1

Ibach, H., Lüth, H. (1991): *Solid-State Physics* (Springer, Berlin). Sects. 5.4, 5.5.2

Ihm, J., Louie, S.G., Cohen, M.L. (1978): Phys. Rev. Lett. **40**, 1208. Sect. 5.6; Table 5.1

Ihm, J., Joannopoulos, J.D. (1982): Phys. Rev. B **26**, 4429. Sect. 3.9.2

Im, H.-J., Kaczer, B., Pelz, J.P., Choyke, W.J. (1998): Appl. Phys. Lett. **72**, 839 and private communication. Sect. 6.1.6, Tables 3.3, 3.4

Im, H-J-, Ding, Y., Pelz, J.P., Choyke, W.J. (2001): Phys. Rev. B **64**, 075310. Sects. 4.4, 4.5.1, 6.1.6; Tables 3.3, 3.4

Jacob, W., Bertel, E., Dose, V. (1987): Europhys. Lett. **4**, 1303. Sect. 1.1.4

Jäger, D., Kassing R. (1977): J. Appl. Phys. **48**, 4413. Sect. 6.1.2, Table 3.3

Jakob, P., Chabal, Y.J., Raghavashari, K., Dumas, P., Christman, S.B. (1993): Surf. Sci. **285**, 251. Sect. 10.1

Jin, M.S., Kim, W.T., Yoon, C.S. (1993): J. Phys. Chem. Solids **54**, 1509. Sect. 6.2.8

Jones, F.E., Wood, B.P., Myers, J.A., Daniels-Hafer, C., Lonergan, M.C. (1999): J. Appl. Phys. **86**, 6431. Sect. 3.5

Jong, A.F. de, Bulle-Lieuwma, C.W.T. (1990): Phil. Mag. A **62**, 183. Sect. 10.2.2

Käckell, P., Wenzien, B., Bechstedt, F. (1994): Phys. Rev. B **50**, 10761. Sect. 6.1.6, Table 6.1

Kahn, A., Stiles, K., Mao, D., Horng, S.F., Young, K. (1989): in *Metallization and Metal–Semiconductor Interfaces,* ed. by I.P. Batra (Plenum, New York), p. 163. Sect. 3.9.2

Kaiser, W.J., Bell, L.D. (1988): Phys. Rev. Lett. **60**, 1406. Sects. 1.1.5, 3.6, 4.1

Kajiyama, K., Mizushima, Y., Sakata, S. (1973): Appl. Phys. Lett. **23**, 458. Sects. 6.1.10, 6.2.3

Kampen, T.U., Mönch, W. (1995): Surf. Sci. **331**, 490. Sects. 3.5, 6.1.2, 10.1; Table 3.3

Kampen, T.U., Mönch, W. (1997): Appl. Surf. Sci. **117**, 388. Sects. 3.5, 6.1.5; Table 3.3

Kampen, T.U., Schmitsdorf, R.F., Mönch, W. (1995): Appl. Phys. A **60**, 391. Sects. 3.5, 10.1

Kampen, T.U., Eyckeler, M., Mönch, W. (1998): Appl. Surf. Sci. **123/124**, 28. Sect. 3.9.2

Karpov, K.A. (1965): *Tables of the function* $w(z) = \exp(-z^2)\,_0\!\int^z \exp(x^2)dx$ *in the complex domain* (Pergamon, Oxford). Sect. 3.1

Karrer, U., Dobner, A., Ambacher, O., Stutzmann, M. (1999): phys. stat. sol. (a) **176**, 163. Sect. 6.1.5, Table 3.3

Kawarada, H., Aoki, M., Sasaki, H., Tsugawa, K., Ohdomari, I. (1994): in *Control of Semiconductor Interfaces,* ed. by I. Ohdomari, M. Oshima, and A. Hiraki (Elsevier, Amsterdam), p. 161. Sect. 10.1

Ke, S., Wang, R., Huang, M. (1994): Phys. Rev. B **49**, 10495. Sect. 8.3

Ke, S.., Zi, J., Zhang, K., Xie, X. (1996a): Phys. Rev. B **54**, 8789. Sect. 6.1.6, Table 6.1

Ke, S., Zhang, K., Xie, X. (1996b): J. Phys.: Condens. Matter **8**, 10209. Sect. 6.6, Table 6.1

Keister, J.W., Rowe, J.E., Kolodziej, J.J., Niimi, H., Madey, T.E. Lucovsky, G. (1999): J. Vac. Sci. Technol. B **17**, 1831. Table 6.4

Khan, M.R.H., Nakayama, H., Detchprom, T., Hiramatsu, K., Sawaki, N. (1997): Solid-State Electron. **41**, 287. Sect. 6.1.10

Khemka, V., Chow, T.P., Gutman, R.J. (1998): J. Electron Mater. **27**, 1128. Sect. 6.1.6, Table 3.3

Kim, J.W., Kim, S., Seo, J.M., Tanaka, S., Kamada, M. (1998): J. Phys.: Condens. Matter **10**, 3731. Sect. 3.11, Table 3.7

King, S.W., Ronning, C., Davis, R.F., Benjamin, M.C., Nemanich, R.J. (1998): J. Appl. Phys. **84**, 2086. Sect. 6.2.6

King, S.W., Davis, R.F., Ronning, C., Nemanich, R.J. (1999): J. Electron. Mater. **28**, L34. Sect. 6.2.6

Klein, A., Pettenkofer, C., Jaegermann, W., Lux-Steiner, M., Bucher, E. (1994): Surf. Sci. **321**, 19. Sect. 6.2.5

Klein, A., Löher, T., Tomm, Y., Pettenkofer, C., Jaegermann, W. (1997): Appl. Phys. Lett. **70**, 1299. Sect. 6.2.4, Table 6.3

Koenigsberger, J., Weiss, J. (1911): Ann. Physik (IV) **35**, 1. Sect. 1.1.1

Kohyama, M., Hoekstra, J. (2000): Phys. Rev. B **61**, 2672. Sect. 7.2, Table 7.1

Koma, A., Sunouchi, K., Miyajima, T. (1985): J. Vac. Sci. Technol. B **3** 724. Sect. 6.2.1

Koma, A., Ueno, K., Saiki, K. (1991): J. Crystal Growth **111**, 1029. Sect. 6.2.1

Kościelniak-Mucha, B., Opanowicz, A. (1992): phys. stat. sol. (a) **141**, K67. Table 6.2

Kościelniak-Mucha, B., Opanowicz, A. (1994): phys. stat. sol. (a) **130**, K55. Table 6.2

Kowalczyk, S.P., Kraut, E.A., Waldrop, J.R., Grant, R.W. (1982a): J. Vac. Sci. Technol. **21**, 482. Sects. 6.2.2, 10.2.3

Kowalczyk, S.P., Schaffer, W.J., Kraut, E.A., Grant, R.W. (1982b): J. Vac. Sci. Technol. **20**, 705. Sect. 6.2.2

Kraut, E.A., Grant, R.W., Waldrop, J.R., Kowalczyk, S.P. (1980): Phys. Rev. Lett. **44**, 1620. Sects. 3.12.2, 6.2.2

Kroemer, H. (1985): in *Molecular Beam Epitaxy and Heterostructures*, ed. by LL. Chang and K. Ploog (M. Nijhoff, Dordrecht), p. 331. Sect. 1.2

Kuech, T.F., McCaldin, J.O. (1980): J. Vac. Sci. Technol. **17**, 891. Sect. 6.1.10

Kuhaimi, S.A. Al (1998): Sol. Energy Mater. Sol. Cells **52**, 69. Sect. 6.2.4, Table 6.3

Kuo, H.C., Kuo, J.M., Wang, Y.C., Lin, C.H., Chen, H., Stillman, G.E. (1997): J. Electron. Mater. **26**, 944. Sect. 6.2.3

Kurtin, S., Mead, C.A. (1969): J. Phys. Chem. Solids **30**, 2007. Sects. 1.1.4, 6.1.8, 6.2.5; Table 3.5

Kurtin, S., McGill, T.C., Mead, C.A. (1969): Phys. Rev. Lett. **22**, 1433. Sects. 1.1.4, 5.1.1, 5.6

Kwon, S.D., Kwon, H.K., Choe, B.-D., Lim, H., Lee, J.Y. (1995): J. Appl. Phys. **78**, 2482. Sects. 6.1.10, 6.2.3

Lambrecht, W.R.L., Segall, B., Andersen, O.K. (1990): Phys. Rev. B **41**, 2813. Sect. 5.1.2

Landmann, I., Jäger, D. (1975): Thesis, Universität Münster (unpublished). Sect. 6.1.2, Table 3.3

Landoldt-Börnstein (2001): *Numerical Data and Functional Relationships in Science and Technology, Vol.* III/41, ed. by O. Madelung, U. Rössler, and M Schulz (Springer, Berlin)

Lang, D.V., Panish, M.B., Capasso, F., Allam, J., Hamm, R.A., Sergent, A.M., Tsang, W.T. (1987): Appl. Phys. Lett. **50**, 736. Sect. 6.2.3

Lang, O., Klein, A., Pettenkofer, C., Jaegermann, W. (1996): J. Appl. Phys. **80**, 3817. Sect. 6.2.5

Langer, J.M., Heinrich, H. (1985): Phys. Rev. Lett. **55**, 1414. Sect. 6.2.8

Langer, J.M., Dobaczewski, L., Missous, M. (1995): in *Proc. 22nd Int'l Conf. on the Physics of Semiconductors*, ed. by D.J. Lockwood (World Scientific, Singapore), p. 588. Sect. 8.2.1

Laridjani, M.R., Masri, P., Majewski, J.A. (2001): Mat. Res. Soc. Symp. Proc. **639**, G11.34. Sect. 7.3.2

Lazzarino, M., Scarel, G., Rubini, S., Bratina, G., Sorba, L., Franciosi, A., Berthod, C., Binggeli, N., Baldereschi, A. (1998): Phys. Rev. B **57**, R9431. Sect. 7.2, Table 7.1

Ledebo, L.-Å., Ridley, (1982): J. Phys. C: Solid State Phys. **15**, L961. Sect. 6.2.8

Lee, J.-K., Cho, Y.-H., Choe, B.-D., Kim, K.S., Jeon, H.I., Lim, H., Razeghi, M. (1997): Appl. Phys. Lett. **71**, 912. Sect. 6.2.3

Lee, S.-K., Zetterling, C.-M., Östling, M. (2001): J. Electron. Mat. **30**, 242. Sect. 4.4, Table 3.3

Lee, S.-K., Zetterling, C.-M., Östling, M., Åberg, I., Magnusson, M.H., Deppert, K., Wernersson, L.E., Samuelson, L., Litwin, A. (2002): Solid-State Electron. **46**, 1433. Sect. 4.4, Table 3.3

Leuther, A., Förster, A., Lüth, H., Holzbrecher, H., Breuer, U. (1996): Semicond. Sci. Technol. **11**, 766. Sect. 6.2.3

Lin, C.L., Chu, P., Kellner, A.L., Wieder, H.H., Rezek, E.A. (1986): Appl. Phys. Lett. **49**, 1593. Sects. 6.1.10, 6.2.3

Lin, D.-S., Miller, T., Chiang, T.-C. (1991a): Phys. Rev. B **44**, 10719. Sect. 3.11, Table 3.7

Lin, D.-S., Miller, T., Chiang. T.-C. (1991b): Phys. Rev. Lett. **67**, 2187. Sect. 3.11, Table 3.7

Lince, J.R., Carré, D.J., Fleischauer, P.D. (1987): Phys. Rev. B **36**, 1647. Sect. 6.2.5

Lindell, A., Pessa, M., Salokatve, A., Bernardini, F., Nieminen, R.M., Paalanen, M. (1997): J. Appl. Phys. **82**, 3374. Sect. 8.1

Linz, R., Clemens, H.J., Mönch, W. (1993): J. Vac. Sci. Technol. B 14, 1591. Sects. 1.1.4, 3.9.2, 6.1.4; Table 3.6

Loenen, E.J. van, Frenken, J.W.M., Veen, J.F. van der, Dev, B.N., Valeri, S. (1985): Phys. Rev. Lett. **54**, 827. Sect. 10.2.2

Löher, T. (1995): *Halbleiter-Heteroübergänge: II–VI/Schichtgitterchalkogenide- und CdS/CuInSe2-Grenzflächen*; Dissertation, Freie Universität Berlin. Sect. 6.2.5

Löher, T., Pettenkofer, C., Jaegermann, W. (1994): in *Proc. 24th IEEE Photovoltaic Specialists Conf.*, p. 295. Table 6.2

Löher, T., Jaegermann, W., Pettenkofer, C. (1995): J. Appl. Phys. **77**, 731. Sect. 6.2.4, Table 6.3

Louie, S.G., Cohen, M.L. (1976): Phys. Rev. B **13**, 2461. Sects. 1.1.3, 5.3.1, 5.4

Louie, S.G., Chelikowsky, J.R., Cohen, M.L. (1977): Phys. Rev. B **15**, 2154. Sects. 1.1.3, 5.6; Table 5.1

Louis, E., Yndurain, F., Flores, F. (1976): Phys. Rev. B **13**, 4408. Sect. 1.1.3; Table 5.1

Lucas, C.A., Loretto, D. (1995): Surf. Sci. **344**, L1219. Sects. 3.5, 10.2.1

Ludeke, R. (1989): in *Metallization and Metal–Semiconductor Interfaces,* ed. by I.P. Batra (Plenum, New York), p. 39. Sect. 9.1

Ludeke, R. (2002): J. Non-Cryst. Solids **303**, 159. Table 6.4

Ludeke, R., Jezequel, G., Taleb-Ibrahimi, A. (1988a): J. Vac. Sci. Technol. B **6**, 1277. Sect. 9.1

Ludeke, R., Jezequel, G., Taleb-Ibrahimi, A. (1988b): Phys. Rev. Lett. **61**, 601. Sect. 9.1

Ludeke, R., Prietsch, M., Samsawar A. (1991): J. Vac. Sci. Technol. B **9**, 2342. Sect. 3.6

Ludwig, M., Heymann, G., Janietz, P. (1986): J. Vac. Sci. Technol. B **4**, 485. Sect. 1.1.3

Maeda, K., Kitahara, E. (1998): Appl. Surf. Sci. **130**, 925. Sects. 3.5, 6.1.2; Table 3.3

Maffeis, T.G.G., Simmonds, M.C., Clark, S.A., Peiro, F., Haines, P., Parbrook, P.J. (2000): J. Phys. D: Appl. Phys. **33**, L115 and private communication. Sect. 6.1.5, Table 3.3

Magnusson, K.O., Wiklund, S., Dudde, R., Reihl, B. (1991): Phys. Rev. B **44**, 5657. Sect. 3.11, Table 3.7

Majewski, J.A., Städele, M., Vogl, P. (1997): Mat. Res. Soc. Symp. Proc. **449**, 917. Sect. 7.3.2

Majewski, J.A., Vogl, P. (1998): MRS Internet J. Nitride Semicond. Res. **3**, 21. Sect. 6.1.6, Table 6.1

Mamy, R., Zaoui, X., Barrau, J., Chevy, A. (1990): Revue Phys. Appl. **25**, 947. Sects. 1.1.4, 6.2.5

Manago, T., Miyanishi, S., Akinaga, H., Roy, W. van, Roelfsema, R.F.B., Sato, T., Tamura, E. (2000): J. Appl. Phys. **88**, 2043. Sect. 6.1.3, Table 3.3

Marée, P.M.J., Nakagawa, K., Mulders, F.M., Veen, J.F. van der (1987): Surf. Sci. **191**, 305. Sect. 3.9.2

Margaritondo, G., Katnani, A.D., Stoffel, N.G., Daniels, R.R., Zho, T.-X. (1982): Solid State Commun. **43**, 163. Sect. 6.2.6

Martin, G., Strite, S.C., Botchkarev, A., Agarwal, A., Rockett, A., Lambrecht, W.R.L., Segall, B., Morkoç, H. (1994): Appl. Phys. Lett. **65**, 610. Sect. 6.2.6

Mattern, B., Bassler, M., Pensl, G., Ley, L. (1998): Mater. Sci. Forum **264-268**, 375. Table 6.5

Maue, A.W. (1935): Z. Physik **94**, 717. Sects. 1.1.3, 1.1.5, 5.3.1, 5.4

McCafferty, P.G., Sellai, A., Dawson, P., Elabd, H. (1996): Solid-State Electron. **39**, 583. Sect. 6.1.2, Table 3.3

McCaldin, J.O., McGill, T.C., Mead, C.A. (1976): J. Vac. Sci. Technol. **13**, 802. Sect. 1.1.4

McLean, A.B., Williams, R.H. (1988): J. Phys. C: Solid State Phys. **21**, 783. Sect. 1.1.3

McLean, A.B., Dharmadasa, I.M., Williams, R.H. (1986a): Semicond. Sci. Technol. **1**, 137. Table 3.3

McLean, A.B., Evans, D.A., Williams, R.H. (1986b): Semicond. Sci. Technol. **2**, 547. Sect. 1.1.3

Mead, C.A. (1965): Phys. Lett. **18**, 218. Sect. 6.2.5

Mead, C.A. (1966): Solid-State Electron. **9**, 1023. Sect. 1.1.4

Mead, C.A., McGill, T.C. (1976): Phys. Lett. **58A**, 249. Sect. 10.1, Table 3.5

Mead, C.A., Spitzer, W.G. (1964): Phys. Rev. **134**, A713. Sect. 6.2.5, Tab.3.5

Mezey, L.Z., Giber, J. (1982): Japan. J. Appl. Phys. **21**, 1569. Sect. 3.9.2

Michaelson, H.B. (1977): J. Appl. Phys. **48**, 4729. Sect. 5.2

Miedema, A.R. (1978): Z. Metallkde. **69**, 287. Sect. 3.9.2

Miedema, A.R., Boer, F.R. de, Châtel, P.F. de (1973): J. Phys. F **3**, 1558. Sect. 5.2

Miedema, A.R., de Chätel, P.F., de Boer, F.R. (1980): Physica **100**B, 1. Sect. 5.2

Missous, M., Rhoderick, E.H. (1986): Electron. Lett. **22**, 477. Sect.3.4

Missous, M., Truscott, W.S., Singer, K.E. (1990): J. Appl. Phys. **68**, 2239. Sect. 6.1.10

Miura, Y., Fujieda, S., Hirose, K. (1994): Phys. Rev. B **50**, 4893. Sect. 6.1.2, Table 3.3

Miyazaki, S. (2001): J. Vac. Sci. Technol. B **19**, 2212. Table 6.4

Mönch, W. (1969): phys. stat. sol. **36**, 9. Sect. 3.6

Mönch, W. (1970): phys. stat. sol. **40**, 257. Sect. 3.10

Mönch, W. (1983): Surf. Sci. **132**, 92. Sect. 1.1.4

Mönch, W. (1986): in *Festkörperprobleme* (Adv. Solid State Physics) Vol. 26, ed. by P. Grosse (Vieweg, Braunschweig), p. 67. Sects. 1.1.4, 1.1.5, 1.2, 5.6

Mönch, W. (1987): Phys. Rev. Lett. **58**, 1260. Sect. 1.1.5

Mönch, W. (1988a): Phys. Rev. B **37**, 7129. Sect. 1.1.5

Mönch, W. (1988b): J. Vac. Sci. Technol. B **6**, 1270. Sect. 1.1.5

Mönch, W. (1994a): in *Control of Semiconductor Interfaces*, ed. by I. Ohdomari, M. Oshima, and A. Hiraki (Elsevier, Amsterdam), p. 169. Sect. 6.1.6

Mönch, W. (1994b): Europhys. Lett. **27**, 479. Sect. 10.1

Mönch, W. (1996a): Appl. Surf. Sci. **92**, 367. Sects. 1.1.5, 5.2, 5.6

Mönch, W. (1996b): J. Appl. Phys. **80**, 5076. Sects. 1.1.5, .5.5.1, 5.5.2, 6.1.5, 6.1.10, 6.2.3; Table 5.1

Mönch, W. (1996c): J. Vac. Sci. Technol. B **14**, 2985. Sect. 6.1.11

Mönch, W. (1998a): Solid-State Electron. **42**, 470. Sect. 6.1.10

Mönch, W. (1998b): Appl. Phys. Lett. **72**, 1899. Sects. 6.1.8, 6.2.5

Mönch, W, (1999): J. Vac. Sci. Technol. B **17**, 1867. Sects. 1.1.3, 1.1.5

Mönch, W. (2001): *Semiconductor Surfaces and Interfaces*, 3rd Edn. (Springer, Berlin). Sects. 1.1.3, 1.1.5, 1.2, 2.1, 3.11, 10.1, 10.2.1

Mönch W. (2002): at *Int'l Workshop on Electronic, Structural, and Chemical Properties of Cu-Chalcopyrite Surfaces and Interfaces*, Triberg, June 2002 (unpublished). Sects. 6.2.4, 6.2.8

Mönch, W., Koenders, L. (1985): in Proc. *17th Int'l Conf. on the Physics of Semiconductors,* ed. by J.D. Chadi and W.A. Harrison (Springer, New York), p. 85. Sect. 8.1

Morgan, B.A., Talin, A.A., Bi, W.G., Kavanagh, K.L., Williams, R.S., Tu, C.W., Yasuda, T., Yasui, T., Segawa, Y. (1996a): Mater. Chem. Phys. **46**, 224. Sects. 4.4, 6.1.2; Table 3.4

Morgan, B.A., Ring, K.M., Kavanagh, K.L., Talin, A.A., Williams, R.S., Yasuda, T., Yasui, T., Segawa, Y. (1996b): J. Appl. Phys. **79**, 1532. Sects. 6.1.2, 6.1.7, 6.2.5; Table 3.4

Morkel, M., Weinhardt, L., Lohmüller, B., Heske, C., Umbach, E., Riedl, W., Zweigart, S., Karg, F. (2001): Appl. Phys. Lett. **79**, 4482. Sect. 6.2.4, Table 6.3

Morrison, S.R. (1953): J. Phys. Chem. **57**, 860. Sect. 1.1.3

Mott, N.F. (1938): Proc. Camb. Philos. Soc. **34**, 568. Sect. 1.1.2

Mott, N.F., Gurney, R.W. (1940): *Electronic Processes in Ionic Crystals* (Oxford University Press, Oxford), p. 174. Sect. 3.8

Muller, D.A., Shashkow, D.A., Benedek, R., Yang, L.H., Silcox, J., Seidman, D.N. (1998): Phys. Rev. Lett. **80**, 4741. Sect. 6.1.12

Nadenau, V., Braunger, D., Hariskos, D., Schock, H.W. (1998): in *Ternary and Multinary Compounds* (Inst. of Phys., Bristol), p. 955. Sect. 6.2.4, Table 6.3

Nanda, A., Hafich, M.J., Woods, L.M., Patrizi, G.A., Robinson, G.Y. (1992): Appl. Phys. Lett. **61**, 81. Sect. 6.1.10

Narayanamurti, V., Kozhevnikov, M. (2001): Phys. Rep. **349**, 447. Sect. 3.6

Nathan, M., Shoshani, Z., Ashkinazi, G., Meyler, B., Zolotarevski, O. (1996): Solid-State Electron. **39**, 1457. Sect. 6.1.3, Table 3.3

Neamen, D.A., Grannemann, W.W. (1971): Solid-State Electron. **14**, 1319. Sect. 6.1.10

Needs, R.J., Charlesworth, J.P.A., Godby, R.W. (1994): Europhys. Lett. **25**, 31. Table 7.1

Nelson, A.J., Gebhard, S., Kazmerski, L.L., Engelhardt, M., Höchst, H. (1991a): J. Vac. Sci. Technol. A **9**, 978. Table 6.2

Nelson, J.S., Wright, A.F., Fong, C.Y., (1991b): Phys. Rev. B **43**, 4908. Sect. 7.3.1

Nelson, A.J., Schwerdtfeger, C.R., Wie, S.-H., Zunger, A., Rioux, D., Patel, R., Höchst, H. (1993): Appl. Phys. Lett. **62**, 2557. Sect. 6.2.4, Table 6.3

Nelson, A.J., Niles, D.W., Schwerdtfeger, C.R., Wei, S.-H., Zunger, A., Höchst, H. (1994): J. Electron Spectrosc. Relat. Phenom. **68**, 185. Sect. 6.2.5

Neuhold, G., Horn, K., Magnusson, K.O., Evans, D.A. (1995): J. Vac. Sci. Technol. A **13**, 666. Sect. 6.2.2

Newman, N., Spicer, W.E., Kendelewicz, T., Lindau, I. (1986a): J. Vac. Sci. Technol. B **4**, 931. Sect. 1.1.3

Newman, N., Schilfgaarde, M. van, Kendelewicz, T., Williams, M.D., Spicer, W.E. (1986b): Phys. Rev. B **33**, 1146. Sect. 8.1

Niedermann, P., Quattropani, L., Solt, K., Kent, A.D., Fischer, Ø. (1992): J. Vac. Sci. Technol. B **10**, 580. Sect. 6.1.2, Table 3.4

Nicolini, R., Vancetti, L., Mula, G., Bratina, G., Sorba, L., Franciosi, A., Peressi, M., Baroni, S., Resta, R., Baldereschi, A., Angelo, J.E., Gerberich, W.W. (1994): Phys. Rev. Lett. **72**, 294. Sect. 10.2.3

Niles, D.W., Tang, M., McKinley, J., Zanoni, R., Margaritondo, G. (1988): Phys. Rev. B **38**, 10949. Sect. 6.2.2

Niles, D.W., Contreras, M., Ramanathan, K., Noufi, R. (1996): in *Conf. Record of the 25th IEEE Photovoltaic Specialists Conf.* (IEEE, New York, NY), p. 833. Sect. 6.2.4, Table 6.3

Nuhoğlu, Ç., Ayyíldíz, E., Sağlam, M., Türüt, A. (1998): Appl. Surf. Sci. **135**, 350 and private communication. Sect. 6.1.3, Table 3.3

Nur, O., Sardela, M.R. jr., Willander, M., Turan, R. (1995): Semicond. Sci. Technol. **10**, 551. Sect. 6.1.2, Table 3.3

Oh, J, Nemanich, R.J. (2002): J. Appl. Phys. **92**, 3326. Sect. 6.1.2, Table 3.3

Ohdomari, I., Aochi, H. (1987): Phys. Rev. B **35**, 682. Sect. 4.1

Ohdomari, I., Tu, K.N., d'Heurle, F.M, Kuan, T.S., Petersson, S. (1978): Appl. Phys. Lett. **33**, 1028. Sect. 1.1.5, 3.5, 6.1.2; Table 3.3

Ohler, C., Moers, J., Förster, A., Lüth, H. (1995): J. Vac. Sci. Technol. B **13**, 1728. Sect. 8.3

Okabayashi, J., Ono, K., Mano, T., Mizuguchi, M., Horiba, K., Nakamura, K., Fujimori, A., Oshima, M (2002): Appl. Phys. Lett. **80**, 1764. Sect. 3.12.2

Okamoto, K., Wood, C.E.C., Eastman, L.F. (1981): Appl. Phys. Lett. **38**, 636. Sect. 6.1.10

Okuda, T., Shigeoka, H., Daimon, H., Suga, S., Kinoshita, T., Kakizaki, A. (1994): Surf. Sci. **321**, 105. Sect. 3.11, Table 3.7

Olbrich, A., Vaneca, J., Kreupl, F., Hoffmann, H. (1997): Appl. Phys. Lett. **70**, 2559. Sect. 4.1

Olbrich, A., Vaneca, J., Kreupl, F., Hoffmann, H. (1998): J. Appl. Phys. **83**, 358. Sect. 4.1, Table 3.4

Olmstead, M.A., Bringans, R.D. (1990): Phys. Rev. B **41**, 8420. Sect. 3.11, Table 3.7

Olmstead, M.A., Bringans, R.D., Uhrberg, R.I.G., Bachrach, R.Z. (1987): Phys. Rev. B **35**, 7526. Sect. 3.11, Table 3.7

Opanowicz, A., Kościelniak-Mucha, B., Boroś, G. (1988): phys. stat. sol. (a) **106**, K197. Table 6.2

Ospelt, M., Menz, J., Flepp, L., Känel, H. von (1988): Appl. Phys. Lett. **52**, 227. Sect. 10.2.2

Ossicini, S., Bisi, O., Bertoni, C.M. (1990): Phys. Rev. B **42**, 5735. Sect. 10.2.2

Ottaviani, G., Tu, N.K., Mayer, J.W. (1980): Phys. Rev. Lett. **44**, 284. Sect. 1.1.4

Padovani, F.A., Stratton, R. (1966): Solid-State Electron. **9**, 695. Sect. 11

Paggel, J.J., Neuhold, G., Haak, H., Horn, K. (1998): Surf. Sci. **414**, 221. Sect. 6.1.2, Table 3.6

Palm, H., Arbes, M., Schulz, M. (1993): Phys. Rev. Lett. **71**, 2224. Sect. 6.1.2, Table 3.4

Palmstrøm, C.J., Fimland, B.-O., Sands, T., Garrison, K.C., Bartynski, R.A. (1989): J. Appl. Phys. **65**, 4753. Sect. 6.1.10

Park, C.Y., An, K.S., Kim, J.S., Park, R.J., Chung, J.W., Kinoshita, T., Kakizaki, A., Ishii, T. (1995): Phys. Rev. B **52**, 8198. Sect. 3.11, Table 3.7

Parkes, J., Tomlinson, R.D., Hampshire, M.J. (1973): Solid-State Electron. **16**, 773. Table 6.2

Pauling, L.N. (1939/1960): *The Nature of the Chemical Bond* (Cornell Univ., Ithaca, NY). Sects. 1.1.5, 3.11, 5.2

Penn, D.R. (1962): Phys. Rev. **128**, 2093. Sects. 1.1.5, 5.5.1, 5.6

People, R., Wecht, K.W., Alavi, K., Cho, A.Y. (1983): Appl. Phys. Lett. **43**, 118. Sect. 6.2.3

Phatak, P., Newman, N., Dreszer, P., Weber, E.R. (1995): Phys. Rev. B **51**, 18003. Sect. 8.2.1

Phillips, J.C. (1963): Solid State Commun. **12**, 861. Sect. 5.6

Pickard, G.W. (1906): U.S. Patent No. 836,531. Sect. 1.1.1

Pickett, W.E., Cohen, M.L. (1978): Phys. Rev. B **18**, 939. Sect. 5.3.2

Pickett, W.E., Louie, S.G., Cohen, M.L. (1978): Phys. Rev. B **17**, 815. Sect. 5.3.2

Picozzi, S., Continenza, A., Massidda, S., Freeman, A.J. (1998): Phys. Rev. B **57**, 4849. Sects. 5.3.1, 7.2; Table 5.1

Picozzi, S., Continenza, A., Satta, G., Massidda, S., Freeman, A.J. (2000): Phys. Rev. B **61**, 16736. Sects. 5.3.1, 7.2; Table 7.1

Picozzi, S., Profeta, G., Continenza, A., Massidda, S., Freeman, A.J. (2002): Phys. Rev. B **65**, 165316. Sect. 7.2, Table 7.2

Poganski, S. (1952): Z. Elektrochemie **56**, 193. Preface

Poganski S. (1953): Z Physik 134, 469. Preface

Prasad, J.J.B., Rao, D.K., Sridevi, D., Majhi, J., Reddy, K.V., Sobhanadri, J. (1985): Solid-State Electron. **28**, 1251. Table 6.2

Prietsch, M. (1995): Phys. Rep. **253**, 163. Sect. 3.6

Prietsch, M., Domke, M., Laubschat, C., Mandel, T., Xue, C., Kaindl, G. (1989): Z. Phys. B, Condensed Matter **74**, 21. Sect. 3.11

Qiao, D., Yu, L.S., Lau, S.S., Redwing, J.M., Lin, J.Y., Jiang, H.X. (2000): J. Appl. Phys. **87**, 801. Sect. 6.1.5, 6.1.10; Table 3.5

Qiao, D.J., Lau, S.S. (2003): private communication of the original data

Qteish, A., Heine, V., Needs, R.J. (1992): Phys. Rev. B **45**, 6534. Sect. 6.1.6, Table 6.1

Radermacher, K., Schüppen, A., Mantl, S. (1994): Solid-State Electron. **37**, 443. Sect. 6.1.2, Table 3.3

Raether, H. (1980): *Excitation of Plasmons und Interband Transitions by* Electrons, Springer Tracts Mod. Phys., Vol. 88 (Springer, Berlin). Sect. 5.5.1

Ramesh, P.P., Uthanna, S., Naidu, B.S., Reddy, P.J. (1997): Thin Solid Films **292**, 290. Table 6.2

Rao, D.K., Prasad, J.J.B., Sridevi, D., Reddy, K.V., Sobhanadri, J. (1986): phys. stat. sol. (a) **94**, K193. Table 6.2

Rao, M.A., Caine, E.J., Kroemer, H., Long, S.I., Babic, D.I. (1987): J. Appl. Phys. **61**, 643. Sect. 6.2.3

Rau, U., Güttler, H.H., Werner, J.H. (1992): Mat. Res. Soc. Symp. Proc. **260**, 245. Sect. 4.1

Revva, P., Langer, J.M., Missous, M., Peaker, A.R. (1993): J. Appl. Phys. **74**, 416. Sect. 8.1. 10.1.10

Rhoderick, E.H. (1972): J. Phys. D: Appl. Phys. **5**, 1920. Sect. 3.4

Rich, D.H., Samsavar, A., Miller, T., Lin, H.F., Chiang, T.-C., Sundgren, J.-E., Greene, J.E. (1987): Phys. Rev. Lett. **58**, 579. Sect. 3.11, Table 3.7

Rich, D.H., Miller, T., Samsawar, A., Lin, H.F., Chiang, T.-C. (1988): Phys. Rev, B **37**, 10221. Sect. 3.11, Table 3.7

Rich, D.H., Miller, T., Chiang, T.-C. (1990): Phys. Rev. B **41**, 3004. Sect. 3.11, Table 3.7

Rizzi, A., Lantier, R., Monti, F., Lüth, H., Della Sala, F., Di Carlo, A., Lugli, P. (1999): J. Vac. Sci. Technol. B **17**, 1674. Sects. 6.2.2, 6.2.6

Robinson, P., Wilson, J.I.B. (1977): Inst. Phys. Conf. Ser. **35**, 229. Sect. 6.2.4, Table 6.3

Robinson, I.K., Tung, R.T., Feidenhans'l, R. (1988): Phys. Rev. B **38**, 3632. Sect. 10.2.2

Rohlfing, M., Krüger, P., Pollmann, J. (1993): Phys. Rev. B **48**, 17791. Sects. 5.5.1, 7.2

Rohlfing, M., Krüger, P., Pollmann, J. (1995): Phys. Rev. Lett. **75**, 3489. Sect. 5.5.1

Rohlfing, M., Krüger, P., Pollmann, J. (1996): private communication. Sect. 5.5.1

Rotelli, B., Tarricone, L., Gombia, E., Mosca, R., Perotin, M. (1997): J. Appl. Phys. **81**, 1813. Table 3.5

Rozkov, V.A., Trusova, A.Yu., Berezhnoy, I.G. (1998): Thin Solid Films **325**, 151. Table 6.4

Ru, G.-P., Qu, X.-P., Zhu, S.-Y., Li, B.-Z., Detavernier, C., Meirhaeghe, R.L. van, Cardon, F., Donaton, R.A., Maex, K. (2000): J. Vac. Sci. Technol. B **18**, 1942. Sect. 6.1.2, Table 3.4

Rubini, S., Pelucchi, M., Lazzarino, M., Kumar, D., Franciosi, A., Berthod, C., Binggeli, N., Baldereschi, A. (2001): Phys. Rev. B **63**, 235307 and private communications. Sects. 7.2, 6.1.7; Tables 3.3, 7.1

Ruini, A., Resta, R., Baroni, S. (1997): Phys. Rev. B **56**, 14921. Table 7.1

Sadwick, L.P., Kim, C.W., Tan, K.L., Streit, D.C. (1991): IEEE Electron Device Lett. **12**, 626. Sects. 6.1.10, 6.2.3

Sakaki, J., Chang, L.L., Ludeke, R., Chang, C.-A., Sai-Halasz, G.A., Esaki, L. (1977): Appl. Phys. Lett. **31**, 211. Sect. 6.2.2

Sakamoto, K., Sakamoto, T., Nagao, S., Hashiguchi, G., Kunisyoshi, K., Bando, Y. (1987): Japan. J. Appl. Phys. **26**, 666. Sect. 3.9.2

Samsavar, A., Miller, T., Chiang, T.-C. (1988): Phys. Rev. B **38**, 9889. Sect. 3.11, Table 3.7

Sands, D., Scott, C.G. (1995): J. Appl. Phys. **77**, 3295. Table 3.3

Sanz-Maudes, J., Jiménez-Leube, F.J., Clement, M. (1999): J. Vac. Sci. Technol. B **17**, 397. Sect. 6.1.2, Table 3.3

Savage, D.E., Lagally, M.G. (1986): J. Vac. Sci. Technol. B **4**, 943. Sect. 3.9.2

Sawada, T., Ito, Y., Imai, K., Suzuki, K., Tomozawa, H., Sakai, S. (2000): Appl. Surf. Sci. **159/160**, 449. Sect. 3.5, 6.1.5; Table 3.3

Schilfgaarde, M. van (1990): J. Vac. Sci. Technol. B **8**, 990. Sect. 4.5.1

Schlaf, R., Lang, O., Pettenkofer, C., Jaegermann, W., Armstrong, N.R. (1997): J. Vac. Sci. Technol. A **15**, 1365. Sect. 6.2.5

Schlüter, M. (1978): Phys. Rev. B **17**, 5044. Sects. 1.1.4, 5.6

Schlüter, M. (1982): Thin Solid Films **93**, 3. Sects. 1.1.4, 5.6

Schmitsdorf, R. (1993): Diploma Thesis, Universität Duisburg (unpublished). Sects. 2.2, 3.4

Schmitsdorf, R.F., Mönch, W. (1999): Eur. Phys. J. B **7**, 457 Sects. 3.5, 4.3, 6.1.2, 10.1; Table 3.3

Schmitsdorf, R., Kampen, T.U., Mönch, W. (1995): Surf. Sci. **324**, 249. Sects. 1.1.5, 3.4, 3.5, 6.1.2; Table 3.3

Schoen, K.J., Woodall, J.M., Cooper, J.A., Melloch, M.R. (1998): IEEE Trans. **ED-45**, 1595. Sect. 6.1.6, Table 3.3

Schottky, W. (1914): Physik. Zeitschr. **15**, 872. Sects. 1.1.5, 2.3

Schottky, W. (1938): Naturwissenschaften **26**, 843. Sect. 1.1.2

Schottky, W. (1939): Z. Physik **113**, 367. Sect. 1.1.1

Schottky, W. (1940): Physik. Zeitschr. **41**, 570. Sects 1.1.1, 1.1.2

Schottky, W. (1942): Z. Physik **118**, 539. Sects. 1.1.1, 1.1.5, 2.1, 2.2, 4.5.1

Schottky, W., Deutschmann, W. (1929): Physik. Zeitschr. **30**, 839. Sects. 1.1.1, 2.2

Schottky, W., Spenke, E. (1939): Wiss. Veröff. Siemens-Werke **18**, 225. Sect. 1.1.2, 3.1

Schramm, C., Bach, H.G., Künzel, H., Praseuth, J.P. (1991): J. Electrochem. Soc. **138**, 2808. Sects. 6.1.10, 6.2.3

Schröder, C., Heiland, W., Held, R., Loose, W. (1996): Appl. Phys. Lett. **68**, 1957. Table 3.3

Schultz, W. (1954): Z. Physik **138**, 598. Sect. 3.1

Schuster, A. (1874): Phil. Mag. (4) **48**, 251. Sect. 1.1.1

Schwarz, G. (2002): unpublished. Sect. 8.2

Schweickert, H. (1939): Verhandl. D. Phys. Gesell. **3**, 99. Sects. 1.1.1, 1.1.2

Shan, W., Li. M.F., Yu, P.Y., Hansen, W.L., Walukiewicz, W. (1988): Appl. Phys. Lett. **53**, 974. Sect. 8.2.1

Shenai, K. (1991): IEEE Electron Dev. Lett. **12**, 169. Sect. 6.1.2, Table 3.3

Shigiltchoff, O., Kimoto, T., Hobgood, D., Neudeck, P.G., Porter, L.M., Devaty, R.P., Choyke, W.J. (2002): Mater. Sci. Forum **389-393**, 921. Sect. 6.1.6, Table 3.5

Shigiltchoff, O., Bai, S., Devaty, R.P., Choyke, W.J., Kimoto, T., Hobgood, D., Neudeck, P.G., Porter, L.M. (2003): Mater. Sci. Forum **433-436**, 705. Sect. 6.1.6, Table 3.3

Shive J.N. (1940): see Bardeen, J. (1947). Sect. 1.1.2

Shockley, W. (1939): Phys. Rev. **56**, 317. Sect. 1.1.3

Shockley, W. (1949): Bell System Tech. J. **28**, 435. Sect. 3.1

Shockley, W., Pearson, G.L. (1948): Phys. Rev. **74**, 232. Sect. 1.1.3

Siemens, W. (1875): Monatsber. Berliner Akad. Wissensch. p. 259. Sect. 1.1.1

Sirringhaus, H., Lee, E.Y., Känel, H. von (1994): Phys. Rev. Lett. **73**, 577. Sect. 6.1.2, Table 3.4

Sirringhaus, H., Lee, E.Y., Kafader, U., Känel, H. von (1995): J. Vac. Sci. Technol. B **13**, 1848. Sect. 6.1.2, Table 3.4

Sirringhaus, H., Meyer, T., Lee, E.Y., Känel, H. von (1996): Phys. Rev. B **53**, 15944. Sects. 3.6, 4.5.2

Skromme, B.J., Luckowski, E., Moore, K., Bhatnagar, M., Weitzel, C.E., Gehoski, T., Ganser, D. (2000): J. Electron. Mater. **29**, 376. Sects. 4.4, 6.1.6; Table 3.3

Sobolewski, M.A., Helms, C.R. (1989): Appl. Phys. Lett. **54**, 638. Sect. 6.1.11

Soukiassian, P., Kendelewicz, J., Hurych, Z.D. (1989): Phys. Rev. B **40**, 12570. Sect. 3.11, Table 3.7

Sousa Pires, J. de, Ali, P., Crowder, B., D'Heurle, F., Petersson, S., Stolt, L., Tove, P.A. (1979): Appl. Phys. Lett. **35**, 202. Sect. 6.1.2, Table 3.5

Spicer, W.E., Gregory, P.E., Chye, P.W., Babaola, J.A., Sukegawa, T. (1975): Appl. Phys. Lett. **27**, 617. Sects. 1.1.5, 3.9.2, 6.1.3; Table 3.6

Spicer, W.E., Chye, P.W., Skeath, P.R., Su, C.Y., Lindau, I. (1979): J. Vac. Sci. Technol. **16**, 1422. Sects. 1.1.4, 9.3

Spicer, W.E., Lilienthal-Weber, Z., Weber, E.R., Newman, N., Kendelewicz, T., Cao, R., McCants, C., Miyano, K., Mahowald, P.H., Lindau, I. (1988): J. Vac. Sci. Technol. B **6**, 1245. Sect. 4.5.2

Spitzer, W.G., Mead, C.A. (1964): Phys. Rev. **133**, A872. Sect. 6.1.10

Städele, M., Majewski, J.A., Vogl. P. (1997): Phys. Rev. B **56**, 6911. Sect. 7.3.2

Stadler, R., Vogtenhuber, D., Podloucky, R. (1999): Phys. Rev. B **60**, 17112. Sect. 7.2, Table 7.1

Stadler, R., Podloucky, R. (2000): Phys. Rev. B **62**, 2209. Sect. 7.2, Table 7.1

Stampfl, C., Van de Walle, C.G. (1998): Phys. Rev. B **57**, R15052. Sects. 6.1.5, 6.1.6, 7.2; Table 6.1

Stockman, L., Kempen, H. van (1998): Surf. Sci. **408**, 232. Sect. 6.1.3, Table 3.4

Stranski, I.N., Krastanow, L. von (1939): Akad. Wiss. Lit. Mainz, Math.-Natur. Kl. IIb, **146**, 797. Sect. 3.9.2

Suezaki, T., Kawahito, K., Hatayama, T., Uraoka, Y., Fuyuki, T. (2001): Jpn. J. Appl. Phys. **40**, L43. Sect. 6.1.6, Table 3.3

Sullivan, J.P., Tung, R.T., Pinto, M.R., Graham, W.R (1991): J. Appl. Phys. **70**, 7403. Sect. 4.1

Surkova, T.P., Giriat, W., Godlewski, M., Kauczor, P., Permogorov, S.A., Tenishev, L.N. (1996): J. Cryst. Growth **159**, 471. Sect. 6.2.8

Surkova, T.P., Giriat, W., Godlewski, M., Permogorov, S. (1997): Acta Physica Pol. 92, 1009. Sect. 6.2.8

Sze, S.M. (1981): *Physics of Semiconductor Devices,* 2nd Edn. (Wiley, New York). Sect. 1.1.4

Sze, S.M., Crowell, C.R., Kahng, D. (1964): J. Appl. Phys. **35**, 2534. Sect. 2.3

Takayanagi, K., Tanishiro, Y., Takahashi, M., Takahashi, S. (1985a): J. Vac. Sci. Technol. A **3**, 1502. Sect. 3.4

Takayanagi,. K., Tanishiro, Y., Takahashi, M., Takahashi, S. (1985b): Surf Sci. **164**, 367. Sect. 3.4

Talin, A.A., Williams, R.S., Morgan, B.A., Ring, K.M., Kavanagh, K.L. (1994): J. Vac. Sci. Technol. B **12** , 2634. Sects. 3.7, 6.1.2; Table 3.4

Tamm, I.E. (1932): Physikal. Z. Sowjetunion **1**, 733. Sect. 1.1.3

Tejedor, C., Flores, F. (1978): J. Phys. C **11**, L19. Sects. 1.1.3, 1.2, 5.3.3

Tejedor, C., Flores, F., Louis, E. (1977): J. Phys. C **10**, 2163. Sect. 1.1.3, Table 5.1

Tersoff, J. (1984a): Phys. Rev. Lett. **52**, 465. Sect. 1.1.5, 5.5.2; Table 5.1

Tersoff, J. (1984b): Phys. Rev. B **30**, 4874. Sect. 1.2

Tersoff, J. (1985): Phys. Rev. B **32**, 6968. Sect. 5.6

Tersoff, J. (1986a): Surf. Sci. **168**, 275. Sect. 5.5.2, Table 5.1

Tersoff, J. (1986b): Phys. Rev. Lett. **56**, 2755. Table 5.1

Tersoff, J. (1986c): J. Vac. Sci. Technol. B **4**, 1066. Sects. 1.1.5, 5.5.1, 5.5.2, 6.1.10, Table 5.1

Thurmond, C.D. (1975): J. Electrochem. Soc. **122**, 1133. Sect. 8.1

Topping, J. (1927): Proc. Roy. Soc. A **114**, 67. Sects. 3.11, 5.1.1

Troost, D., Strathausen, A.K., Mönch, W. (1994): in *Proc. 4th Int'l Conf. on the Formation of Semiconductor Interfaces,* ed. by B. Lengeler, H. Lüth, W. Mönch, and J. Pollmann (World Scientific, Singapore), p. 305. Sect. 10.1

Truchsess, M. von, Latussek, V., Becker, C.R., Batke, E. (1996): J. Crystal Growth **159**, 1128 Sect. 8.1

Tsaur, B.-Y., Geis, M.W., Fan, J.C.C., Gale, R.P. (1981): Appl. Phys. Lett. **38**, 779. Sect. 3.9.2

Tsunoda, M., Fukutani, K., Murata, Y., Yamashita, H., Komaki, K., Kobajashi, K. (1994): in *Proc. of the 4th Int'l Conf. on the Formation of Semiconductor Interfaces*, ed. by B. Lengeler, H. Lüth, W. Mönch, and J. Pollmann (World Scientific, Singapore), p. 162. Sects. 3.5, 10.1

Tung, R.T. (1984a): Phys. Rev. Lett. **52**, 461. Sects. 1.1.5, 10.2.2

Tung, R.T. (1984b): J. Vac. Sci. Technol. B **2**, 465. Sects. 1.1.5, 6.1.2, 10.2.2; Table 3.3

Tung, R.T. (1989): J. Vac. Sci. Technol. A **7**, 598. Sect. 10.2.2

Tung, R.T. (1992): Phys. Rev. B **45**, 13509. Sects. 1.1.5, 4.1, 4.2, 4.3, 4.4, 4.5.2

Tung, R.T., Gibson, J.M., Bean, J.C., Poate, J.M., Jacobson, D.C. (1982): Appl. Phys. Lett. **40**, 684. Sect. 10.2.2

Tung, R.T., Gibson, J.M., Poate, J.M. (1983a): Phys. Rev. Lett. **50**, 429. Sect. 10.2.2

Tung, R.T., Gibson, J.M., Poate, J.M. (1983b): Appl. Phys. Lett. **42**, 888. Sect. 10.2.2

Tung, R.T., Levi, A.F.J., Gibson, J.M., Ng, K.K., Chantre, A. (1986): Mat. Res. Soc. Symp. Proc. **54**, 458. Sects. 6.1.2, 10.2.2; Table 3.3

Tung, R.T., Levi, A.F.J., Sullivan, J.P., Schrey, F. (1991): Phys. Rev. Lett. **66**, 72. Sect. 10.2.2

Turan, R., Akman, N. (1993): Semicond. Sci. Technol. **8**, 1999. Table 3.3

Turan, R., Aslan, B., Nur, O., Yousif, M.Y.A., Willander, M. (2001): Appl. Phys. A **72**, 587. Sects. 3.8, 6.1.2; Tables 3.3, 3.5

Ubara, H., Imura, T., Hiraki, A. (1984): Solid State Commun. **50**, 673. Sect. 10.1

Ueno, K., Abe, A., Saiki, K., Koma, A. (1991): Jpn. J. Appl. Phys. Part 2 **30**, L1352. Sect. 6.2.1

Unewisse, M.H., Storey, J.W. (1993): J. Appl. Phys. **73**, 3873. Sect. 6.1.2, Table 3.3

Vandenberg, J.M., Panish, M.B., Temkin, H., Hamm, R.A. (1988): Appl. Phys. Lett. **53**, 1920. Sect. 8.3

Vandenberg, J.M., Panish, M.B., Hamm, R.A., Temkin, H. (1990): Appl. Phys. Lett. **56**, 910. Sect. 8.3

Varshni, Y.P. (1967): Physica **34**, 149. Sect. 8.1

Vlieg, E., Fischer, A.E.M.J., Veen, J.F. van der, Dev, B.N., Materlik, G. (1986): Surf. Sci. **178**, 36. Sect. 10.2.2

Vogl, P. (1985): in *Festkörperprobleme* (Adv. Solid State Physics) Vol. 25, ed. by P. Grosse (Vieweg, Braunschweig), p. 671. Sect. 6.2.8

Volmer, M., Weber, E. (1926): Z. Phys. Chem. **119**, 277. Sect. 3.9.2

Vrijmoeth, J., Veen, J.F. van der, Heslinga, D.R., Klapwijk, T.M. (1990): Phys. Rev. B **42**, 9598. Sects. 1.1.4, 10.2.2

Waldrop, J.R. (1984): J. Vac. Sci. Technol. B **2**, 445. Sect. 1.1.3

Waldrop, J.R. (1985a): J. Vac. Sci. Technol. B **3**, 1197. Sect. 10.1

Waldrop, J.R. (1985b): Appl. Phys. Lett. **47**, 1301. Sect. 10.1

Walukiewicz, W. (1988): Phys. Rev. B **37**, 4760. Sect. 1.1.4

Wang, X.J., He, L. (1998): J. Electron. Mater. **27**, 1272. Sect. 6.1.5, Table 3.3

Wang, W.I., Stern, F. (1985): J. Vac. Sci. Technol. B **3**, 1280. Sects. 1.2, 6.2.2, 6.2.3

Wang, L.P., Yang, J.R., Hwang, J. (1993): J. Appl. Phys. **74**, 6251. Sect. 6.1.2, Table 3.3

Wang, R.-Z., Ke, S.-H., Huang, M.-C. (1992): J. Phys., Condens. Matter **4**, 8093. Sect. 8.3

Wang, Z.-G., Ledebo, L.-Å., Grimmeis, H.G. (1984): J. Appl. Phys. **56**, 2762. Sect. 6.2.8
Wasserstrom, E., McKenna, J. (1970): Bell System Tech. J. **49**, 853. Sect. 4.1
Watanabe, M.O., Ohba, Y. (1987): Appl. Phys. Lett. **50**, 906. Sect. 6.2.3
Weber, E. R., Ennen, H., Kaufmann, U., Windscheif, J, Schneider, J. (1982): J. Appl. Phys. **53**, 6140. Sect. 1.1.4
Wei, S.-H., Zhang, S.B. (2000): Phys. Rev. B **62**, 6944. Sect. 6.1.6, Table 6.1
Wei, S.-H., Zunger, A. (1993): Appl. Phys. Lett. **63**, 2549. Sect. 6.2.4
Wei, S.-H., Zunger, A. (1995): J. Appl. Phys. **78**, 3846. Sect. 6.2.4
Weilmeier, M.K., Rippard, W.H., Buhrman, R.A. (1999): Phys. Rev. B **59**, R2521. Sect. 3.6
Weitering, H.H., Chen, J., DiNardo, N.J., Plummer, E.W. (1993): Phys. Rev. B **48**, 8119. Sect. 3.11, Table 3.7
Weitering, H.H., Shi, X, Erwin, S.C. (1996): Phys. Rev. B **54**, 10585. Sect. 3.11, Table 3.7
Welker, H. (1943): see Handel, K.Chr. (1999*): Anfänge der Halbleiterforschung und – entwicklung*;, Dissertation, RWTH Aachen. Sect. 1.1.2
Werner, J.H. (1989): Appl. Phys. Lett. **54**, 1528. Sect. 8.2.1
Werner, J.H., Güttler, H.H. (1991): J. Appl. Phys. **69**, 1522. Sect. 6.1.2, Table 3.3
Werner, J.H., Güttler, H.H. (1993): J. Appl. Phys. **73**, 1315. Sect. 8.2.1
Werner, J., Levi, A.F.J., Tung, R.T., Anzlowar, M., Pinto, M. (1988): Phys. Rev. Lett. **60**, 53. Sect. 10.2.2
Werner, P., Jäger, W., Schüppen, A. (1993): J. Appl. Phys. **74**, 3846. Sects. 3.5, 6.1.2, 10.2.2; Table 3.3
Weyers, S.J., Janzen, O., Mönch, W. (1999): J. Phys., Condens. Matter **11**, 8489. Sects. 1.1.4, 3.9.2, 3.10, 6.1.2,; Table 3.6
Williams, R. (1965): Phys. Rev. **140**, A569. Table 6.4
Williams, R., Bube, H.R. (1960): J. Appl. Phys. **31**, 968. Sect. 3.8
Wilson, A.H. (1931): Proc. Roy. Soc. A **133**, 458 and **134**, 277. Sect. 1.1.1
Wittmer, M. (1990): Phys. Rev. B **42**, 5249. Sect. 6.1.2, Table 3.3
Wood, E.A. (1964): J. Appl. Phys. **35**, 1306. Sects. 1.1.5, 3.4
Wörz, M., Pschorr-Schoberer, E., Flierl, R., Preis, R., Gebhardt, W. (1998): J. Appl. Phys. **84**, 2871. Sect. 6.2.4, Table 6.3

Yablonovitch, E., Allara, D.L., Chang, CC., Gmitter, T., Bright, T.B. (1986): Phys. Rev. Lett. **57**, 249. Sect. 10.1
Yeh, C.N., McNeil, L.E., Nahory, R.E., Bhat, R. (1995): Phys. Rev. B **52**, 14682. Sect. 8.3
Yeom, H.W., Abukawa, T., Takakuwa, Y., Nakamura, M., Kimura, M., Kakizaki, A., Kono, S. (1996a): Surf. Sci. **365**, 328. Sect. 3.11, Table 3.7
Yeom, H.W., Abukawa, T., Takakuwa, Y., Mori, Y., Shimatani, T., Kakizaki, A., Kono, S. (1996b): Phys. Rev. B **54**, 4456. Sect. 3.11, Table 3.7
Yu, L.S., Qiao, D.J., Xing, Q.J., Lau, S.S., Boutros, K.S., Redwing, J.M. (1998): Appl. Phys. Lett. **73**, 238. Table 3.3

Zachau, M., Helgesen, P., Kux, A., Koch, F., Grützmacher, D., Meyer, R., Jürgensen, H., Balk., P. (1989): Superlatticies Microstruct. **5**, 19. Sect. 6.2.3
Zahn, D.R.T., Esser N., Müller, C., Richter, W., Stephens, C., Whittle, R., McGovern, I.T., Kulkarni, S., Braun, W. (1992): Appl. Surf. Sci. **56-58**, 228. Sect. 10.1
Zakharov, O., Rubio, A., Blase, X., Cohen, M.L., Louie, S.G. (1994): Phys. Rev. B **50**, 10780. Sect. 7.1

Zegenhagen, J., Huang, K.-G., Hunt, BD., Schowalter, L.J. (1987): Phys. Rev. Lett. **51**, 1176. Sect. 10.2.2

Zegenhagen, J., Huang, K.-G., Gibson, W.M., Hunt, B.D., Schowalter, L.J. (1989): Phys. Rev. B **39**, 10254. Sect. 10.2.2

Zhang, S.B., Cohen, M.L., Louie, S.G. (1985): Phys. Rev. B **32**, 3955. Sect. 9.1

Zhang, S.B., Cohen, M.L., Louie, S.G. (1986): Phys. Rev. B **34**, 768. Sect. 3.11, Table 7.1

Zhang, S.B., Cohen, M.L., Louie, S.G., Tománek, D., Hybertsen, M.S. (1990): Phys. Rev. B **41**, 10058. Sect. 7.3.1

Zhang, X.S., Fan, C.Y., Xu, Y.B., Sui, H., Bao, S., Xu, S.H., Pan, H.B., Xu, P.S. (1996): J. Phys., Condens. Matter **8**, 699. Sect. 3.11, Table 3.7

Zhu, S., Meirhaeghe, R.L. van, Detavernier, C., Ru, G.-P., Li, B.-Z., Cardon, F. (1999): Solid State Commun. **112**, 611. Sect. 6.1.2, 8.1; Table 3.4

Zhu, S., Detavernier, C., Meirhaeghe, R.L. van, Cardon, F., Blondeel, A., Clauws, P., Ru, G.-P., Li, B.-Z. (2001): Semicond. Sci. Technol. **16**, 83. Sects. 8.1, 6.1.2; Tables 3.3, 3.4

Zubkov, V.I., Mel'nik, M.A., Solomonov, A.V., Pikhtin, A.N., Bugge, F. (1999): Semiconductors **33**, 858. Sects. 3.12.1, 6.2.3, 8.1

Zunger, A. (1981): Phys. Rev. B **24**, 4372. Sect. 3.9.2

Zur, A., McGill, T.C., Smith, D.L. (1983): Phys. Rev. B **28**, 2060. Sect. 9.1

Subject Index

Acceptor-bonds, interfacial 161
Acceptor-like interface states 9
Ab-initio calculations *see* First-principles calculations
Adatom-induced core-level shifts 74ff.
Anderson rule 18
Atomic term values A.2 (table)
Average-bond-energy model 198

Baldereschi's concept of mean-value *k*-points 14, 122
Ballistic-electron-emission microscopy 18, 56
– barrier-height 56, 89
– – distribution 100, 101. 190
– – dependence on donor density 101
– – temperature dependence 190
– – voltage dependence 32
– current-voltage characteristic 59
– laterally inhomogeneous contacts 90
– lateral profiling 18, 61, 90
Band-edge discontinuities *see* Band-edge offsets of heterostructures
Band-edge offsets of heterostructures 19, 109ff.
– between different modifications of semiconductors 144 (table)
– I–III–VI_2 chalcopyrites 168
– III–nitrides 187
– determination by
– – IPEYS 80
– – XPS 81
– extrinsic dipole contribution 162
– insulator–semiconductor interfaces 173ff., 174 (table)
– intrinsic dipole contribution 110
– Langer–Heinrich rule 176
– lattice-matched III–V alloys 163ff., 186, 196
– lattice strain, dependence on 200
– layered semiconductors 170
– metamorphic heterostructures 172
– non-polar (110) heterostructures 162
– polar heterostructures 161
– pressure dependence 196
– pseudomorphic heterostructures 197
– silicon carbide heterostructures 187
– slope parameter 119
– temperature dependence 193
– types of 164
Band gap of semiconductors A.1 (table)
– temperature dependence 190
Bardeen case 115
Barrier height of Schottky contacts 5
– BEEM barrier height 59
– – distribution of 100, 101, 190
– determination from
– – *I*/*V* characteristics 33ff.
– – BEEM I_{coll}/V_{tip} characteristics 59
– – IPEYS 65
– – XPS 73
– effective 45
– – correlation with ideality factor 16, 45, 47ff., 97, 98, 192
– electronegativity difference, dependence on 16
– extrapolated 16, 47
– extrinsic dipole contributions 209ff.
– flat-band 26
– image-force lowering of 28
– – studied by BEEM 31
– – studied by IPEYS 31
– interface-structure effects 17, 18, 48, 49, 214ff., 217ff.
– intrinsic dipole contribution 107ff.
– laterally homogeneous contacts 47
– – tables 54, 63, 67, 74
– lateral inhomogeneities

– – artificial origin 97
– – extrinsic origin 104
– – intrinsic or natural origin 17, 98, 103
– MIGS-and-electronegativity concept 14
– orientation of semiconductors 51
– pressure dependence 193
– Schottky –Mott rule 5
– slope parameter 108, 131
– voltage-dependence of 30
– zero-bias 38
– zero-charge-transfer 108, 111
Bauer's wetting criterion 71
BEEM *see* Ballistic-electron-emission microscopy
Bethe's criterion 37
Branch point 9
Branch-point energy 14
– of binary compounds 130 (table)
– of different modifications of semiconductors 141, 144
– of elemental semiconductors 130 (table)
– of ternary I –III –IV_2 compounds 130 (table)
– of ternary III –V alloys 149ff.
– temperature dependence of 192
Bulk properties of semiconductors A.1 (table)

Capacitance–voltage characteristics of Schottky contacts 26
– determination of flat-band barrier height 26
Charge neutrality, condition of, at interfaces 7, 111, 205, 212
Charge neutrality level *see* Branch-point energy
Chemical shifts *see* Adatom-induced core-level shifts, Core-level shifts, Metal adatom-induced core-level shifts
Clean Si surfaces prepared by
– HF dips 42, 209
– decomposition of SiO_2 46
CNL *see* Charge neutrality level, Branch point
Common-anion rule 11
Complex band structure 119ff., 122ff.
Contact potential 113
Contact resistance
– definition 228
– thermionic-emission *I*/*V* curves 228
– tunnel contacts 230
Core-level photoemission spectroscopy 68ff.
Core-level shifts, metal-induced 74ff., 77 (table)
Covalent bonds, ionicity of 76
Covalent radii of atoms A.2 (table)
Current–voltage characteristics of
– ideal Schottky contacts 33ff.
– – diffusion current over the barrier 35
– – thermionic-emission current over the barrier 35, 228
– – tunneling through the barrier 230
– *patchy* Schottky contacts 91ff.
– real Schottky contacts 42, 46
– – influence of clean-surface preparation (silicon)
– – – HF dips 42
– – – decomposition of SiO_2 46
– – standard relation 45
– – thermionic emission over the barrier 44, 45
C/*V* characteristics *see* Capacitance–voltage characteristics of Schottky contacts

DAS model *see* Dimer-adatom-stacking fault model
Dawson function 36
Debye length 5, 22
Defect-induced gap states 13
Defect-induced changes of barrier heights 203ff.
Defects at metal–semiconductor interfaces
– GaP Schottky contacts 206
– influence on barrier heights 203ff.
– Unified defect model 12, 208
Density-functional theory 181
Depletion layer 4, 21ff.
– capacitance of 25
– electric-field strength in 23, 24
– shape of potential of 24
– space charge of 23
– width of 24

DFT *see* Density functional theory
Diamond Schottky contacts 211
Dielectric band gap 14, 121, 130 (table)
Dielectric constants of semiconductors A.1 (table)
Dielectric mid-gap level *see* Branch point
Dielectric work function 114
DIGS *see* Defect-induced gap states
Dimer-adatom-stacking fault model of Si(111)-7×7 reconstruction 42ff.
Donor-bonds, interfacial 161
Donor-like interface states 9

Effective barrier height
– correlation with patches 16, 92, 95, 97
– correlation with ideality factor 16, 45, 47ff., 97, 98
– definition of 45
Effective work-function model 12
Einstein relation 34
Electron affinity (definition) 5
Electron escape depth 69
Electronegativity 14
– average value of semiconductors 108
– of binary compounds 108
– and bond polarity 76
– correlation between Pauling and Miedema scales 113
– correlation with work function 10, 113
– ionicity of bonds and 76
– Miedema values A.3 (table)
– Pauling values A.3 (table)
Electronegativity concept of charge transfer 76
Electrostatic interaction of adatom dipoles *see* Topping model
Empirical tight-binding approximation 125ff.
– matrix elements 127
– – inter-atomic 127
– – inter-orbital 127
ETB approximation *see* Empirical tight-binding approximation
Extrinsic dipole contributions to
– barrier heights 209ff.
– valence-band offsets 224

First-principles calculations of
– Schottky barrier heights 181ff., 214
– valence-band offsets 144, 186
Flat-band barrier height of Schottky contacts 26
Frank–van der Merve growth mode 69

Gallium arsenide
– escape depth of electrons 69
– Schottky barrier heights 9, 16, 139, 183, 194
Gallium chalcogenides
– Schottky barrier heights 146
Gallium nitride
– Schottky barrier heights 141, 183
Gallium phosphide
– interface defects 206
– escape depth of electrons 69
– Schottky barrier heights 140
Germanium
– escape depth of electrons 69
– metal-adatom-induced core-level shifts 76, 77 (table)
Growth modes of metals on semiconductors 69ff.
– Bauer's wetting criterion 71
– Frank–van der Merve growth mode 69
– Stranski–Krastanov growth mode 70
– Volmer–Weber growth mode 69
GW correction 181

Hartee–Fock term values of atoms A.2 (table)
Heterostructures 160ff.
– I–III–VI_2 chalcopyrites 168, 169 (table)
– determination of band offsets
– – by IPEYS 80
– – by XPS 81
– insulators 173, 174 (table)
– Langer–Heinrich rule 176
– lattice-matched 160, 163
– layered semiconductor compounds 162, 170
– metamorphic 160, 172
– non-polar (110) interfaces 162

– polar interfaces 161
– pseudomorphic 160
– types of band-structure lineup 164
HF dips of Si surfaces 43, 209

Ideality factor
– correlation with effective barrier heights 16, 45, 47ff., 98, 192
– definition of 30
– image-force-determined 31
– patch-induced 92
– of real Schottky contacts 45,47
– temperature dependence 50
IFIGS *see* Interface-induced gap states
IFIGS-and-electronegativity concept 136ff.
Impurity levels in binary compounds and ternary alloys, alignment of 176ff.
Image-force effect 16, 27ff.
– barrier-height lowering caused 28
– ideality factor determined by 31
– position of potential maximum 28
– voltage dependence of 30
Inhomogeneities, lateral, of Schottky barrier heights *see* Patches
Insulator interfaces
– metal–insulator contacts 154
– – barrier heights 154
– semiconductor–insulator interfaces 173ff.
– – valence-band offsets 174 (table), 176 (table)
Interface bonds 79
– acceptor type 161
– donor type 161
Interface charge neutrality 7, 111, 205, 212
Interface defects *see* Defects at metal–semiconductor interfaces
Interface dielectric constant 131ff.
Interface dipoles in heterostructures 160
– extrinsic
– – non-polar interfaces 161
– – polar interfaces 224
– intrinsic 109
– – polar interfaces 161
Interface dipoles in Schottky contacts
– extrinsic 17, 209ff.
– interface doping 209
– interface structure
– intrinsic 9, 17, 19, 107
Interface-induced gap states 18, 115ff., 135
– branch-point energy 125ff., 128, 130 (table)
– decay length 129, 130 (table)
– density of states of, at branch point 130 (table)
Interface doping 209
Interface-molecule concept 75, 107ff., 212
Interface states 7, 14ff., 111ff.
Interface structure 17, 214ff.
Internal photoemission yield spectroscopy 62
– threshold photonenergy of Schottky contacts 64
– determination of band offsets in heterostructures 80
– temperature dependence of barrier heights 192
Intrinsic interface states *see* MIGS and IFIGS
Ionicity of covalent bonds and electronegativity 76
Ionization energy (definition) 5
– temperature dependence 192
IPEYS *see* Internal photoemission yield spectroscopy
I/*V* characteristics *see* Current–voltage characteristics

Langer–Heinrich rule 176
Lattice-matched heterostructures 163ff., 186, 196
– strain in 198
Lattice mismatch 160, 200
Lattice parameter of semiconductors A.1 (table)
Layered semiconductor compounds 162, 170

Matrix elements *see* ETB matrix elements
Mean-value $\underline{k}$-point 14, 122
– bulk band gap at 124

Metal-induced gap states 115ff.
– branch-point energy 128
– concept 8
– decay length 129, 130 (table), 223
– density of states of, at branch point 130 (table)
– observation 158
Metal adatom-induced core-level shifts 74ff., 77 (table)
Metal–insulator contacts 154
Metals on semiconductors, growth mode 69
Metamorphic heterostructures 172
MIGS *see* Metal-induced gap states
MIGS-and-electronegativity concept 14
Misfit dislocations 73, 160

Non-polar heterostructures 161

Offset *see* Band-edge offset
Ohmic contacts
– natural 152, 153
– tunnel barriers 228

Patches of lower barrier height
– artificial 97
– circular shape 86ff.
– – potential distribution 85ff.
– – saddle point of potential 85ff.
– – – barrier height 89, 92, 94
– – – position 88
– condition for pinch-off 87, 90
– effective area 92, 95
– effective barrier height 92, 95
– ideality factor 92
– linear stripes 105
– origin of
– – extrinsic nonuniformities 98, 104
– – intrinsic or natural nonuniformities 17, 99, 103
– patch function 94
– patch parameter 89, 94
– space-charge parameter 89
Penn's model 14, 122, 132
Pinning of the Fermi level 12, 115
Plasmon energy of valence electrons of semiconductors 123, A.1 (table)
– average value of semiconductors 132
Point-charge model of interface bonds 76, 107, 212
Poisson's equation 21, 84
Polar heterostructures 162
Poly(pyrrole)/*n*-InP(001) Schottky contacts 53
Pressure dependence of
– Schottky barrier heights 193
– valence-band offsets 196
Pseudomorphic heterostructures 197ff.

Quasi-Fermi level 33
Quasi-particle energies *see* First-principles calculations

Restatoms 42, 215
Richardson constant 35
– corrected effective, of semiconductors 41
– effective, of semiconductors 35, 39, 41 (table)

sp^3 hybrids 127
– ETB energy of 128
Saddle point of potential in front of patches 16
Schottky barrier 21ff.
– capacitance of 25ff.
Schottky barrier height *see* Barrier height
Schottky contacts
– $CuInSe_2$ 148
– current transport 33ff.
– diamond 211
– edge effects 84
– layered semiconductor compounds 170
– laterally homogeneous 47ff.
– – barrier heights (tables) 54, 63, 67
– – GaAs 9, 16, 139, 194
– – GaN 141, 183
– – GaP 140
– – Ga–VI 146
– – Se 6
– – Si 107, 136, 185, 194, 214
– – SiC 142, 183
– – ZnSe 145, 182

– quaternary III–V alloys 163ff.
– ternary III–V alloys 148
Schottky effect *see* Image-force effect
Schottky–Mott rule 5, 115
Scissors operator 14
Screening length *see* Debye length, Thomas–Fermi length
Silicide–silicon Schottky contacts 137, 185, 217
– interface structure 217ff.
– pressure dependence 194
– temperature dependence 192
Silicon
– escape depth of electrons 69
– insulator heterostructures 174 (table)
– influence of surface preparation on *I*/*V* characteristics
– – HF dips 42, 209
– – decomposition of SiO_2 46
– Schottky barrier heights 107, 136, 185, 194, 214
– metal-adatom-induced core-level shifts 76, 77 (table)
Si(111)-(1×1)^i interface structure 17
Si(111)-(7×7)^i interface structure 17, 42, 214ff.
Silicon carbide
– Schottky barrier heights 142, 183
– valence-band offsets 187
Si_3N_4
– metal contacts 156
– Si heterostructure 174 (table)
SiO_2
– branch-point energy 175
– metal contacts 154
– semiconductor heterostructures 174 (table), 175, 176 (table)
Slope parameter of
– heterostructures 110, 119
– Schottky contacts 6, 11, 14, 108, 131
– – dependence on density of states of interface states 115, 131ff.
– – dependence on width of dielectric band gap 132
Space-charge layer *see* Depletion layer, Schottky barrier
Stacking fault 42, 216
– dipole moment of 216
Stranski–Krastanov growth mode 70
Surface-molecule concept 75

Tailing length of MIGS 129, 130 (table), 223
Term values (Hartree–Fock) of atoms A.2 (table)
Thermionic-emission *I*/*V* characteristics
– of ideal Schottky contacts 33ff., 228
– of real Schottky contacts 42, 45, 46
Thomas–Fermi screening length 5, 133
Topping model 78, 107
Tunneling parameter 229

Valence-band offset *see* Band-edge offsets of semiconductors
– between different modifications of semiconductors 144 (table)
– I–III–VI_2 chalcopyrites 168
– III–nitrides 187
– extrinsic dipole contribution 162
– intrinsic dipole contribution 10
– insulator–semiconductor heterostructures 173ff.
– Langer–Heinrich rule 176
– lattice-matched III–V alloys 163ff., 186
– – pressure dependence 196
– – strain at 198
– lattice strain, dependence on 200
– layered semiconductors 170
– metamorphic heterostructures 172
– non-polar (110) heterostructures 162
– polar heterostructures 161
– pressure dependence 196
– pseudomorphic heterostructures 197
– silicon-carbide heterostructures 187
– slope parameter of 119
– temperature dependence 193
– types of 164
ViGS *see* Virtual gap states
Virtual crystal approximation 149, 169
Virtual gap states
– branch-point energy 125, 130 (table)
– decay length 129, 130 (table)

– one-dimensional model 119ff.
– – density of states 122
– three-dimensional model 122ff.
Volmer–Weber growth mode 69

Unified defect model 12, 208

Wetting criterion 71
Work function
– definition 5
– dielectric 114
– electronegativity, correlation with 10, 113

XPS *see* X-ray photoemission spectroscopy

X-ray photoemission spectroscopy, determination of
– band offsets in heterostructures 81
– Schottky barrier heights 73
– metal-induced core-level shifts 74

Zero-bias barrier height 38
Zero-bias image-force lowering 30, 39
Zero-charge-transfer barrier height 108, 111
Zinc selenide, Schottky contacts 145, 182

SPRINGER SERIES IN SURFACE SCIENCES

1 **Physisorption Kinetics**
By H. J. Kreuzer, Z. W. Gortel

2 **The Structure of Surfaces**
Editors: M. A. Van Hove, S. Y. Tong

3 **Dynamical Phenomena at Surfaces, Interfaces and Superlattices**
Editors: F. Nizzoli, K.-H. Rieder, R. F. Willis

4 **Desorption Induced by Electronic Transitions, DIET II**
Editors: W. Brenig, D. Menzel

5 **Chemistry and Physics of Solid Surfaces VI**
Editors: R. Vanselow, R. Howe

6 **Low-Energy Electron Diffraction**
Experiment, Theory and Surface Structure Determination
By M. A. Van Hove, W. H. Weinberg, C.-M. Chan

7 **Electronic Phenomena in Adsorption and Catalysis**
By V. F. Kiselev, O. V. Krylov

8 **Kinetics of Interface Reactions**
Editors: M. Grunze, H. J. Kreuzer

9 **Adsorption and Catalysis on Transition Metals and Their Oxides**
By V. F. Kiselev, O. V. Krylov

10 **Chemistry and Physics of Solid Surfaces VII**
Editors: R. Vanselow, R. Howe

11 **The Structure of Surfaces II**
Editors: J. F. van der Veen, M. A. Van Hove

12 **Diffusion at Interfaces: Microscopic Concepts**
Editors: M. Grunze, H. J. Kreuzer, J. J. Weimer

13 **Desorption Induced by Electronic Transitions, DIET III**
Editors: R. H. Stulen, M. L. Knotek

14 **Solvay Conference on Surface Science**
Editor: F. W. de Wette

15 **Surfaces and Interfaces of Solids**
By H. Lüth*)

16 **Atomic and Electronic Structure of Surfaces**
Theoretical Foundations
By M. Lannoo, P. Friedel

17 **Adhesion and Friction**
Editors: M. Grunze, H. J. Kreuzer

18 **Auger Spectroscopy and Electronic Structure**
Editors: G. Cubiotti, G. Mondio, K. Wandelt

19 **Desorption Induced by Electronic Transitions, DIET IV**
Editors: G. Betz, P. Varga

20 **Scanning Tunneling Microscopy I**
General Principles and Applications to Clean and Adsorbate-Covered Surfaces
Editors: H.-J. Güntherodt, R. Wiesendanger
2nd Edition

21 **Surface Phonons**
Editors: W. Kress, F. W. de Wette

22 **Chemistry and Physics of Solid Surfaces VIII**
Editors: R. Vanselow, R. Howe

23 **Surface Analysis Methods in Materials Science**
Editors: D. J. O'Connor, B. A. Sexton, R. St. C. Smart
2nd Edition

24 **The Structure of Surfaces III**
Editors: S. Y. Tong, M. A. Van Hove, K. Takayanagi, X. D. Xie

25 **NEXAFS Spectroscopy**
By J. Stöhr

26 **Semiconductor Surfaces and Interfaces**
By W. Mönch
3rd Edition

27 **Helium Atom Scattering from Surfaces**
Editor: E. Hulpke

28 **Scanning Tunneling Microscopy II**
Further Applications and Related Scanning Techniques
Editors: R. Wiesendanger, H.-J. Güntherodt
2nd Edition

29 **Scanning Tunneling Microscopy III**
Theory of STM and Related Scanning Probe Methods
Editors: R. Wiesendanger, H.-J. Güntherodt
2nd Edition

30 **Concepts in Surface Physics**
By M. C. Desjonquères, D. Spanjaard*)

31 **Desorption Induced by Electronic Transitions, DIET V**
Editors: A. R. Burns, E. B. Stechel, D. R. Jennison

32 **Scanning Tunneling Microscopy and Its Applications**
By C. Bai
2nd Edition

33 **Adsorption on Ordered Surfaces of Ionic Solids and Thin Films**
Editors: H.-J. Freund, E. Umbach

34 **Surface Reactions**
Editor: R. J. Madix

35 **Applications of Synchrotron Radiation**
High-Resolution Studies of Molecules and Molecular Adsorbates on Surfaces
Editor: W. Eberhardt

36 **Kinetics of Metal-Gas Interactions at Low Temperatures: Hydriding, Oxidation, Poisoning**
By E. Fromm

37 **Magnetic Multilayers and Giant Magnetoresistance**
Fundamentals and Industrial Applications
Editor: U. Hartmann*)

*) Available as a textbook

Printing: Saladruck Berlin
Binding: Lüderitz&Bauer, Berlin

Printed in the United Kingdom
by Lightning Source UK Ltd.
118106UK00001BB/17